NANOSTRUCTURE, NANOSYSTEMS, AND NANOSTRUCTURED MATERIALS

Theory, Production, and Development

NANOSTRUCTURE, NANOSYSTEMS, AND NANOSTRUCTURED MATERIALS

Theory, Production, and Development

Edited by
P. M. Sivakumar, PhD, Vladimir I. Kodolov, DSc
Gennady E. Zaikov, DSc and A. K. Haghi, PhD

Apple Academic Press

TORONTO   NEW JERSEY

Apple Academic Press Inc. | Apple Academic Press Inc.
3333 Mistwell Crescent | 9 Spinnaker Way
Oakville, ON L6L 0A2 | Waretown, NJ 08758
Canada | USA

©2014 by Apple Academic Press, Inc.

First issued in paperback 2021

Exclusive worldwide distribution by CRC Press, a member of Taylor & Francis Group

No claim to original U.S. Government works

ISBN 13: 978-1-77463-279-6 (pbk)
ISBN 13: 978-1-926895-49-9 (hbk)

Library of Congress Control Number: 2013945479

Library and Archives Canada Cataloguing in Publication

Nanostructure, nanosystems, and nanostructured materials: theory, production, and development/edited by Prof. P.M. Sivakumar, Dr. Vladimir Ivanovitch Kodolov, Dr. Gennady E. Zaikov, and Dr. A.K. Haghi.

Includes bibliographical references and index.
ISBN 978-1-926895-49-9
1. Nanocomposites (Materials). 2. Nanostructured materials.
3. Carbon composites. 4. Metallic composites. 5. Nanochemistry.
I. Sivakumar, P. M., writer of preface, editor of compilation

TA418.9.N35N42 2013   620.1'18   C2013-904998-3

Apple Academic Press also publishes its books in a variety of electronic formats. Some content that appears in print may not be available in electronic format. For information about Apple Academic Press products, visit our website at **www.appleacademicpress.com** and the CRC Press website at **www.crcpress.com**

ABOUT THE EDITORS

P. M. Sivakumar, PhD

P. M. Sivakumar, PhD, is a Foreign Postdoctoral Researcher (FPR) at RIKEN, Wako Campus, in Japan. RIKEN is Japan's largest comprehensive research institution renowned for high-quality research in a diverse range of scientific disciplines. He received his PhD from the Department of Biotechnology, Indian Institute of Technology Madras, India. He is a member of the editorial boards of several journals and has published papers in international peer-reviewed journals and professional conferences. His research interests include bioinformatics and computational approaches for drug design, small molecule synthesis and their biological evaluation, and bionanotechnology and biomaterials.

Vladimir I. Kodolov, DSc

Vladimir I. Kodolov, DSc, is Professor and Head of the Department of Chemistry and Chemical Technology at M. I. Kalashnikov Izhevsk State Technical University in Izhevsk, Russia, as well as Chief of Basic Research at the High Educational Center of Chemical Physics and Mesoscopy at the Udmurt Scientific Center, Ural Division at the Russian Academy of Sciences. He is also the Scientific Head of Innovation Center at the Izhevsk Electromechanical Plant in Izhevsk, Russia.

Gennady E. Zaikov, DSc

Gennady E. Zaikov, DSc, is Head of the Polymer Division at the N. M. Emanuel Institute of Biochemical Physics, Russian Academy of Sciences, Moscow, Russia, and professor at Moscow State Academy of Fine Chemical Technology, Russia, as well as professor at Kazan National Research Technological University, Kazan, Russia. He is also a prolific author, researcher, and lecturer. He has received several awards for his work, including the the Russian Federation Scholarship for Outstanding Scientists. He has been a member of many professional organizations and on the editorial boards of many international science journals.

A. K. Haghi, PhD

A. K. Haghi, PhD, holds a BSc in urban and environmental engineering from University of North Carolina (USA); a MSc in mechanical engineering from North Carolina A&T State University (USA); a DEA in applied mechanics, acoustics and materials from Université de Technologie de Compiègne (France); and a PhD in engineering sciences from Université de Franche-Comté (France). He is the author and editor of 65 books as well as 1000 published papers in various journals and conference proceedings. Dr. Haghi has received several grants, consulted for a number of major corporations, and is a frequent speaker to national and international audiences. Since 1983, he served as a professor at several universities. He is currently Editor-in-Chief of the *International Journal of Chemoinformatics and Chemical Engineering* and *Polymers Research Journal* and on the editorial boards of many international journals. He is also a faculty member of University of Guilan (Iran) and a member of the Canadian Research and Development Center of Sciences and Cultures (CRDCSC), Montreal, Quebec,Canada.

CONTENTS

LIST OF CONTRIBUTORS

L. F. Akhmetshina
Basic Research—High Educational Centre of Chemical Physics and Mesoscopy, Udmurt Scientific Centre, Ural Division, Russian Academy of Sciences, Izhevsk
OJSC "Izhevsk Electromechanical Plant—Kupol"

I. I. Blagodatskikh
M. T. Kalashnikov Izhevsk State Technical University

A. Yu. Bondar
M. T. Kalashnikov Izhevsk State Technical University

M. A. Chashkin
Basic Research—High Educational Centre of Chemical Physics and Mesoscopy, Udmurt Scientific Centre, Ural Division, Russian Academy of Sciences, Izhevsk
OJSC "Izhevsk Electromechanical Plant—Kupol"

Siva Chidambaram
Division of Nanoscience and Technology, Anna University-BIT Campus, Thiruchirapalli, 620024, India

Ramasamy Jayavel
Centre for Nanoscience and Technology, Anna University, Chennai, 600025, India

Park Jinsub
Department of Electronic and Computer Engineering, Hanyang University, Seoul 133791, Korea

Jayaraman Kandasamy
Defence Research and Development Organization (DRDO), New Delhi, India

Jin Kawakita
Electronics Material Centre, National Institute for Materials Science, Namiki, Tsukuba 305-0044, Japan

N. V. Khokhriakov
Basic Research—High Educational Centre of Chemical Physics and Mesoscopy, Udmurt Scientific Centre, Ural Division, Russian Academy of Sciences, Izhevsk
Izhevsk State Agricultural Academy

V. I. Kodolov
M. T. Kalashnikov Izhevsk State Technical University
Basic Research—High Educational Centre of Chemical Physics and Mesoscopy, Udmurt Scientific Centre, Ural Division, Russian Academy of Sciences, Izhevsk

G. A. Korablev
Izhevsk Agricultural Academy
Basic Research—High Educational Centre of Chemical Physics and Mesoscopy, Udmurt Scientific Centre, Ural Division, Russian Academy of Sciences, Izhevsk

Ganesan Mohan Kumar
Department of Electronic and Computer Engineering, Hanyang University, Seoul 133791, Korea, E-mail: mobhuu@gmail.com
Centre for Nanoscience and Technology, Anna University, Chennai, 600025, India

V. I. Ladyanov
Ural Division, Physical Technical Institute, Russian Academy of Sciences, Izhevsk

A. M. Lipanov
Basic Research—High Educational Centre of Chemical Physics and Mesoscopy, Ural Division, Udmurt Scientific Centre, Russian Academy of Sciences, Izhevsk
Institute of Mechanics, Ural Division, Russian Academy of Sciences, Izhevsk

L. G. Makarova
Ural Division, Physical Technical Institute, Russian Academy of Sciences, Izhevsk

E. A. Naimushina
Ural Division, Physical Technical Institute, Russian Academy of Sciences, Izhevsk

Sreevidya Narasimhan
Women Scientist –C, Lakshmi Kumaran and Sridharan Attorneys, No. 2, Wallace garden 2nd street, Nungambakkam, Chennai, TamilNadu India

Kasi Nehru
Department of chemistry, Anna University-BIT Campus, Thiruchirapalli 620024, India

R. M. Nikonova
Ural Division, Physical Technical Institute, Russian Academy of Sciences, Izhevsk

Ramesh Chandra Pandey
Ph. D., Department of Pediatric Pneumology and Neonatology, Hannover Medical School, Hannover 30625, Germany, E-mail: rcpandey.vs@gmail.com

Yu. V. Pershin
M.T. Kalashnikov Izhevsk State Technical University
Basic Research—High Educational Centre of Chemical Physics and Mesoscopy, Ural Division, Udmurt Scientific Centre, Russian Academy of Sciences, Izhevsk

Ya. A. Polyotov
M.T. Kalashnikov Izhevsk State Technical University
Basic Research—High Educational Centre of Chemical Physics and Mesoscopy, Udmurt Scientific Centre, Ural Division, Russian Academy of Sciences, Izhevsk

V. I. Ryabova
Ural Division, Physical Technical Institute, Russian Academy of Sciences, Izhevsk

G. V. Sapozhnikov
Ural Division, Physical Technical Institute, Russian Academy of Sciences, Izhevsk

Vijay Kumar Saxena
M. VSc., Central Sheep and Wool Research Institute, Avikanagar, Rajasthan, India
Dr. An A. SeongSoo
Associate Professor, Ph. D., Department of Bionano Technology, Gachon University, Sungnam, Korea, E-mail: seongaan@gachon.ac.kr

I. N. Shabanova
Ural Division, Physical Technical Institute, Russian Academy of Sciences, Izhevsk
Basic Research—High Educational Centre of Chemical Physics and Mesoscopy, Ural Division, Ud-murt Scientific Centre, Russian Academy of Sciences, Izhevsk

Vishwas Sharma
M. Sc., Hannover Medical School, Hannover 30625, Germany

Muthusamy Sivakumar
Division of Nanoscience and Technology, Anna University-BIT Campus, Thiruchirapalli 620024, In-dia, E-mail: siva.may3@gmail.com,muthusiva@gmail.com

N. S. Terebova
Ural Division, Physical Technical Institute, Russian Academy of Sciences, Izhevsk
Basic Research—High Educational Centre of Chemical Physics and Mesoscopy, Ural Division, Ud-murt Scientific Centre, Russian Academy of Sciences, Izhevsk

Rajangam Thanavel
M. pharm, (Ph.D candidate), Department of Bionano Technology, Gachon Bionano Research Institute,

Gachon University, Sungnam, Korea, E-mail address: thana.vel2009@gmail.com

V. V. Trineeva
Basic Research—High Educational Centre of Chemical Physics and Mesoscopy, Ural Division, Ud-murt Scientific Centre, Russian Academy of Sciences, Izhevsk
Institute of Mechanics, Ural Division, Russian Academy of Sciences, Izhevsk

M. A. Vakhrushina
Basic Research—High Educational Centre of Chemical Physics and Mesoscopy, Udmurt Scientific Centre, Ural Division, Russian Academy of Sciences, Izhevsk
OJSC "Izhevsk Electromechanical Plant—Kupol"

Yu. M. Vasil'chenko
Basic Research—High Educational Centre of Chemical Physics and Mesoscopy, Ural Division, Ud-murt Scientific Centre, Russian Academy of Sciences, Izhevsk
M. T. Kalashnikov Izhevsk State Technical University
OJSC "Izhevsk Electromechanical Plant—Kupol"

G. E. Zaikov
N. M. Emanuel Institute of Biochemical Physics, Russian Academy of Sciences, Moscow

LIST OF ABBREVIATIONS

AA	Acetylacetone
Adv-hBMP-2	Adenovirus-mediated human BMP-2 gene
AEM	Aryl ethynylene macrocycle
AES	Auger-electron spectroscopy
AFM	Atomic force microscopy
AP	Ammonium perchlorate
APP	Ammonium polyphosphate
APPh	Ammonium polyphosphate
ARDB	Aeronautical Research and Development Board
AcAc	Acetylacetone
bFGF	Basic fibroblast growth factor
BHK	Baby-hamster kidney
BMP	Bone morphogenic protein
BMSC	Bone marrow stromal cell
BSSE	Basis sets of complex molecules
CHEC	Cold-hardened epoxy composition
CTAB	Cetyltrimethyl ammonium bromide
CVs	Cyclic voltammograms
DAcA	Diacetonealcohol
DRDO	Defence Research and Development Organisation
DSSC	Dye sensitized solar cells
EC	Endothelial cell
ECM	Extracellular matrix
ED	Electron diffraction
EDAX	Energy Dispersive Analysis by X-Rays
EDR	Epoxy diane resin
EGF	Epidermal growth factors
ER	Epoxy resin
eV	Electron-volts
FDA	Food and Drug Administration
FET	Field-effect transistor
FGF	Fibroblast growth factor
FN	Fibronectin
FS	Fine suspensions
GATET	Gas Turbine Enabling Technology Initiative

GN	Gadolinium nitrate
GTRE	Gas Turbine Research and Establishment
HA	Hydrogen bonding agent
HBBA	2-(4′-Hydroxybenzeneazo) benzoic acid
hES	Human embryonic stem cells
IEP	Isoelectric point
IR MDCIR	Infrared microscopy of multiply disturbed complete inner reflection
KGF	Keratinocyte growth factor
LED	Light-emitting diode
LFE	Linear dependencies of free energies
LMMA	Laser microprobe mass-analysis
MPC	Monolayer protected cluster
MRI	Magnetic resonance imaging
MRSA	Methicillin-resistant S. aureus
MSs	Mesenchymal stem cell
nAl	Nano Al particle
NGF	Nerve growth factors
NS	Nanostructures
OFET	Organic field effect transistors
OLED	Organic light-emitting diode
P3HT	Poly (3-hexylthiophene)
PAGA	Poly (-[4-aminobutyl]-l-glycolic acid)
PAMAM	Poly (amido amine)
PAP	Polyammonium-phosphate
PC	Polycarbonate
PCM	Polymeric composite materials
PEG	Polyethylene glycol
PEPA	Polyethylene polyamine
PFR	Phenol-formaldehyde resins
PGA	Polyglycolic acid
PHA	Polyhydroxyalkanoate
PHB	Polyhydroxybutyrate
PHBV	Polyhydroxy-co-valerate
PLA	Poly-lactic acid
PLGA	Polyglycolic-co-lactic acid
PMMA	Polymethyl methacrylate
PPF	Polypropylene fumarate
PPI	Poly(propylene-imine)
PPy	Polypyrrole
PS-PVP	Polystyrene-b-polyvinylpyridine
PTCDI	Perylene tetracarboxylic diimide

PV	Photovoltaic
PVA	Polyvinyl alcohol
PVAc	Polyvinyl acetate
PVC	Polyvinyl chloride
rhVEGF	Recombinant human vascular endothelial growth factor
RS	Raman spectroscopy
SAM	Self-assembled monolayer
Sc	Coked surface
SC	Schwann cell
SCD	Spectroscopy of combination dissipation
SED	Spectroscopy of ionic dissipation
SEM	Scanning electron microscopy
SEP	Spatial-energy parameter
SFD	Squeeze Film Damper
SIMS	Secondary ionic mass-spectrometry
siRNA	Small interference RNA
SMD	Sauter mean diameter
SREM	Scanning raster electron microscopy
TEM	Tunnel electron microscopy
TEMEMD	Transmission electron microscopy with electron microdiffraction
TG-DTA	Thermo-gravimetric and differential thermal analyze
TGF	Transforming growth factor
TiO2	Titanium dioxide
TU	Technical condition
UPS	Ultraviolet photoelectron spectroscopy
UPSIO	Ultra small particulars of iron oxide
UVES	Ultraviolet electron spectroscopy
WF	Work function
XPS	X-Ray photoelectron spectroscopy
ZN	Zirconyl nitrate
ZPVE	Zero-point vibration energy

PREFACE

This volume accumulates the most important information about new trends in nanochemistry and also in the science about materials modified by nano-structures.

The editors selected papers, including reviewed articles, in the field of chemical physics of metal/carbon nanocomposites.

The book includes information on the new classes of metal/carbon nanocomposites and their new production methods in the nanoreactors of polymeric matrixes. In recent years, the chemistry in nanoreactors of polymeric matrixes has successfully been developed. For synthesis of metal containing nanophases in carbon or polymeric shells, it is expedient to evaluate the possibilities of redox reactions with the participation of metal containing phases and organic (or polymeric) compounds (matrixes).

The book raises discussion of the following topics:

1. The place of metal/carbon nanocomposites besides other nanostructures.
2. Main notions and characteristics of metal/carbon nanocomposites and nanosystems including them.
3. Nanochemistry principles for nanostructures synthesis in nanoreactors of polymeric matrixes.
4. Modeling of processes for obtaining nanocomposites, nanosystems and nanostructured materials.
5. Dependence of nanocomposites activity on their composition, sizes, forms and synthesis methods.
6. Experimental features of metal/carbon nanocomposite redox synthesis.
7. Types of fine dispersed suspensions and the interaction of metal/carbon nanocomposites with different media.
8. Material modification methods by super small quantities of metal/carbon nanocomposites.
9. The investigation of properties of materials modified by fine dispersed suspensions of metal/carbon nanocomposites.

The specific features of this book include:

- Computer prognosis, including quantum chemical modeling, for metal/carbon nanocomposites synthesis processes as well as fine dispersed suspensions obtaining processes and material modification processes
- Determination of new notions for metal/carbon nanocomposites and estimation of their activity
- Redox synthesis of metal/carbon nanocomposites in nanoreactors of polymeric matrixes
- Application of Avrami equations to determine the conditions for processes of metal/carbon nanocomposites synthesis and material modification
- Mechanic-chemical functionalization methods for increasing nanocomposite surface energy electronic component
- Modification methods for changing different materials properties

This book is unique and important because the new trends in nanochemistry and new objects of nanostructures are discussed. Its appeal to potential readers consists of its full information about new perspective nanostructures, new methods of synthesis in nanoreactors of polymeric matrixes and also different material modification by super small quantities of metal/carbon nanocomposites. Researchers, professors, post and undergraduates, students and other readers will find a lot of interesting information and obtain new knowledge in chemical physics of metal/carbon nanocomposites.

The editors and contributors will be happy to receive comments from readers, which we can use in our research and studies in future.

— **P. M. Sivakumar, Vladimir I. Kodolov,**
Gennady E. Zaikov, and A. K. Haghi

INTRODUCTION

A lot of scientific information in the field of nanotechnology and nanomaterial science has appeared recently. Since the Nobel prize winner R. Smalley [1] defined the range from 1 to 1,000 nm as the interval of nanostructures with the control of self-organization processes as the main feature, papers appeared [2-4] in which the upper value of nanostructures was limited by the size of 100 nm. At the same time, there was definitely insufficient substantiation taken from [5]. But nanostructures differing in size and shape found in the range given by Smalley are rather active in self-organization processes and differ by a special set of properties. In some papers [6-8] the problems of nanostructure activity and the influence of their supersmall quantities on active media structuring were discussed.

The present time can be defined as the time of quiet revolution in the concepts of matter and substance due to the discoveries in the field of nanotechnology. A lot of nanoparticles (nanostructures) of various shapes, sizes and element composition are known. Such terms as "fullerenes", "nanotubes", "graphene" have become widely known in popular scientific literature, though the shapes of nanostructures found are not limited by them and their names grow in number exponentially but not in arithmetic progression.

Carbon, silicide, metal, metal oxide, boron nitride and metal/carbon nanostructures are distinguished by composition but this list of classes is far from being complete. Carbon nanostructures have the most of shapes, though such nanostructures with peculiar shapes as nanorotors, nanonails, nanowalls and nanowires are known for zinc oxide nanostructures.

Naturally, shapes and sizes of nanostructures are stipulated by their composition and formation conditions. In turn, the nanostructure characteristics mentioned define the originality of nanostructure properties and possible fields of their application.

As usual, when new phenomena and new investigation objects appear, breakthrough in certain fields is expected. Such expectation "to obtain superconductors" was felt when conductive single nanotubes were discovered; however the corresponding result has not been obtained so far on nanotubes filled with metals. Despite recent developments and achievements in nanoelectronics and nanopower engineering, the center of attention for nanostructure application is shifting to nanomedicine and is becoming most widely represented in nanomaterial science. However, although "the fashion" on carbon nanostructures of different shapes still remains, the preferences in investigations depending on nanostructure shape are changing. First, there was a rush and everybody wrote about fullerenes. Then there was a turn of carbon singlewall nanotubes. Some time later it was found out that multiwall nanotubes were the most convenient to give definite properties to the materials. Finally it was the turn of graphene. However there was information [9] on obtaining hydrocarbon polymeric films (graphane) by reducing graphene.

The main feature of all nanostructures obtained is a considerable excess of a number of surface atoms over the atom number in volume. This excess increases with the nanostructure size decrease. Therefore the nanoparticle size is its main feature.

In the majority of nanotechnology standards and programs [10] the range of nanostructure existence is defined by sizes 0.1– 100 nm. The authors of [11] demonstrated that the interval in which the nanostructure activity changes depends on nanostructure nature and shape. However, if the energy of nanoparticle field is comparable with the energy of electromagnetic radiation, when significant changes take place due to chemical reactions in substances under the action of radiation upon them in the wavelength interval indicated, the nanoparticle activity in size interval up to 100 nm will be considerable.

It should be pointed out that atoms on the surface of nanostructures are in energy non-compensated state. In general, this results in the growth of nanoparticle energy which can be presented as the total of atom energies on the particle surface. It is quite obvious that the freedom of movement of surface atoms is limited except for the possibilities of oscillatory motion and movement of electrons. Both these movement forms are interconnected since the shifts of electron clouds in atoms inevitably result in changes of oscillation frequencies of corresponding bonds with the atom participation. In turn, the changes in the position of valence electrons in bonds result in the change of bond polarity and the so-called "supramolecule" [12]. Here the electron shift to higher energy levels is possible.

In this regard, metal/carbon nanostructures are the most interesting objects of investigation. Since metal clusters in these nanostructures are associated with the carbon envelope protecting them from the environment, they were called metal/carbon nanocomposites.

In contrast with carbon nanostructures, metal and metal oxide clusters, metal/carbon nanocomposites, apart from metal and covalent bonds, have a significant share of coordination bonds contributing to their self-organization. At the same time, the bonds indicated can arise changes in the electron structure of d metals contributing to the increase in the number of unpaired electrons and growth of the number of atom magnetic momentum.

Metal/carbon nanocomposites are interesting for the investigation with the help of X-ray electron spectroscopy and electron paramagnetic resonance.

It is amazing that such informatively powerful method as X-ray electron spectroscopy is envisaged for the investigation of the electron structure of atoms in surface layers is insufficiently applied in the investigation of nanostructures and nanostructurized materials, especially polymers.

Therefore, the editors put down the discussions of the investigation results of metal/carbon nanocomposites with the application of X-ray photoelectron spectrometers with magnetic focusing in certain chapters of the book. A special place is provided for computer modeling of processes with the participation of metal/carbon nanocomposites. The development of prognosticating methods for the obtaining of nanocomposites is very important for the assessment of their activity in different media and modification of polymeric materials.

The authors and editors pay much attention to the application of metal/carbon nanocomposites in the form of fine suspensions and sols in the corresponding media for the modification of polymeric organic and inorganic materials.

Since, despite of a large number of theoretical and experimental works, there is no unified concept of the formation of nanostructures and nanocomposites, as well as nanostructurized materials with predetermined properties, there have been attempts in a number of papers to close this gap or, at least, to come closer to solving the problem.

REFERENCES

1. Smalley, R. E. & Cole, R. *Initiatives in Nanotechnology*. 1995. http// pcheml.rice.edu/nanoinit.html. Дата обращения - февраль 1997г.

2. Pomogailo, A. D., Rozenberg, A. S., & Uflyand, I. E. (2000) *Metal nanoparticles in polymers*. M.: Himiya . 672p.

3. Gusev, A. I. (2005) *Nanomaterials, nanostructures, nanotechnologies*. M.:Fizmatlit. 412p.

4. Suzdalev, I. P. (2005) *Nanotechnology. Physic-chemistry of nanoclusters, nanostructures and nanomaterials*. M.: KomKniga, 589p.

5. Furstner, A. (Ed.) (1996) *Active Metals*. VCH: Weinheim . 233p

6. Kodolov, V. I., Khokhriakov, N. V., Kuznetsov, A. P. et al. (2007) Perspectives of the application of nanostructures and nanosystems when producing composites with predictable behavior // In the book: Space challenges in 21 century. V. 3. *Novel materials and technologies for space rockets and space development*. M.: Torus press. P. 201–205.

7. Krutikov V.A., Didik A.A., & Yakovlev G.I. et al. (2005) Composite material with nanoreinforcement//*Alternative power engineering and ecology*. № 4(24). P. 36–41.

8. Kodolov ,V. I., Khokhriakov, N. V., & Kuznetsov, A. P. (2006) To the issue on the mechanism of the nanostructure influence on structurally changing media at the formation of "smart" composites//*Nanotehnika*. № 3(7). P. 27–35.

9. Elias, D. C. et al. (2009) *Science*, . N 323. P. 610.

10. Alfimov, M. V., Gohberg, A. M., & Fursov, K. S. (2010) *Nanotechnologies: definitions and classification*. // Rossiyskie nanotehnologii. V. 5. № 7–8. P. 8–15.

11. Kodolov, V. I., Khokhriakov, N. V., Trineeva, V. V., & Blagodatskikh, I. I. (2008) Activity of nanostructures and its display in nanoreactors of polymeric matrixes and active media. //*Chemical physics and mesoscopy*. V. 10. № 4. P. 448–460.

12. Rusanov, A. I. (2006)Nanothermodynamics: chemical approach. //Russian Chemical Journal. V. 1. № 2. P. 145–151.

CHAPTER 1

FUNDAMENTAL DEFINITIONS FOR DOMAIN OF NANOSTRUCTURES AND METAL/CARBON NANOCOMPOSITES

V. I. KODOLOV and V. V. TRINEEVA

CONTENTS

1.1   DETERMINATION OF EXISTENCE AREA OF NANOSTRUCTURES AND NANOSYSTEMS BASIC NOTIONS IN NANOCHEMISTRY AND NANOTECHNOLOGY NANOSIZED INTER VAL

In every new field of knowledge specific notions and terms or "language" are created limiting and isolating this field. However, first existence boundaries of new notions and ideas corresponding to the above field of knowledge are determined.

The appearance of "nanoscience" and "nanotechnology" stimulated the burst of terms with "nano-" prefix. Historically the term "nanotechnology" appeared before and it was connected with the appearance of possibilities to determine measurable values up to 10^{-9} of known parameters: 10^{-9}m–nm (nanometer), 10^{-9}s–ns (nanosecond), 10^{-9} degree (nanodegree, shift condition). Nanotechnology and molecular nanotechnology comprise the set of technologies [1] connected with transport of atoms and other chemical particles (ions, molecules) at distances contributing the interactions between them with the formation of nanostructures with different nature. Although Nobel laureate Richard Feyman (1959) showed the possibility to develop technologies on nanometer level, Eric Drexler due to his emotional book "Future without boundaries—revolution of nanotechnology" is considered [1] to be the founder and ideologist of nanotechnology. When scanning tunnel microscope was invented by Nobel laureates Rorer and Binig (1981) there was an opportunity to influence atoms of a substance thus stimulating the work in the field of probe technology, which resulted in substantiation and practical application of nanotechnological methods in 1994. With the help of this technique it is possible to handle single atoms and collect molecules or aggregates of molecules, construct various structures from atoms on a certain substrate (base). Naturally, such a possibility cannot be implemented without preliminary computer designing of so-called "nanostructures architecture". Nanostructures architecture assumes a certain given location of atoms and molecules in space that can be designed on computer and afterwards transferred into technological program of nanotechnological facility.

The term "chemical assembly" appeared together with the development of chemistry and physics of surface after the birth of "electron spectroscopy for chemical analysis" founded by Nobel laureate K. Zigban [2].

In Russia, chemical assembly comprising the interaction of chemical particles with the surface and "grafting" of functional groups to the surface or interface boundary "gas–solid" has been developed by the school of V. B. Aleskovsky [3].

The paper with the participation of Nobel laureates (Nobel Prize in 1995) Croto, Smalley and Carl on synthesis of fullerenes by graphite evaporation was published in 1985. At the same time in nanoproduct obtained fullerene C_{60} predominates by its content, which represents an enclosed cluster of 60 carbon atoms. This cluster had a stable and symmetrical structure. Further, a specialist in electron microscopy Iijima discovered nanotubes in 1991. Afterwards, the investigations in the field of nanoparticles and nanosystems started spreading all over the world.

The development of these trends predetermined the appearance and development of so-called "nanoscience". In the same way as "nanotechnology" and "nanoscience" is determined as a combination of scientific knowledge from various disciplines, such as physics, chemistry, biology, mathematics, programming, and so on, adapted to nanostructures and nanosystems. Nanoscience comprises fundamental and applied scientific knowledge. Therefore, it is possible to speak about nanochemistry, nanometallurgy, nanoelectronics, nanomachine-building, science of nanomaterials, and similar disciplines. If we think of the world from the point that the nature is unified and different disciplines in the science were created by people for the convenience of perception and understanding of the world around, the appearing areas of nanoscience closely connected with nanotechnology represent a vast aggregation of disciplines the list of which will still be incomplete analogously too [4]. Let us be restricted to definitions connected with science of nanomaterials.

The notion "science of nanomaterials" assumes scientific knowledge for obtaining, composition, properties and possibilities to apply nanostructures, nanosystems and nanomaterials. A simplified definition of this term can be as follows: material science dealing with materials comprising particles and phases with nanometer dimensions. To determine the existence area for nanostructures and nanosystems it is advisable to find out the difference of these formations from analogous material objects.

From the analysis of literature the following can be summarized: the existence area of nanosystems and nanoparticles with any structure is between the particles of molecular and atomic level determined in picometers and aggregates of molecules or permolecular formations over micron units. Here, it should be mentioned that in polymer chemistry particles with nanometer dimensions belong to the class of permolecular structures, such as globules and fibrils by one of parameters, for example, by diameter or thickness. In chemistry of complex compounds clusters with nanometer dimensions are also known.

The notion "cluster" assumes energy-wise compensated nucleus with a shell, the surface energy of which is rather small, as a result under given conditions the cluster represents a stable formation.

In chemical literature a cluster is equated with a complex compound containing a nucleus and a shell. Usually a nucleus consists of metal atoms combined with metallic bond, and a shell of ligands. Manganese carbonyls $[(Co)_5MnMn(Co)_5]$

and cobalt carbonyls $[Co_6(Co)_{18}]$, nickel pentadienyls $[Ni_6(C_5H_6)_6]$ belong to elementary clusters.

In recent years, the notion "cluster" has got an extended meaning. At the same time the nucleus can contain not only metals or not even contain metals. In some clusters, for instance, carbon ones there is no nucleus at all. In this case their shape can be characterized as a sphere (icosahedron, to be precise)—fullerenes, or as a cylinder—fullerene tubules. Surely a certain force field is formed by atoms on internal walls inside such particles. It can be assumed that electrostatic, electromagnetic, and gravitation fields conditioned by corresponding properties of atoms contained in particle shells can be formed inside tubules and fullerenes. If analyze papers recently published, it should be noted that a considerable exceeding of surface size over the volume and, consequently, a relative growth of the surface energy in comparison with the growth of volume and potential energy is the main feature of clusters. If particle dimensions (diameters of "tubes" and "spheres") change from 1 up to several hundred nanometers, they would be called nanoparticles. In some papers, the area of nanoclusters existence is within 1–10 nm [5].

Based on classical definitions, in given paper, metal nanoparticles and nanocrystals are referred to as nanoclusters. Apparently, the difference of nanoparticles from other particles (smaller or larger) is determined by their specific characteristics. The search of nanoworld distinctions from atomic-molecular, micro- and macroworld can lead to finding analogies and coincidences in colloid chemistry, chemistry of polymers, and coordination compounds. Firstly, it should be noted that nanoparticles usually represent a small collective aggregation of atoms being within the action of adjacent atoms, thus conditioning the shape of nanoparticles. A nanoparticle shape can vary depending upon the nature of adjacent atoms and character of formation medium. Obviously, the properties of separate atoms and molecules (of small size) are determined by their energy and geometry characteristics, the determinative role being played by electron properties. In particular, electron interactions determine the geometry of molecules and atomic structures of small size and mobility of these chemical particles in media, as well as their activity or reactivity.

When the number of atoms in chemical particle exceeds 30, a certain stabilization of its shape being also conditioned by collective influence of atoms constituting the particle is observed. Simultaneously, the activity of such a particle remains high but the processes with its participation have a directional character. The character of interactions with the surroundings of such structures is determined by their formation mechanism.

During polymerization or co-polymerization the influence of macromolecule growth parameters changes with the increase of the number of elementary acts of its growth. According to [6], after 7–10 acts the shape or geometry of nanoparticles formed becomes the main determinative factor providing the further growth of macromolecule (chain development). A nanoparticle shape is usually determined not only by its structural elements but also by its interactions with surrounding chemical particles.

From the aforesaid, it can be concluded that the possibility of self-organization of nanoparticles with the formation of corresponding nanosystems and nanomaterials is the main distinction of nanoworld from pico-, micro-, and macroworld. Recently, much attention has been paid to synergetics or the branch of science dealing with self-organization processes since these processes, in many cases, proceeds with small energy consumption and, consequently, is more ecologically clear in comparison with existing technological processes.

In turn, nanoparticle dimensions are determined by its formation conditions. When the energy consumed for macroparticle destruction or dispersion over the surface increases, the dimensions of nanomaterials are more likely to decrease. The notion "nanomaterial" is not strictly defined. Several researchers consider nanomaterials to be aggregations of nanocrystals, nanotubes, or fullerenes. Simultaneously, there is a lot of information available that nanomaterials can represent materials containing various nanostructures. The most attention researchers pay to metallic nanocrystals. Special attention is paid to metallic nanowires and nanofibers with different compositions.

Here are some names of nanostructures:

1) fullerenes, 2) gigantic fullerenes, 3) fullerenes filled with metal ions, 4) fullerenes containing metallic nucleus and carbon (or mineral) shell, 5) one-layer nanotubes, 6) multi-layer nanotubes, 7) fullerene tubules, 8) "scrolls", 9) conic nanotubes, 10) metal-containing tubules, 11) "onions", 12) "Russian dolls", 13) bamboo-like tubules, 14) "beads", 15) welded nanotubes, 16) bunches of nanotubes, 17) nanowires, 18) nanofibers, 19) nanoropes, 20) nanosemi-spheres (nanocups), 21) nanobands and similar nanostructures, as well as various derivatives from enlisted structures. It is quite possible that a set of such structures and notions will be enriched.

In most cases nanoparticles obtained are bodies of rotation or contain parts of bodies of rotation. In natural environment there are minerals containing fullerenes or representing thread-like formations comprising nanometer pores or structures. In the first case, it is talked about schungite that is available in quartz rock in unique deposit in Prionezhje. Similar mineral can also be found in the river Lena basin, but it consists of micro- and macro-dimensional cones, spheroids, and complex fibers [7]. In the second case, it is talked about kerite from pegmatite on Volyn (Ukraine) that consists of polycrystalline fibers, spheres, and spirals mostly of micron dimensions, or fibrous vetcillite from the state of Utah (USA); globular anthraxolite and asphaltite.

Diameters of some internal channels are up to 20–50nm. Such channels can be of interest as nanoreactors for the synthesis of organic, carbon, and polymeric substances with relatively low energy consumption. In case of directed location of internal channels in such matrixes and their inner-combinations the spatial structures of certain purpose can be created. Terminology in the field of nanosystems existence is still being developed, but it is already clear that nanoscience obtains qualitatively new knowledge that can find wide application in various areas of human practice thus, significantly decreasing the danger of people's activities for themselves and environment.

The system classification by dimensional factor is known [1], based on which we consider the following:

- Microobjects and microparticles 10^{-6}–10^{-3}m in size;
- Nanoobjects and nanoparticles 10^{-9}–10^{-6}m in size;
- Picoobjects and picoparticles 10^{-12}–10^{-9}m in size.

Assuming that nanoparticle vibration energies correlate with their dimensions and comparing this energy with the corresponding region of electromagnetic waves, we can assert that energy action of nanostructures is within the energy region of chemical reactions. System self-organization refers to synergetics [4]. Quite often, especially recently, the papers are published, for example, by Malinetsky [8], in which it is considered that nanotechnology is based on self-organization of metastable systems. As assumed [9], self-organization can proceed by dissipative (synergetic) and continual (conservative) mechanisms. Simultaneously, the system can be arranged due to the formation of new stable ("strengthening") phases or due to the growth provision of the existing basic phase. This phenomenon underlies the arising nanochemistry. Below is one of the possible definitions of nanochemistry.

Nanochemistry is a science investigating nanostructures and nanosystems in metastable ("transition") states and processes flowing with them in near-"transition" state or in "transition" state with low activation energies.

To carry out the processes based on the notions of nanochemistry, the directed energy action on the system is required, with the help of chemical particle field as well, for the transition from the prepared near-"transition" state into the process product state (in our case–into nanostructures or nanocomposites). The perspective area of nanochemistry is the chemistry in nanoreactors. Nanoreactors can be compared with specific nanostructures representing limited space regions in which chemical particles orientate creating "transition state" prior to the formation of the desired nanoproduct. Nanoreactors have a definite activity which predetermines the creation of the corresponding product. When nanosized particles are formed in nanoreactors, their shape and dimensions can be the reflection of shape and dimensions of the nanoreactor [10].

In the last years a lot of scientific information in the field of nanotechnology and science of nanomaterials appeared. Nobel laureate R. Smalley [11] defined the interval from 1 to 1000nm as the area of nanostructure existence, the main feature of which is to regulate the system self-organization processes. However, later some scientists [12–14] limited the upper threshold at 100nm. At the same time, it was not well-substantiated and taken from [15]. Now many nanostructures varying in shapes and sizes are known. These nanostructures have sizes that fit into the interval, determined by Smally, and are active in the processes of self-organization, and also demonstrate specific properties.

Problems of nanostructure activity and the influence of nanostructure super small quantities on the active media structural changes are explained [16–18].

The molecular nanotechnology ideology is analyzed [19]. In accordance with the development tendencies in self-organizing systems under the influence of nanosized excitations the reasons for the generation of self-organization in the range 10^{-6}–10^{-9}m should be determined.

Based on the law of energy conservation the energy of nanoparticle field and electromagnetic waves in the range 1–1,000nm can transfer, thus corresponding to the range of energy change from soft X-ray to near IR radiation. This is the range of energies of chemical reactions and self-organization (structuring) of systems connected with them.

Apparently the wavelengths of nanoparticle oscillations near the equilibrium state are close or correspond to their sizes. Then based on the concepts of ideologists of nanotechnology in material science the definition of nanotechnology can be as follows:

- Nanotechnology is a combination of knowledge in the ways and means of conducting processes based on the phenomenon of nano-sized system self-organization and utilization of internal capabilities of the systems that results in decreasing the energy consumption required for obtaining the targeted product while preserving the ecological cleanness of the process.

1.1.1　Theoretical substantiation of the approaches proposed

The activity of nanostructures in self-organization processes is defined by their surface energy thus corresponding to the energy of their interaction with the surroundings. It is known [10] that when the size of particles decreases, their surface energy and particle activity increase. The following ratio is proposed to evaluate their activity:

$$a = \varepsilon_s / \varepsilon_V \tag{1.1}$$

where, ε_s—nanoparticle surface energy, ε_v—nanoparticle volume energy. Naturally in this case $\varepsilon_s \gg \varepsilon_v$ conditioned by the greater surface "defectiveness" in comparison with nanoparticle volume. To reveal the dependence of activity upon the size and shape we take ε_s as $\varepsilon_s^0 \cdot S$, and $\varepsilon_v = \varepsilon_v^0 \cdot V$, where, ε_s^0—average energy of surface unit, S—surface, ε_v^0—average energy of volume unit, V—volume, then the equation (1) is converted to:

$$a = d \cdot \varepsilon_s^0 / \varepsilon_v^0 \; S/V \qquad (1.2)$$

Substituting the values of S and V for different shapes of nanostructures, we see that in general form the ratio S/V is the ratio of the number whose value is defined by the nanostructure shape to the linear size connected with the nanostructure radius or thickness. The equation (2) can be given as

$$a = d \cdot \varepsilon_s^0 / \varepsilon_v^0 \; N/r(h) = \varepsilon_s^0 / \varepsilon_v^0 \; 1/B \qquad (1.3)$$

where, B equals r(h)/N, r—radius of bodies of revolution including hollow ones, h—film thickness depending upon its "distortion from plane", N—number varying depending upon the nanostructure shape. Parameter d characterizes the nanostructure surface layer thickness, and corresponding energies of surface unit and volume unit are defined by the nanostructure composition. For the corresponding bodies of revolution the parameter B represents an effective value of the interval of nanostructure linear size influencing the activity at the given interval r from 1 to 1,000nm (Table 1.1). The table shows spherical and cylindrical bodies of revolution. For nanofilms the surface and volume are determined by the defectiveness and shape of changes in conformations of film nanostructures depending upon its crystallinity degree. However, the possibilities of changes in nanofilm shapes at the changes in the medium activity are higher in comparison with nanostructures already formed. At the same time, the sizes of nanofilms formed and their defectiveness (disruptions and cracks on the surface of nanofilms) play are important.

TABLE 1.1   Changes in interval B depending upon the nanoparticle shape.

Nanostructure shape	Internal radius as related to the external radius	Interval of changes B, nm
Solid sphere	–	0.33(3)–333.(3)
Solid cylinder	–	0.5–500
Hollow sphere	8/9	0.099–99

TABLE 1.1   *(Continued)*

Hollow sphere	9/10	0.091–91
Hollow cylinder	8/9	0.105–105
Hollow cylinder	9/10	0.095–95

Proposed the parameter called the nanosized interval (B) may be used to demonstrate the nanostructures activity. Depending on the structure and composition of nanoreactor internal walls, distance between them, shape and size of nanoreactor, the nanostructures differing in activity are formed. The correlation between surface energy, taking into account the thickness of surface layer, and volume energy was proposed as a measure of the activity of nanostructures, nanoreactors and nanosystems [1].

It is possible to evaluate the relative dimensionless activity value (A) of nanostructures and nanoreactors through relative values of difference between the modules of surface and volume energies to their sum:

$$A = (\varepsilon_s - \varepsilon_v)/(\varepsilon_s + \varepsilon_v) = [(\varepsilon_s^{\,0}d)S - \varepsilon_v^{\,0}V]/[(\varepsilon_s^{\,0}d)S + \varepsilon_v^{\,0}V]$$

$$= [(\varepsilon_s^{\,0}d/\varepsilon_v^{\,0})S - V]/[(\varepsilon_s^{\,0}d/\varepsilon_v^{\,0})S + V] \qquad (1.4)$$

If $\varepsilon_s \gg \varepsilon_v$, A tends to 1.

If $\varepsilon_s^{\,0}d/\varepsilon_v^{\,0} \approx 1$, the equation for relative activity value is simplified as follows:

$$A \approx [(S - V)/(S + V)] = [(1 - B)/(1 + B)] \qquad (1.5)$$

If we accept the same condition for a, the relative activity can be expressed via the absolute activity:

$$A = (a - 1)/(a + 1) \qquad (1.6)$$

At the same time, if $a \gg 1$, the relative activity tends to 1.

Nanoreactors represent nanosized cavities, in some cases nanopores in different matrixes that can be used as nanoreactors to obtain desired nanoproducts. The main task for nanoreactors is to contribute to the formation of "transition state" of activated complex being transformed into a nanoproduct practically without any losses for activation energy. In such case, the main influence on the process progress and direction is caused by the entropic member of Arrhenius equation connected either with the statistic sums or the activity of nanoreactor walls and components participating in the process.

The surface energy of nanostructures represents the sum of parts assigned forward motion (ε_{fm}), rotation (ε_{rot}), vibration (ε_{vib}), and electron motion (ε_{em}) in the nanostructure surface layer:

$$\varepsilon_S = \Sigma(\varepsilon_{fm} + \varepsilon_{rot} + \varepsilon_{vib} + \varepsilon_{em}) \tag{1.7}$$

The assignment of these parts on values depends on nature of nanostructure and medium. The decreasing of nanostructure sizes and their quantity usually leads to the increasing of the surface energy vibration part, if the medium viscosity is great. When the nanostructure size is small, the stabilization of electron motion takes place and the energy of electron motion is decreased. Also the possibility of coordination reaction with medium molecules is decreased. In this case the vibration part of surface energy corresponds to the total surface energy.

The nanostructures formed in nanoreactors of polymeric matrixes can be presented as oscillators with rather high oscillation frequency. It should be pointed out that according to references [1] for nanostructures (fullerenes and nanotubes) the absorption in the range of wave numbers 1,300–1,450cm^{-1} is indicative. These values of wave numbers correspond to the frequencies in the range 3.9–4.35·10^{13}Hz, i.e. in the range of ultrasound frequencies.

If the medium into which the nanostructure is placed blocks its translational or rotational motion giving the possibility only for the oscillatory motion, the nanostructure surface energy can be identified with the vibration energy:

$$\varepsilon_S \approx \varepsilon_v = m\upsilon_v^2/2, \tag{1.8}$$

where, m—nanostructure mass, a υ_κ—velocity of nanostructure vibrations. Knowing the nanostructure mass, its specific surface and having identified the surface energy, it is easy to find the velocity of nanostructure vibrations:

$$\upsilon_v = \sqrt{2\varepsilon_v/m} \tag{1.9}$$

If only the nanostructure vibrations are preserved, it can be logically assumed that the amplitude of nanostructure oscillations should not exceed its linear nanosize, that is $\lambda < r$. Then the frequency of nanostructure oscillations can be found as follows:

$$v_v = \upsilon_v/\lambda \tag{1.10}$$

Therefore, the wave number can be calculated and compared with the experimental results obtained from IR spectra.

However with the increasing of nanostructures numbers in medium the action of nanostructures field on the medium is increased by the inductive effect.

The reactivity of nanostructure or the energy of coordination interaction may be represented as:

$$\Sigma \varepsilon_{coord} = \Sigma[(\mu_{ns} \cdot \mu_m)/r^3] \qquad (1.11)$$

and thus, the activity of nanostructure:

$$a = \{[\varepsilon_{vib} + \Sigma[(\mu_{ns} \cdot \mu_m)/r^3]\}/\varepsilon_V \qquad (1.12)$$

When, $\varepsilon_{vib} \rightarrow 0$, the activity of nanostructures is proportional product $\mu_{ns} \cdot \mu_m$, where μ_{ns}—dipole moment of nanostructure, μ_m—dipole moment of medium molecule.

1.2 NANOCOMPOSITES AND NANOMATERIALS. WHAT IS NANO-COMPOSITE AND IT'S UNLIKE NANOMATERIAL NANOSTRUCTURED MATERIALS (MATERIALS MODIFIED BY NANOSTRUCTURES)

At present the precise definition of nanomaterials and nanocomposites creates certain difficulties. Prefix "nano-" assumes small-dimensional characteristics of materials, and the notion "material" has a macroscopic meaning. In accordance with [20], material is a common substance used to produce objects and articles. Unlike substance, material has heterogeneous or heterophase composition. What is nanomaterial? Formally, it is a nanoparticle or aggregation of nanoparticles with properties necessary to produce several articles, for instance, in nanoelectronics or nanomachine–building. It is been assumed that obtained rather active nanoparticles, which can be interconnected in certain succession with the help of UV-laser radiation to produce spatial articles. Naturally, the swarm of nanoparticles in certain surroundings should be compacted accordingly in pulse electrostatic or electromagnetic field to afterwards treat the "phantom" formed with several laser beams successively connecting nanoparticles in accordance with computer software. In this case, the corresponding nanoparticles can actually be called nanomaterial further used to produce articles. Truly, the material formed from nanoparticles will not completely repeat the characteristics of nanoparticles constituting it. The material obtained from nanoparticles is also called [21] nanomaterial or nanocomposite if nanoparticles differ by their nature. Simultaneously, the notion nanocomposite or composition material has a wider meaning, under composites

[20], any macroheterophase materials consisting of two or more heterogeneous components with different physical or mechanical properties. In general case, composition material [22] represents multi-component and multi-phase material formed from the composition hardened when simultaneously obtaining material or article and containing mineral and organic, often polymeric materials, as a rule, with predominance of one of the components, for instance, a mineral one, thus allowing obtaining unique properties of composite formed. The definition of nanocomposites represents similar complexity as the definition of nanomaterials. In the paper by A. L. Ivanovsky [23], it is mentioned that fullerenes and tubules filled with various materials got to be called nanocomposites. At the same time, it was mentioned that effects for filling nanotubes with various substances are also interesting from the point of studying capillar properties of tubules.

Intercalated nanotubes or threads-like bunches of nanotubes, ordered layers of nanotubes in combination with different matrixes are referred to nanotubular composites. The notion "nanocomposite" in classical variant should also contain heterogeneous nanostructures or nanostructures with encapsulated nanoparticles or nanocrystals. At the same time under nanoparticles we understand nanoformations of different character without strict internal order. For gigantic tubules and fullerenes consisting of several spheres or tubes filled with other microphases and with various shapes, the corresponding verbal designations or terms, for instance "onions" or "beads", can appear. In case of formation of extensive structures, such as tubules or nanofibers their mutual interlacing with the formation of nets or braids is possible, between which particles of other components and other phases are located. Here, it should be mentioned that the mixture of phases with different shapes and structures can be referred to nanocomposites, although their composition remains the same. However, when the shape and ordering system change, the properties of nanoparticle change as well. Its surface energy and, consequently, interaction potentials between the particles of different shapes will be different in comparison with corresponding interaction potentials of nanoparticles homogeneous in composition, dimensions and shape. To some extent such conclusions follow from well-known definition of regularities for polymeric, ceramic, and metallic composites. It is known that multi-phase material with unique properties can be created under mechanical-chemical effect (monoaxial pulling from polymer melt), in which crystalline regions with different crystallization degree and amorphous regions will alternate. At the same time, in such material the formation of self-organizing reinforcing phases is possible. The availability of other component in the material, which can also form different nanophases, increases the abilities of nanocomposite formed.

What is nanophase?

By analogy with common definition of the phase it can be said that nanophase is a homogeneous part of nanosystem, which is isolated by physical boundaries on nanometer level from other similar parts. However, the difference in properties of nanophases can be insignificant. This is conditioned by the tendency to decrease total energy of nanosystem.

A considerable number of recent researches are dedicated to nanocrystals. When we extend the notion "cluster" adding particles with nuclei, substances consisting of non metallic components and shells different in energy and composition to this class, there appeared papers [24, 25] where the stability of clusters, nanoclusters to be more exact, is provided by their shape or surface on which these particles are located. Besides, in several cases nanoclusters identify with nanocrystals for which there are no corresponding protective shells, therefore nanoclusters and nanocrystals can be classified based on the type of crystalline structure, composition, sizes, and form of cluster structure, if a certain crystalline structure is preserved. At the same time, it should not be forgotten that cluster is translated as a group or bunch, and with nanoclusters this notion transforms into a group of chemical particles. Now, it is possible to produce clusters with a precisely determined number of atoms. Due to final sizes these small particles have structures and properties differing from "volumetric" characteristics of crystals of bigger sizes. At the same time, properties of nanoclusters change considerably even when one atom is removed from the cluster. It should be mentioned that a cluster can contain aggregates or groups of 2 up to 10^4 atoms. Therefore, aggregates of weakly bound condensed molecules can be presented as clusters. Thus nanocrystals or quasicrystals can be classified based on their sizes. Is there any difference between nanoclusters and nanocrystals? From the aforesaid translation of the word "cluster", it can be seen that a group or bunch of atoms or molecules in this chemical formation differs from the corresponding nanocrystals or crystals by weaker bonds. At the same time, the availability of crystalline lattice in nanocrystals assumes stronger organization of the system and increase in the strength of chemical bonds between the particles constituting nanocrystals. By their composition nanoclusters with a certain order, that can be an element of crystalline structure, can consist of the element of one type, for instance, carbon, metallic, silicon, and so on particles. It is interesting to note that nanoclusters of "noble" metals have fewer atoms in comparison with the clusters formed from more active elements. In Figure 1.1 there are three nanostructures of elements differing by electronegativity and electrochemical potentials.

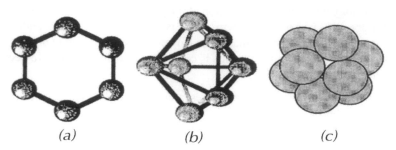

      (a)   *(b)*   *(c)*

FIGURE 1.1 Nanoclusters of elements given according to the decrease of electronegativity of metals: (a) gold, (b) silicon, (c) aluminum.

Based on the increase of their activity in redox processes these elements can be arranged as follows: Au < Si < Al. The least number of atoms is in nanocluster of gold—6 (χ_{Au} = 2.54, by Polling), in nanocluster of silicon—7 atoms ($_{Si}$ = 1.9), and in nanocluster of aluminum—13 atoms ($_{Al}$ = 1.61). Atoms of gold form six-term ring, atoms of silicon—pentagonal pyramid, and nanocluster of aluminum has a shape of icosahedron. However, the least amount of atoms in a nanocluster is also determined, apparently, by the sizes of components and possibilities of their aggregation into certain initial geometrical formations. Let us point out that the structures given in the Figure refer to the third and sixth periods. The activity of elements of the second period is higher therefore the least number of particles in clusters of these elements can be expected to increase. For example, for carbon a six-term ring with π-electrons distributed between the atoms is initially the most stable, and the least number of atoms in carbon clusters is determined as 32 [26–28], that was obtained. However, in fullerenes with the shape of icosahedron five-term rings are also found. Silicon being an analog of carbon in chemical properties can form relatively stable nanostructures from three-term rings, and phosphorus atom can form four-term rings. Clusters are classified by the number of atoms or valence electrons participating in cluster formation. For instance, for atoms of group 1 in the Periodic Table it is pointed out [29] that an even number of valence electrons always participate in their cluster formation, starting from 2 and further 8, 18, 20, and 40. The stability of clusters, on the one hand, is conditioned by the stability of their electron composition, and, on the other hand, by the corresponding external potential of the medium surrounding particles. To increase the stability neutral clusters can be transferred into ionic form. Therefore, clusters and nanocrystals can also be classified by their charge. Apart from neutral clusters, clusters-anions and clusters-cations are known. Such clusters are investigated with the help of photoelectron spectroscopy [30, 31].

For metallic cluster-anions stable electron shells appear at odd number of atoms in a cluster. For example, for Cu_7^- p-orbital of the cluster is completely filled, similar situation of complete filling of d-orbitals is noted for cluster Cu_{17}^-. For clusters-cations the increase in stability at odd number of atoms in a cluster is seen [32]. For instance, spherically asymmetrical particles of single-charged silver cations have the following numbers of atoms 9, 11, 15, and 21. The stability of nanoclusters and nanocrystals increases when the changes decreasing the cluster surface energy are introduced into the electron structure, this can be done when other atoms interacting with atoms-"hosts" are introduced into the cluster, or the protective shell is formed on the cluster*, or the cluster is precipitated on a certain substrate that stabilizes the cluster. Thus, clusters and nanocrystals can be classified by the stabilization method. Formation of clusters and nanocrystals on different surfaces is of great interest for researches. There is the phenomenon of epitaxy was mentioned, when the formation of or change in the structure of clusters proceed under the influence of active centers of surface layer located in certain order, thus contributing to the minimization of surface energy. The introduction of a certain structure of nanocrystals into the substrate surface layer results in the changes of nanoparticle electron structure and energy of corresponding active centers of surface layer providing the decrease of surface energy of the system being formed. As nanocrytsals have a rather active surface, their application to the substrate should proceed under definite conditions that could not considerably change their energy and shape.

However, depending upon the energy accumulated by a nanocrystal (nanocluster), even at identical element composition different results in morphology are achieved. The influence of substrate increases due to the dissipation energy brought by the cluster onto the substrate surface (Figure 1.2, [33]).

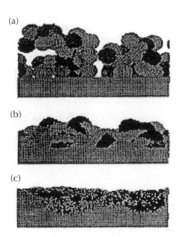

FIGURE 1.2   Morphology of cluster-assembled films created on Mo (100) surface as a function of the cluster energy: (a) 0.1eV/atom; (b) 1eV/atom; (c) 10eV/atom.

As seen from the Figure the elevation of nanoparticle energy from 0.1 eV up to 10 eV results in considerable changes in thin films being formed. The elevation of energy up to 10 eV leads to finer mixing of structures of Mo cluster on molybdenum surface (100). The given type of morphology is observed on many other systems, such as manganese clusters on clean and covered with fullerenes C_{60} silicon surface (111) [34], кластеры железа на поверхности кремния (111) [35], clusters of gold on gold surface (111) [36], as well as clusters of cobalt and nickel on glass surface. In some cases, nanocrystals applied or formed on the surface are joined with the formation of relatively stable dendrite or star-like structures comprising hundreds of atoms. In Figure 1.3, there are pictures of such nanostructures of antimony on graphite surface [37].

FIGURE 1.3 TEM images of islands formed by similar doses of Sb clusters on HOPG at 298K (a), 373K (b).

"Island" structure is produced. The clusters antimony formed contain 2,300 atoms [38]. Such nanostructures are less expressed on amorphous carbon and depend upon the substrate material and number of atoms in the cluster.

When the number of atoms in antimony cluster goes down [39], the nanostructures formed join to a lesser extent and represent "drop-like" formations.

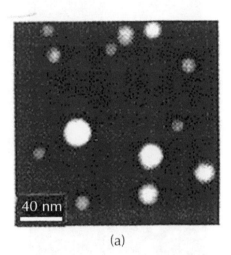

(a)

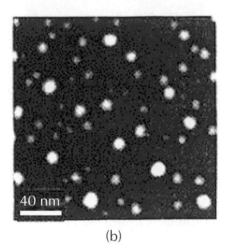

(b)

FIGURE 1.4   *(Continued)*

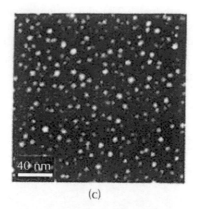

(c)

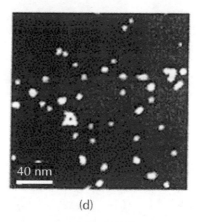

(d)

FIGURE 1.4   STEM images of islands formed by the deposition of different-sized Sb clusters on amorphous carbon (a) Sb4; (b) Sb60; (c) Sb150; (d) Sb2200.

"Island" morphology changes from spherical to the aforesaid shapes for the sizes from 4 up to 2,200 atoms. For transition metals the critical size for the formation of branched nanostructures is much less than for alkaline metals. Thin films from clusters of silver containing less than 200 atoms also have certain joints.

The morphology of structures formed can keep on changing in comparison with the initial one. This is explained, by the analogy with terminology from polymeric chemistry, by Ostwald's maturation when nanoparticles "merge" into larger formations.

The investigations of metallic or shifted nanocrystals are closely connected with the studies on carbon and carbon-containing (metal carbides) nanoclusters (nanocrystals). Therefore, nanocrystals are split into one-, two-, and multi-element structures. The stability of nanocrystals and nanoclusters increases if they comprise two or more elements. Simultaneously, the combination of elements active and passive by redox potentials contributes to the stabilization of multi-element clusters being formed. Clusters containing Co/Cu, (Fe-Co)/Ag, Fe/Au, Co/Pt are known [7]. Several clusters have a stabilizing shell in the form of oxides, carbides or hydrides, for example, Al_2O_3, Sm_2O_3, SiO_2, CO, Met_xC_y, $B_{2n}H_{2n}$ [24, 25, 40, 41].

Thus, clusters and nanocrystals can be classified based on the composition of the shell and nucleus, alongside with their evaluation based on size and shape.

In turn, nanocrystals are classified by the type of "superlattice" [42]. Overview [40] contains the classification of nanoclusters and nanosystems based on their production method. This paper describes the ways for organizing nanostructures from nanoclusters and mentions the new properties appearing during the merge of nanoclusters into a nanosystem.

Classification of carbon clusters (fullerenes and tubules) with metals or their compounds intercalated in them has a special position. The most complete idea was obtained during theoretical and experimental investigations of fullerene C_{60} filled with alkaline metals [29].

The shape of nanocluster or nanocrystal and their sizes are determined by the size and shape of the shell that often consists of ligands of different nature. Here, metal carbonyls are studied best of all, although some factors have to be explained. For instance, the nucleus of cluster Ni_4 is surrounded by the shell of ten ligands CO [43] and the nucleus of cluster Pt_4—only eight ligands CO. This is probably connected with the stabilization of electron shells and transition into a "neutral" cluster.

By their definition nanoreactors are energetically saturated nanosized regions (from 10 nm up to 1 cm), intended for conducting directed chemical processes, in which nanostructures or chemical nanoproducts of certain application are formed. The term "nanoreactor" is applied when considering nanotechnological probe devices. In these devices a nanoreactor is the main unit and is formed between the probe apex and substrate under the directed stream of active chemical particles into the clearance "probe—substrate" [44]. Therefore, nanoreactors are classified by energy and geometrical characteristics, by shape and origin of the matrix in which nanoreactors are formed, possibilities to apply for definite reactions. One of the types of reactions in nanoreactors are topochemical ones proceeding with the participation of solids and localized on the surface of splitting of solid phases of reagent and product. Gaseous and liquid phases can participate in these reactions. Here belong topotaxic polymerization that consists in obtaining macromolecules under low temperatures from monomers located in crystal channels under the radiation and thermal action [45, 46].

Thus, the reactions for producing nanostructures—macromolecules in the channels being nanoreactors were proposed long time ago. The classical example is the topotaxic polymerization of butadiene and its derivatives in the crystals of thiourea and urea. Brown and White [38], as well as Chatani et al [39, 40] studied the polymerization of 2,3-dimethylbutadiene and 2,3-dichlorbutadiene in clathrate complexes of thiourea. The molecules of these substances are arranges in the complexes as shown in Figure 1.5.

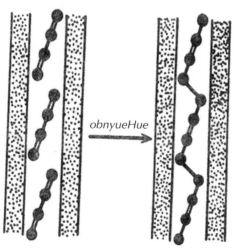

obnyueHue

FIGURE 1.5　Polymerization scheme of dienes in the channel of thiourea molecules of included butadiene (a); Macromolecule of polybutadiene (b).

Molecules of butadiene—flat and fill the rectangle 0.76×0.5nm (Figure 1.5). In a complex crystal the distance between the centers of molecules of butadiene derivatives mentioned equals 0.625 nm. Molecules are arranged obliquely that allows optimizing the interaction. At the same time, inner diameters of the channel (nanoreactor) in thiourea crystals (or clathrate complexes of thiourea and urea) are a little bit bigger than transverse size of butadiene molecule and equal 0.7 nm. The temperature of monomer polymerization reaction under radioactive impact equals approximately -80°C. Since the monomer polymerization process flows in a condensed state, these reactions can be referred to topochemical reactions. However, stereoregulator macromolecules or nanostructures of certain regularity are rarely obtained, in other words, there are only a few topotaxic processes among topochemical reactions. The corresponding orientation of nanoparticles obtained in nanoreactors is determined, based on the results of topotaxic polymerization, by the interaction level of reacting chemical particles with the walls of nanoreactor. Depending upon the value of the interaction energy of chemical particles located inside the nanoreactors with the walls of corresponding containers, it is advisable to separate nanoadsorbers and so-called nanocontainers from nanoreactors. Nanoadsorbers are applied to adsorb different substances, mainly toxic, but nanocontainers—to store certain products. However, the same nanostructures can be nanoadsorbers, nanocontainers, and nanoreactors. For instance, carbon nanotubes are applied to remove dioxins [41]. These nanostructures are used to clean industrial wastes from sulfurous gas [42]. Nanotubes are also applied to adsorb hydrogen as nanocontainers or hydrogen storage [43]. All these containers should have active atoms in their walls to adsorb the corresponding chemical particles.

Nanoadsorbers and nanoreactors are evaluated and classified by the adsorption value that, in turn, means the number of particles (N) or volume (v*) of adsorbed particles absorbed by the unit of adsorption space (w) of nanoadsorber.

Nanoreactors can have cavities dimensionally corresponding to the pores from 0.7 nm up to hundreds of nanometers. The pores are classified by their radii [44]: micropores (under 0.7 nm), supermicropores (from 0.7 up to 1.5 nm), mesopores (from 1.5 up to 100nm), macropores (over 100nm). If we limit the size of the pores by the radius of 1cm, all the pores in porous adsorbers can be considered potential nanoadsorbers and nanoreactors.

Mono- and polymolecular layers are formed in nanoadsorbers with equivalent radius over 1.5nm during the sorption process, thus making the conditions to conduct condensation or chemical transformations. If the equivalent radius is less and when nanoadsorbers are filled completely, the equation by M. M. Dubinin written down in accordance with the theory of volumetric filling of micropores [53] is proposed:

$$a = (w_0/v^*) \exp[-(A/\beta E_0)^n], \qquad (1.13)$$

where, a—value of equilibrium adsorption, w_0—ultimate volume of adsorption space, E_0—characteristic adsorption energy (or interaction energy); v*—volume of adsorbed particles; β—similarity coefficient approximated by the ratio of parachors of the substance adsorbed to the substance taken as a standard (e.g., benzene); n—equation parameter (for the majority of coals equals 2); A—differential work of. Work of adsorption equals

$$A = RT\ln(p_s/p); \qquad (1.14)$$

where, T—adsorption temperature; p_s and p—pressure of the substance saturated steam, approximately, this ratio equals n_s/n, in which n_s—number of particles participating in the interaction, n—equilibrium number of adsorbed particles.

The sense of "a" for nanoadsorbers means an ultimate number of chemical particles adsorbed in nanoadsorber till the equilibrium is reached under the given conditions of the process. The dimensionalities of corresponding parameters will differ from classical Dubinin's equation but the external form of the equation will be preserved.

As mentioned before, materials modified with introduced nanoparticles or substances in which due to the formation conditions self-organizing processes with the formation of certain nanostructures are initiated can be referred to nanocomposites, in this variants for the formation of nanostructures the appearance of nanoparticles in the form of separate fragments of bodies of rotation. Images of bodies of rotation and their separate fragments realized in nanostructures as cones, cylinders, spheres, ellipsoids, slit ellipsoids, slit spheres, hyperboloids, paraboloids, sphere, and elliptical segments and sectors, barrel-like nanostructures. These nanostructures can be described by the formulas of analytical geometry using the corresponding terms.

The diversity of shapes of nanoparticles of one component results in the response reaction of another. If corresponding capillaries are available and caps closing the ends of nanotubes are absent, it is possible to fill them with atoms and ions of another component. An increased activity of one component leads to the dispersion and activity growth of another one with the formation of new nanophases.

Since, a considerable number of interface layers can be found in nanocomposites, the production of composites to be destructed with the formation of corresponding nanophases in certain media at given local pulse impacts, for instance, laser beam can be perspective. Such materials are of interest for their further utilization after the exploitation resource or functional depreciation of articles produced from these composites. The production of nanocomposites of such level is perspective and contributes to the ecological purity of the environment. However, to solve the tasks connected with the problem of environmental pollution by the products of human vital activity it is necessary to change engineering psychology and education of manufacturers of materials, articles, various structures, machines and mechanisms. At present any engineer after getting the task to produce the corresponding article does not think as for the destiny of this product after its life span. This is already resulted in "littering" the Earth and space around it. To clean the environment from the garbage having been produced by the people for many years the task of present and further generations of mankind.

In nanocomposite and nanomaterial terminology the terms taken from adjacent disciplines, such as coordination chemistry, play an important role. Since nanoparticles and nanoclusters are stabilized in particular surroundings, their transition into different medium results in decreasing their stability and changing their "architecture". This increases interactions in the composition consisting of nanostructures (nanoparticles) and can be equated with "re-coordination" reactions. The possibility of such processes can be characterized by corresponding constants of instability or stability of nanoparticles in the medium or composite. If certain nanoparticles obtained in different medium are introduced into the composition, the change in their surface energy results in the particle shape distortion or its destruction with the formation of new bonds and shapes.

When nanoparticles (clusters) are stabilized in the surroundings, such notions as "conformation" and "conformation energy" can be applied. In this case conformation reflects spatial location of separate components of nanosystem. Since any nanoparticles in nanosystems or nanocomposites are highly effective "traps" of energy coming from their habitat, this energy can be conditionally divided into energy consumption connected with reactions of polycoordination or re-coordination (self-organizing energy) and consumption conditioned by orientation processes in the habitat (conformation energy). The increase in the number of nanoparticles can result in the destruction of their shells, changes in the nanoparticle shapes and sizes with the transition to micron level, thus finally leading to coagulation and separation of phases. Let us mention that coagulation means the adhesion of particles of disperse phase in colloid systems and it is conditioned by the system tendency to decrease its free energy. Mainly the process of phase separation in nanocompositions or nanosystems composed of a mixture of particles of different components (nanocomponents) depends upon the "architecture" of nanostructures and physical structure of nanocomposites. In turn, the decrease in concentration of nanoparticles in nanosystem can result in destruction (depending upon the medium activity) or decrease of nanoparticle dimensions, that is, dissolution of nanoparticles in the medium. Since the stability of nanosystems in different media is connected with dissolution processes of physical-chemical nature, such parameters as the degree of nanocomposite filling with nanoparticles, nanocomposite density in general and interface layers in comparison with nanophase density, nanocomposition effective viscosity require great importance. The determination of dissolution phase diagrams is of interest for investigating nanosystems and nanocomposites. In this regard the terminology widely used when studying polymeric systems and compositions seems appropriate.

Why is it possible to use this terminology and notions of polymer chemistry but not chemistry of metals with more advanced study of phase diagrams? It can be explained that nanocomponents are more active due to increasing their surface energy. This, in turn, contributes to the increase of interactions between nanocomponents and their mutual solubility with the formation of new strengthening nanophases.

Such notions as compatibility and operational stability of systems are applied. Interaction parameters (critical on nanophase boundaries, enthalpy, and entropy) are calculated based on experimental data. Upper and lower critical mixing points are determined and phase diagrams with nodes, coexistence curves and spinodals are drawn. Such notions as spinodal decay and critical embryo are used in phase separation [54]. Such notion as "compatibility" defines the ability to form in certain conditions a stable system comprising dispersed (fine) components or phases. In case of nanomaterials or nanocomposites the compatibility is determined by the interaction energy of nanophases and, consequently, the density of interface layer being formed between them. The time period during which the changes in material characteristics are within the values permissible by operational conditions is defined as operational stability. When this time period is exceeded the changes in phase energy are possible, this can result in their destruction, coagulation, and, finally, phase separation and material destruction. The nanocomposite stability and compatibility of nanophases are conditioned by the values of interaction parameters ($\chi^{\text{кр}}$) or forces (energies) of interactions.

Many notions and terms in the field under consideration are taken from chemistry and physics of surface.

1.3 PHYSICS AND CHEMISTRY OF SURFACE: DEFINITIONS AND NOTIONS

1.3.1 The Comparison of Surface Energies of Carbon Nanostructures with Corresponding Energies of Metal or Carbon Nanocomposites

In physics and chemistry of surface of materials the basic notions still raising debates are the following: "surface," "interface layers", "surface", "boundary" layers. Round-table discussions, workshops, and conferences are dedicated to estimating these notions and assessing the rationality of their use.

First, when studying the surface, what surface do we mean? There is purely mathematical notion of the surface as geometrical space of points dividing the phases. A geometrical surface is simplistically shown as a line. However, everyone understands that this is an abstract conception and a surface cannot be presented as a line. Irregularities are always observed on the boundary of gas and liquid separation, liquid and solid separation, gas and solid separation caused by energy fluctuations. Physics of surface indicates that chemical particles on the surface are, on the one hand, in the action field of the particles of solid or liquid, and, on the other hand, in the action field of molecules of gas or liquid being in the contact with this body. Therefore, it is better to estimate properties, structure and composition of surface layers, the dimensions of which change in thickness from 1 nm up to 10 nm depending upon the nature of material being investigated (conductor or dielectric) and depth of surface influence on material inner layers. The relief of surface layer or morphology of material surface is determined by particular features of its formation and nature (composition and structure) of material.

Energy characteristics of surface of solids are usually estimated by curved angles of wetting. In turn, in authors' opinion [55], molecules from the layer about 1nm thick (from the side of solid or liquid phase volume) contribute to the surface energy, and then the influence of deeper layers decreases. Changes in the surface layer energy are determined by its chemical composition and structure, as well as the aggregate of chemical particles surrounding this surface layer and belonging to material and medium in which this material is placed. We are going to consider a surface layer of any body as a boundary separating the body and surroundings. At the same time, surface energy always tends to get balanced with the energy of surroundings (liquid or gas medium).

The notion "interface layer" has a wider interpretation and can refer to "surface layer", and "boundary layer". The notion "surface layer" is used to determine the boundaries: gas–solid, liquid–gas; liquid–solid. The notion "boundary layer" is used when phase boundaries in a solid, suspension and emulsion are considered.

However, interface layers are more often considered in complex compositions or composites containing a big number of components. The notion "boundary layer" is more often used for multilayer materials with a clearly lamellar structure, or for coatings, materials of solids and liquids located on a certain substrate. At the same time "boundary layers" can define the boundaries between phases in solid materials. Naturally, it is difficult to imagine quick jumps when transferring from one phase to another. As was already mentioned, surface energy tends to get balanced with chemical particles surrounding it. Here the notion "surface energy" will be used both for surface and interface and boundary layers of materials.

When investigating polymer films on different substrates (metal substrates), it is mentioned [56] that intermolecular interactions on the boundary of phase separation results in the appearance of structural heterogeneity on molecular and permolecular levels in polymer films. Simultaneously, the structure defective and heterogeneous by its film thickness is formed. The influence of nature and structure of substrate spreads to over 400cm. The more intermolecular interactions in a polymer or on a substrate boundary are revealed, the greater distance the substrate influences to film structure changes [57].

The surface characteristics determine many practicable macroscopic properties of materials. Such characteristics are revealed in the surface chemical and physical structure and morphology, that is, are immediately connected with surface chemical functional composition, crystallinity degree, shape, and roughness. Chemical composition of material surface represents complex characteristics determining their properties and reactivity expressed via the rate of chemical interaction of the surface of material being considered with other adjoining materials or media. Under the material surface chemical composition we understand its chemical composition and availability of molecule fragments or separate atoms on the surface that enhance the surface activity or, on the contrary, passivate the surface. The surface of solids always contains the layers of gases physically and chemically sorbed [53]. The layer of adsorbed gases or impurities (usually hydrocarbonaceous) is about several monolayers thick. The layer of chemisorbed molecules can be found under the layer of molecules physically adsorbed. The thickness of such layer depends upon chemical activity of centers on the material surface. The material surface layer can be represented as adsorbed particles and the layer of surface atoms and molecules of material itself. Not only chemical composition but also surface geometry influence gas and impurity adsorption. It is possible to present the models reflecting the influence of surface geometry and various roughnesses of the surface in a simple way, when the active center or heteroatom is surrounded by different numbers of surface layer atoms (Figure 1.6).

a   *b*   *c*

FIGURE 1.6   Model of various surface roughness (a–c heteroatom × is surrounded by different numbers of other atoms).

At the same time, on ledges the activity of heteroatoms on roughnesses increases in comparison with other surrounding atoms. The activity of neighboring atoms with chemical particles of environment increases. The surface peculiarities result in local elevations of surface energy or surface potential and influence the formation of adsorbed layers and their thickness on the surface.

It is known that [53] on the surface of monocrystals active centers are located on crystallographic steps or at the points where dislocations intersect the surface. Then an adsorbed particle can interact with several lattice atoms, as a result, the total interaction increases. For instance, the "adhesion" of oxygen molecule to silicon stepped surface is 500 times more probable than to the smooth surface.

Surface layers can also be found in the volume of material with defective regions such as micro- and macro-cracks, pores. Here it is possible that surface layers in one and the same material have different chemical composition. For instance, for cellular plastics chemical composition of such layers is practically the same (surfaces of material and pore), since pores are channels connected with material surface. For foam plastics, in which there are basically closed pores not connected with material surface, the surfaces of material and pore differ in chemical composition since gaseous medium in foam plastics bubble is considerably different from the gaseous medium surrounding the material.

As the majority of permolecular structures in polymers can be referred to nanostructures being investigated and polymeric materials to nanocomposites, in our opinion, several notions from the science of polymers can be referred to chemical physics of nanostructures. The main stage of polymer and nanoparticle formation is the formation of closed shape or their surface determining the existence area of polymeric molecule or nanoparticle. The particle surface formed in non-equilibrium conditions tends to transfer into the state of thermodynamic equilibrium. The rates of relaxation processes mainly depend on the surroundings ("habitat"), temperature and nature of the particle or polymeric material. However, we can distinguish two most important factors influencing the formation of properties and structure (chemical and physical) of the surface of polymers and nanoparticles—mobility of constituent groups of atoms and molecules (taking into account the conformation energy) and surface energy. In equilibrium conditions the surface energy is usually minimal and this minimum is provided by the mobility of chemical particles migrating from the volume to the surface and vice versa or adsorption and hemisorption of chemical particles from the environment onto the surface. This takes place even when "artificial pairs" are formed, for instance, polystyrene covered with polyethylene-oxide. The surface energy of polystyrene is lower the corresponding energy of polyethylene-oxide ($\gamma_{ps} = \gamma_{peo}$) [58]. Therefore, when this material is kept under the temperature above the vitrification temperature of both polymers, the outer layer is enriched with polystyrene. Polystyrene nanoformations in polyethylene-oxide film are 5nm thick. Depending upon the substrate surface energy that can be connected with its polarity, the chemical composition and structure of surface and boundary layers of static and graft copolymers change as well, for example, depending upon substrate material the polarity of boundary layers of vinylchloride and vinylacetate copolymers changes due to the decreased number of polar groups in boundary region copolymer-substrate. The changes in surface energies γ of copolymer and some materials of substrates are given below:

Substrate material	AuNi Al **PTFE***		
Surface energy, γ, mN/m of			
substrate material	4337.3319		
copolymer	5148.4638		

*PTFE—polytetrafluorine ethylene film (sheet)

Based on X-ray photoelectron spectroscopy (XPS) the greatest number of polar acetate groups are present in the layer boundary to substrate of gold that is characterized by the highest surface energy. Depending upon the formation conditions and substrate types the surface energy of polymers changes as well. The change in the surface energy (γ, mN/m) of some copolymers in comparison with polyethylene depending upon their formation conditions are given in Table 1.2.

TABLE 1.2   Values of surface energy for sheets and films obtained on the boundary of different media [58].

Polymer	Surface energy, γ, mN/m		
	Sheets of polyester film obtained by hot pressing	Sheets of PTFE obtained by die casting	Film of PTFE obtained by bulge formation
Non-polar polymers: Low-density polyethylene	32	31 (32)	32
Copolymer of vinylacetate ethylene (86:14)	33	32 (33)	33
Polar polymers:			
Statistic copolymer of ethylene and methacrylic acid (85:15)	44	37	38
Graft copolymer of maleic anhydride (2.1%) to high-density polyethylene	50	33 (33)	-
Graft copolymer of vinyltrime toxisilane (1.1%) to co-polymer of ethylene and vinylacetate (72:28)	39	36	33

Values of surface energy during the mold water-cooling are given in brackets

The results given above were obtained with the help of X-ray photoelectron spectroscopy (XPS) methods and infrared microscopy of multiply disturbed complete inner reflection (IR MDCIR). The essence of investigation methods is given below together with currently widely known nanoscale methods for examining nanostructures, nanosystems, and nanocomposites. To determine the morphology of surface and shape of nanoparticles and nanosystems, various types of electron microscopy, such as scanning, transmission, tunnel, and atom force are widely applied. Diffractometric and spectroscopic methods are used to determine the structure and composition. Table 1.3 contains methods for investigating the surface of materials, boundary layers and nanostructures.

TABLE 1.3   Methods for investigating surface and boundary layers, nanostructures

Method	Information	Depth	Sensitivity, profiling, nm %
X-ray photoelectron spectroscopy (XPS)	Element composition Chemical surroundings Conformation analysis	<10	10^{-1}
Auger-electron spectroscopy (AES)	Element composition	<5	10^{-1}
Ultraviolet electron spectroscopy (UVES)	Chemical surroundings Conformation analysis	~1	10^{-2}
Spectroscopy of ionic dissipation (SED)	Element analysis	<1	10^{-2}
Secondary ionic mass-spectrometry (SIMS)	Element analysis	<1	10^{-3}
Laser microprobe mass-analysis (LMMA)	Element analysis	10	3×10^{-7}
Infrared microscopy of multiply disturbed complete inner reflection (IR MDCIR)	Chemical surroundings Conformation analysis	>1,000	monolayer portions
Spectroscopy of combination dissipation (SCD)	Chemical surroundings	~1000	monolayer portions
Raman spectroscopy (RS)	Conformation analysis Nanostructural analysis	~100	
Atom force microscopy (AFM)	Surface morphology and polarity	<100	monolayer portions

TABLE 1.3 *(Continued)*

Scanning raster electron microscopy (SREM)	Surface morphology	<10	10^{-1}
Transmission electron microscopy with electron microdiffraction (TEMEMD)	Chemical composition and structure	~1	10^{-1}
	Surface morphology		
Tunnel electron microscopy (TEM)	Surface morphology	0.1	10^{-2}

1.4 BASIC NOTIONS IN MODELING NANOSTRUCTURES AND METAL OR CARBON NANOCOMPOSITES

To determine the possibility of nanostructure existence and forecasting the formation of different nanosystems it is necessary to develop the corresponding theoretical and computation apparatus. The creation of such an apparatus is possible based on corresponding ideas of quantum chemistry, molecular mechanics and thermodynamics. Apparently, in developing the overall theory of nanosystems the symbiosis of similar theoretical trends in physics, chemistry, biology, and computer modeling will be found.

The processes of nanostructure formation, the subject investigated by chemical physics, are of great interest. Therefore, apparently, it is more appropriate to discuss the creation of computational apparatus of chemical physics of nanostructure formation processes. In recent years, a lot of investigations connected with the development of different software products have been carried out in quantum chemistry, molecular dynamics and molecular mechanics of cluster and nanosystems. These investigations are fulfilled using semi-empirical and *ab initio* methods. Basically, "paired" interactions were used in all the methods, although there are works discussing "collective" interactions on atomic and molecular levels.

It is advisable to arrange a hierarchical scheme for predicting nanostructure formation. First, the stability of interacting particles and reaction centers should be determined. Under reaction center we understand the group of atoms being changed during the reaction. Potential reactivity centers, that is, groups of atoms able to participate in reactions, are usually called functional groups. This terminology can mainly be found in organic and polymer chemistry. The stability of particles and interaction energy are evaluated with methods of quantum chemistry. After reactivity centers are selected possible reactions with their participation are assumed, this operation for finding reaction options is similar to the isolation of so-called "reaction series" in physical organic chemistry. Here, under reaction series we understand the directivity of processes in reactivity centers or similar nanosystems with the formation of nanoparticles of certain structure and composition. The advantage of one or another reaction series when compared with the others is determined based on energy consumption for performing successive acts of the process and the rate with which these acts are performed. Computational experiment can be carried out using apparatuses of quantum chemistry and molecular dynamics, in some cases in combination with semi-empirical methods of thermodynamics in the frameworks of activated complex theory or spatial-energy concept. At previous two stages of nanosystem modeling and formation single chemical particles, reaction centers or fragments of complex nanocomponents of the systems were discussed. The changes of nanostructures during the action of various fields, such as thermal, electrical, magnetic, field of particles, gravitation upon nanosystems, and nanoparticles are also determined at the aforesaid stages. Therefore, at third and the following stages collective interactions of nanoparticles with the help of apparatuses of molecular mechanics and thermodynamics are discussed. The transitions from one method of computational experiment to another one represent certain terminological and conceptual difficulties. This can be explained not only by the time frames when the corresponding notions, ideas and definitions appeared. In our opinion, the complete positive material accumulated in reactivity theory and reactions of chemical particles with the formation of nanosystems should be used in a new nanoscience. Therefore, modular construction should be introduced into the scheme of basic notions of nanosystem and nanostructure modeling together with hierarchical structure (Figure 1.7).

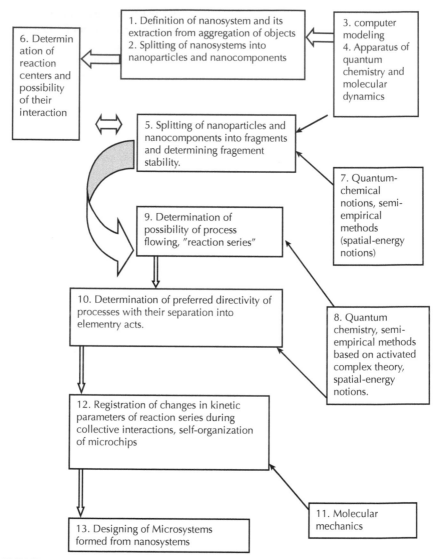

FIGURE 1.7   Scheme of nanosystem consideration.

Assumptions used for computational experiments, especially for transition from one theoretical and computational apparatus are discussed in Part 2.

Therefore, the transition to any trend in thermodynamics—nanothermodynamics—appears natural. The approaches can only be by developing Prigozhin's theorem [60] and synergetics theory. The key term here is entropy production.

The entropy production means the rate of entropy change during the process and is usually marked as $\sigma = ds/d\tau$. At the same time any system (dynamic and isolated) has certain accumulated information being in corresponding structure. If the system is stationary, the total of amount of entropies S and information I is constant [61], equaled to one. It is more correct to consider the changes of these values (ΔS и ΔI). Therefore, it can be written as follows: $\Delta I + \Delta S = 1$. At the same time, when structuring and self-organizing the substances: $\Delta I \to 1$, $\Delta S \to 0$. Actually ΔI mean the change in the record of "information" in the structure determined by structuring energy that consists of the energy of genetic and structural information, according to [61].

Here under genetic information, we understand the energy input of "embryo", substrate or walls of "nanocontainer" into structuring process that proceed with "inheriting" several features of the aforesaid information sources.

In turn, under the energy of structural information, we understand the energy input necessary to form a certain rank of structural formation in accordance with structural hierarchy. In this case, if the dynamic dissipative structure decays, $\Delta I \to 0$, that is, the substance structure decays, for example, into simple components, then $\Delta S \to 1$. The change in entropy grows together with the increase in the number of equiprobable states. The energy dissipates into the environment. The substance self-organizing process is possible, based on Prigozhin's theory, due to the action of energy flow onto the substance. The flow energy is consumed for structuring, but the flow itself is disordered and when the system is removed from equilibrium the system loses stability in flow separation point. Then Prigozhin's theorem can be expressed as follows: under external conditions preventing the system to reach stability, the stationary state corresponds to the production of minimal amount of entropy and maximum amount of information.

Any multi-atom nanoformations with repetitive fragments refer to so-called "permolecules". Academician Rusanov, by analogy with organic compounds, determines the aggregation of such permolecules of one nature and increasing size as homologies, and nanoparticles of one size but different composition as isomers.

Chemical potential is the characteristics of permolecules as well as of the majority of such molecules and their derivatives:

$$\mu_{\{n\}} = G^0_{\{n\}} + kT\ln(C_{\{n\}}\lambda_{\{n\}}^3), \qquad (1.15)$$

Where, μ—chemical potential, $\{n\}$—aggregate of structural units n_i of permolecule, $G^0_{\{n\}}$—Gibbs energy of permolecule taking into account the interactions of permolecule with surroundings and other permolecules present in the system, k—Boltzmann constant, T—temperature, $C_{\{n\}}$—concentration n_i of permolecule in the system, $\lambda_{\{n\}}$—average de Broil wavelength of permolecule [62].

By its sense the value $C_{\{n\}}\lambda_{\{n\}}{}^3$ corresponds to the activity of permolecule directionally influencing the particles surrounding it in space (under spherical wave distribution).

The parameter presented resembles values characterizing the transformation share of one substance state into another in polymers (Zilberberger's equation and its derivatives) [59, 63].

Further Rusanov explains the value $G^0{}_{\{n\}}$ as a total of energies of state, when macromolecule structural units were under vacuum in disconnected state with fixed mass center ($g^0{}_i n_i$), transfer energy of separate structural unit with energy $g^0{}_i$ into permolecule ($w^\lambda{}_i n_i$) and transfer energy of such structural unit onto the permolecule surface ($w^{\lambda\sigma}{}_i b_i n_i{}^{2/3}$). The value $b_i n_i{}^{2/3}$ determines the number of structural units of i-grade on the permolecule surface, and b_i depends upon the permolecule chemical composition and geometry. Values w_i^λ and $w_i^{\lambda\sigma}$ depend upon $\{n\}$ and permolecule size. In turn, $w_i^\lambda < 0$, and $w_i^{\lambda\sigma} > 0$ from general considerations connected with the formation of bonds in the first case and destruction of bonds in the second. If the processes are controlled by chemical affinity (A), then for the process in which from aggregation of $v_{\{n\}}$ moles of one substance we have $v_{\{n\}}'$ moles of another substances we can write down as follows:

$$A = \sum \left(v_{\{n\}\{n\}} - v_{\{n\}}'\mu_{\{n\}}' \right) \tag{1.16}$$

where, μ and μ'—chemical potentials of substances transformed one into another. With proximity or equality of $G_{\{n\}}{}^0$ and $G_{\{n\}}{}^{0'}$, the value of chemical affinity in the process will be determined by the transformation degree of one substance into another. If average de Broil waves of permolecules are close, A is determined only by the correlation of concentrations or equilibrium constant.

As the notions connected with the surface phenomena are closely linked with model concepts based upon the changes in surface energy corresponding to the interaction energy, it is advisable to use schemes in the form of "grid" or "piece" models [60, 22]. In these models a certain order of locating chemical particles, including those of nanodimensions, are specified. In these cases different methods, including electromodeling ones (analog computational complexes), methods from physics of polymers, for instance, Flori–Haggins, Guggengeim, and so on can be considered.

Some definitions from this information area are specified below.

For nanosystems and processes with their participation the computational apparatus based on quantum chemistry or thermodynamic notions can result in inadequate picture of actual processes and its application seems doubtful from the reliability point of results obtained. Unfortunately, a reliable transition from quantum-chemical calculations for picosystems to thermodynamic for macrosystems appears currently unavailable. A unified computational apparatus that can be applied to systems of any size is only being developed as a common network of transformations of picosystems into macrosystems through nanosystems taking into account the transition ways conditioned by the directedness of interactions of system components.

Here, we can use the additivity principle to characterize the system energy, and such notions as "free energy" and "activation energy" to evaluate the energies of system interactions and activity.

Usually the notion "total energy" that is written down in accordance with the law of energy conservation as the total of kinetic and potential energies is used to characterize any system:

$$E = T + U \qquad (1.17)$$

where, **E**—total energy, **T**—kinetic energy, **U**—potential energy.

Similar record is applied when determining Hamiltonian (Hamiltonian function in classical and relativistic mechanics), that is, Hamiltonian function (H) corresponds to:

$$H = T + U \qquad (1.18)$$

Shredinger's equation representing wave equation and reflecting the energy of electrons in the system is written down as follows:

$$H\varphi = E\varphi \text{ или} - (h^2/8\pi^2 m)(\partial^2\varphi/\partial x^2 + \partial^2\varphi/\partial y^2 + \partial^2\varphi/\partial z^2) + U\varphi = E\varphi \qquad (1.19)$$

where, h—Plank's constant, m—electron mass, φ—wave function.

Mainly, the first member of the left-hand part of the equation represents kinetic energy corresponding to the changes in electron density in Cartesian coordinates. The form **Uφ** reflects the potential energy determined via the energy of attraction and repulsion in electron system.

Total energy of the system is found also via the sum of surface and volume energies:

$$\varepsilon_f = \varepsilon_s + \varepsilon_v, \qquad (1.20)$$

where, ε_s—surface energy responsible for the system interaction with the surrounding systems, ε_v—volume energy determining the interactions between the particles constituting the system.

However, if we know the system total energy it is still impossible to evaluate its activity or reactivity and stability. The activity of the system or separate particle is determined by the energy of its interaction with surrounding particles or systems of particles. This energy can be equated with surface or kinetic energy. Note that kinetic energy comprises the energy of progressive, rotary and oscillatory motions. For surface energy of a separate particle such motions are determined by energy fluctuations on the particle surface, changes in electron density in surface layer and particle ionization. When chemical particles or systems are excited, the part of potential energy transfers into kinetic one, this, in turn, results in instability of the aforesaid systems. Therefore, to evaluate the activity of particles and separate fragments of the systems, it is advisable to use the difference between surface and volume energies or kinetic and potential energies:

$$\Delta\varepsilon = \varepsilon_s - \varepsilon_v \text{ or } \mathbf{L} = \mathbf{T} - \mathbf{U} \tag{1.21}$$

where, $\mathbf{L}$—Lagrange function.

The more is the difference between the energies, the more is the activity and reactivity of chemical particles. In the language of relativistic mechanics $\Delta\varepsilon$ and $\mathbf{L}$ are similar and make up the part of energy (free energy) that is consumed for possible interactions with the formation of new products. Since the system kinetic energy decreases during the interaction of particles, the system potential energy goes up; this corresponds to diminishing of Lagrange function. Due to interrelationship of Hamiltonian and Lagrange functions, in relativistic mechanics Lagrange functions can be presented as follows:

$$\mathbf{L}\psi = -\, h^2/8\pi^2 m(\partial^2\psi/\partial x^2 + \partial^2\psi/\partial y^2 + \partial^2\psi/\partial z^2) - \mathbf{U}\psi \tag{1.22}$$

or

$$\mathbf{U}\psi = 1/2\,(\mathbf{H}\psi - \mathbf{L}\psi) \tag{1.23}$$

where, ψ—characterizes the electron density during the formation of molecular orbitals (at bond formation). During the transition to nanosystems and macrosystems Lagrange function basically represents free energy similar to Helmholtz free energy. For the process with constant volume and temperature Lagrangian $\mathbf{L}$—can be compared with isochoric–isothermal potential:

$$\mathbf{L} = \Delta\mathbf{T} - \Delta\mathbf{U} \text{ and } \Delta F = \Delta U - T\,\Delta S, \tag{1.24}$$

where, ΔS—change in entropy, which reflects the changes in progressive, rotary, and oscillatory motions of particles and electrons that corresponds to the changes in kinetic energy, that is, $T = T\Delta S$, T—temperature (heat or other impulse). In other words, in certain cases ΔF can be equaled with L. In this event, during the process for forming new bonds and particles only the changes in forms of particles movement and redistribution of electron densities in the surroundings of groups and separate atoms are taken into account.

Nobel laureate I. Prigozhin proposed to use Lagrange undetermined multipliers to estimate the conditions for maximum reaction output [60]. These multipliers or functions are used when applying the grid model of solution formed by molecules of two or more grades [61] of polymers.

Thus, there is a possibility to evaluate the processes of nanoparticles and nanosystems formation from picosystems, microparticles, and microsystems—from corresponding nanostructures, and, finally, macroparticles and macrosystems—from organized microsystems using Lagrange functions. Unlike quantum-chemical calculations based on the analysis of all electron states, that is, total system energy, when it is necessary to select field potentials for atomic shells to simplify the calculations, the application of Lagrangian reduces the number of interacting electron clouds to the minimum when the processes favorable from energy point take place.

For separate atoms the surface or kinetic energy can be determined via the ionization energy, and volume or potential energy is reflected via the energy of Coulomb attraction of electron clouds to the nucleus. Processes responsible for these energies have different directedness, but they are interrelated. Therefore, the notion of adjusted energy can be introduced. By analogy with adjusted mass it can be expressed as follows:

$$\varepsilon_r = \varepsilon_s \varepsilon_v / (\varepsilon_s + \varepsilon_v), \tag{1.25}$$

where, ε_r—adjusted energy that, when $\varepsilon_s = E_i$, and $\varepsilon_v = q^2/r_i$, is identical to spatial-energy parameter P_i, given in [64].

$$\text{Then } \varepsilon_r = P_i = (q^2/r_i) \times E_i / (q^2/r_i + E_i), \tag{1.26}$$

where q—nucleus effective charge by Clementi, E_i—ionization energy, r_i—orbital radius equal to the distance from atom nucleus to maximum radial electron density. In [64] an alternative formula for characterizing atoms is proposed:

$$P_a = q^2 W_i n_i / (q^2 + W_i n_i r_i), \tag{1.27}$$

where, $q = Z^*/n^*$, Z^*—nucleus effective charge, n^*—effective main quantum number, W_i—orbital energy of electrons, n_i—number of electrons on i-orbital. Corresponding parameters for atoms were tabulated and agree with experimental data by electron densities, as well as with semi-empirical parameters of atoms (Table 1.4).

TABLE 1.4   PE—parameters of atoms and energy of valence electrons (U) for some atoms of metals used when obtaining nanostructures.

Atom	Valence electrons	q2, eVÅ	ri, Å	Wi, eV	ni, E	Pa, eV	U, eV
V	4s2	22.330	1.401	7.50	2	7.730	7.680– 8.151
Cr	4s2	23.712	1.453	7.00	2	7.754	7.013– 7.440
Mn	4s2	25.120	1.278	6.60 (t)	2	7.895	10.890
				7.50	2	10.87	
Fe	4s2	26.570	1.227	8.00	2	9.201	8.598
				7.20 (t)	2	8.647	
Co	4s2	27.980	1.181	8.00	2	10.062– 9.187	9.255– 10.127
				7.50 (t)	2		
Ni	4s2	29.348	1.139	9.00	2	10.60	9.183
				7.70 (t)	2	9.64	
Cu	4s2	30.717	1.191	7.70	2	9.639	9.530
In	5s2	238.30	1.093	11.70	2	21.80	21.30
	4d10	257.23	0.4803	20.00	10	145.80	155.00
Te	5p4	67.28	1.063	9.80	4	24.54	23.59
	5s2	90.577	0.920	19.00	2	27.41	26.54

Bond energy of electrons (W_i) are obtained by: (t)—method Hartree–Fock, the rest—based on optical measurements.

As it is seen from Table 1.4, the values of P_a and U are similar (average deviation does not exceed 2%). By analogy there is a direct bond of corresponding values of spatial-energy parameters with electron density at distance r_i from atom nucleus.

The application of Lagrange functions and P-parameters and notions connected with them will be discussed in Chapter 2.

KEYWORDS

- **Auger-electron spectroscopy**
- **Cartesian coordinates**
- **Laser microprobe mass-analysis**
- **Nanostructures architecture**
- **Plank's constant**
- **Raman spectroscopy**

REFERENCES

1. Kodolov, V. I., & Khokhriakov, N. V. (2009) Chemical physics of the processes of formation and transformation of nanostructures and nanosystems (2 volumes: Vol. 1, p. 360, Vol. 2, p. 415). Izhevsk: Izhevsk State Agricultural Academy.
2. Zigban, K., Nordling, K., Fal'man, A. et al (1971). In Mir M (ed.), Electron Spectroscopy (p. 453). USSR: Elsevier.
3. Aleskovsky, V. B. (1990) Course of permolecular Compounds Chemistry (p. 321). Louisiana: LSU.
4. Melikhov I.V. & Bozhevolnov V.E. (2003) Variability and self-organization in nanosystems. Journal of Nanoparticle Research, 5, 465–472.
5. Binns, C. (2001) Nanoclusters deposited on surface. Surface Science Reports, 44, 1–49.
6. Kodolov, V. I. & Spassky, S. S. (1976) On parameters in Alfrey-Price and Taft equations Vysokomol Soed 18A(9), 1986–1992.
7. Yushkin, N. P. Hydrocarbon crystals as protoorganisms and biological systems predecessors. SPTE Conference on Instruments, Methods, and Missions for Astrobiology, 1998, V. 3441.
8. Malinetsky, G. G. (2010). Designing of the Future and Modernization in Russia (Preprint of M.V. Keldysh Institute of Applied Mechanics), 41, 32.
9. Tretyakov & Yu. D (2003). Self-Organization Processes (Uspekhi Chimii), 72(8), 731–764.
10. Kodolov, V. I., Khokhriakov, N. V., Trineeva, V. V., & Blagodatskikh, I. I. (2008). Activity of nanostructures and its representation in nanoreactors of polymeric matrixes and active media. Chemical physics and mesoscopy, 10(4), 448–460.

11. Kodolov, V. I. & Trineeva, V. V. (2011). Perspectives of Nanosystems self organization development. *Chemical Physics and Mesoscopy, 13*(3), 363–375.

12. Smalley, R. E. & Cole, R. (1995). Initiatives in Nanotechnology. http//pcheml.rice. edu/nanoinit.html. Date of reference February 1997.

13. Pomogailo, A. D., Rozenberg, A. S., & Uflyand, I. E. (2000). *Metal nanoparticles in polymers* (p. 672). Moscow: Himiya .

14. Gusev, A. I. (2005). *Nanomaterials, Nanostructures, Nanotechnologies* (p. 412). Moscow: Fizmatlit.

15. Suzdalev, I. P. (2005). *Nanotechnology. Physico-Chemistry of Nanoclusters, Nanostructures and Nanomaterials* (p. 589). Moscow: KomKniga.

16. Furstner, A. (Ed.) (1996) Active Metals (p. 233). Weinheim: VCH.,

17. Kodolov, V.I., Khokhriakov, N.V., Kuznetsov, A.P. et al. (2007). Perspectives of the application of nanostructures and nanosystems when producing composites with prognosticated behavior *Space Challenges in XXI Century. Novel Materials and Technologies for Space Rockets and Space Development* (Vol. 3, pp. 201–205). Moscow: Torus press.

18. Krutikov, V. A., Didik, A. A., Yakovlev, G. I. et al. (2005). Composite material with nanoreinforcement. *Alternative Energetic and Ecology* (ISJAEE), *4*(24). 36–41.

19. Kodolov, V. I., Khokhriakov, N. V., & Kuznetsov, A. P. (2006). To the question of mechanism of nanostructure influence on structurally changing media when forming "intelligent" composites. *Nanotechnics, 3*(7) 27–35.

20. Smith, R. H. & Hanger, J. (1997) Scott II. Molecular Nanotechnology: Research Funding Sources News New Source for Funding/ European Com. Future and Emerging Technologies/http://www.cordis.lu/ist/fetnid.html. Date of reference February 1997.

21. Tyukaev, V. N. (1983). Composite materials *Chemical Encyclopedia Vocabulary* (pp.270–271). Moscow: Sov. Enc.

22. Kodolov, V. I., Trineeva, V. V., Kovyazina, O. A., & Vasil'chenko, Yu. M. (2012). Production and application of metal/carbon nanocomposites. *The Problems of Nanochemistry for the Creation of New Materials* (pp. 17–22). Torun: IRPMD.

23. Kodolov, V. I. (1992) *Polymeric Composites and the Technology of Rocket-Propelled Vehicle Engine Production* (p. 270). Izhevsk: IMI.

24. Ivanovsky, A. L. (1999). Quantum chemistry in material engineering. *Nanotubules Forms of Substances* (p 173). Ekaterineburg: ICSB UD RAS.

25. Proceedings of Conference. NATO-ASI "Synthesis, functional properties and applications". July-August, 2002.

26. Eberhardt, W (2002). Clusters as new materials. *Surface Science, 500*(1–3), 242–270.

27. Kroto, H. W., Heath, J. R., O'Brien, S. C. et al (1985). C_{60}: *Buckminsterfullerene. Nature, 318,* 162.

28. Krätschmer, W., Lamb L.D., Fostiropoulos, K., & Huffman, D.R. (1990). *Solid C_{60}: a new form of carbon Nature, 347,* 354.

29. Bonard, J. M., Forro, L. et al (1998). Physics and chemistry of nanostructures. *European Chemistry Chronicle, 3,* 9–16.

30. Knight, W. et al. (1984) Electronic shell structure and abundances of sodium clusters. *Physical Review Letters, 52,* 2141.

31. Cheshnovsky, O. et al. (1990) Ultraviolet photoelectron spectra of mass selected cooper clusters: evolution of the 3d band *Physical Review Letters, 64,* 1785.

32. Gantefőr, G. et al (1995). Electronic and geometric structure of small mass selected clusters. *Journal of Electron Spectroscopy 76*, 37.
33. Tiggesbäumber, J. et al (1992). Giant resonances in silver-cluster photo-fragmentation. *Chemical Physics Letters, 42*, 190.
34. Haberland, H., Insepov, Z., & Moseler, M. (1995) Molecular-dynamics simulation of thin-film growth by energetic cluster impact. *Physical Review Letters, 51*, 11061.
35. Upward, M. D., Moriarty, P., Beton, P. H. et al (1997). Measurement and manipulation of Mn clusters on clean and fullerene terminated Si (111) 7×7. *Applied Physics Letters, 7*, 2114.
36. Upward, M. D., Cotier, B. N., Moriarty, P. et al (2000). Deposition of Fe clusters on Si surfaces. *Journal of Vacuum Science and Technology, 18*, 2646–2649.
37. Bardotti, L.,Prevel, B., Melinon, P. et al (2000). Deposition of Fe clusters on Au (111) surfaces. II Experimental results and comparison with simulations. *Physical Review Letters, 62*, 2835.
38. Bardotti, L., Jensen, P., Hoareau, A. et al. (1995). Experimental observation of fast diffusion of large Antimony clusters on Graphite surfaces. *Physical Review Letters, 74*, 4694–4697.
39. Yoon, B., Akulin, V. M., CahuzacPh, et al (1999). Morphology control of the supported islands grown from soft-landed clusters. *Surface Science, 443*(1–2), 76–88.
40. Suzdalev, I. P. & Suzdalev, P.I. (2001). Nanoclusters and nanoclusters systems. Organization, interaction, properties. *USP Chemistry, 70*(3), 203–240.
41. Polukhin, V. A. & Potemkina, E. V. (2000). Structure and stability of Silicon Si$_n$ (8<n<60) nanoclusters under the heating and during the thermal destruction MD experiment. *Izvestiya Chelyab Science C, 2*, 17–22.
42. Pileni, M. P. (2003). Nanocrystals: Fabrication, organization and collective properties. *C. R. Chimie, 6*(8–10), 965–978.
43. Vajda, S., Leisner, T., Wolf, S., & Woste, L. (1999). Reactions of size-selected metal cluster ions. *Philosophical Magazine, 79*, 1353.
44. Hintz, P. A. & Ervin, K. M (1994). Nickel group cluster anion reactions with carbon monoxide rate coefficients and chemisorption efficiency. *The Journal of Chemical Physics, 100*, 5715.
45. Nevolin, V. K. (1988). Nanotechnology Electron Technics. *Microelectronics Series 3, 4*(128), 81.
46. Wunderlikh, B. (1979). In M. Mir (ed.), *Physics of macromolecules* (Vol. 2, p. 574).
47. Brown, J. F. Jr. & White D. M. (1960). *Journal of American Chemical Society, 82*, 5671.
48. Chatani, Y, et al (1970). *Macromolecules, 3*, 481.
49. Chatani, Y. et al (1973). *Journal of Polymer Science and Polymer Physics, 11*, 369.
50. Long, R. Q. & Yang, R. T. (2001). Carbon nanotubes as superior sorbent for Dioxin removal. *Journal of the American Chemical Society, 123*, 2058–2059.
51. Anurov, S. A. (1996). Physico-chemical aspects of sulphur dioxide adsorption by carbon sorbents. *USP Chemistry, 65*(8), 718–732.
52. Simonyan, V. V. & Johnson J.K (2002). Hydrogen storage in carbon nanotubes and graphitic nanofibers. *Journal of Alloys and Compounds, 659*, 330–332.
53 m Keltsev, N. V (1976). *Fundamentals of adsorption technique*. Moscow: Khimiya.

54. Pugachevich, P. P., Beglyarov E. M., & Lavygin I. A (1982). *Surface phenomena in polymers* (p.198). Moscow: Khimiya.
55. Lebedev, E. V. (1986). Colloid-chemical peculiarities of polymer-polymeric compositions Physico-Chemistry of Multi-Component Polymeric Systems (Vol. 2, pp. 121–136). Kiev: Naukova dumka.
56. Lipatov & Yu. S. (1980). Interface phenomenon in polymers (p. 259). Kiev: Naukova Dumka.
57. White J. S. (1972). Related Phenomena. Proceedings 3rd International Conference Nobre Dome, New York-L., 1973, p.81.
58. Gregonis, D. E. & Andrade, J. D. (1985). *Surface and interface aspects of biomed polymers* (Vol.1, pp. 43–75). New York: L.
59. Povstugar, V. I., Kodolov, V. I., & Mihailova, S. S (1988). *Structure and properties of surface of polymeric materials* (p.189). Moscow: Chemistry.
60. Prigozhin, I. & Defay, R.(1966). *Chemical thermodynamics* (p.510). Novosibirsk: Nauka.
61. Aleskovsky, V. B. (2002). Nanostructures from the chemical standpoint *Surface chemistry and synthesis nanosize systems* (p. 122). St. Petersburg: St. Petersburg Technical University.
62. Rusanov, A. I (2002). Nanothermodynamics in chemical approach. *Surface chemistry and synthesis nanosize systems* (p.177). St. Petersburg: St Petersburg Technical University.
63. Bulgakov, V. K., Kodolov, V. I., & Lipanov, A. M (1990). *Modeling of Polymeric Materials Combustion* (p 238). Moscow: Chemistry.
64. Korablev, G. A (2005). *Spatial-energy principles of complex structures formation* (p. 426). Leiden: Brill Academic Publishers.

CHAPTER 2

PROGNOSTIC INVESTIGATIONS OF METAL OR CARBON NANOCOMPOSITES AND NANOSTRUCTURES SYNTHESIS PROCESSES CHARACTERIZATION

N. V. KHOKHRIAKOV, V. I. KODOLOV, G. A. KORABLEV, V. V. TRINEEVA, and G. E. ZAIKOV

CONTENTS

2.1 QUANTUM–CHEMICAL INVESTIGATION OF METAL / CARBON NANOCOMPOSITES REDOX SYNTHESIS

2.1.1 Quantum–Chemical Calculating Experiment for Nanostructures, Containing Metals from Aromatic Hydrocarbons and Metal Chlorides

After carbon nanotubes synthesis in 1991, the interest to these objects is constantly increasing. However, till recently high-temperature evaporation of graphite in arc charge has been the only method of nanotubes synthesis. Significant success has been achieved in the frameworks of this method, but still there is necessity to develop low-temperature synthesis methods with the possibility to control the chemical process, to modify the structure of already prepared tubes and to obtain heterojunctions on their basis.

Thereupon the method, the essence of which consists in the use of fine clusters of transition metals and their salts as stimulators of dehydropolycondensation reaction of aromatic hydrocarbons [1], seems to be perspective. At the same time, phenanthrene and anthracene are used as initial products for the following polymerization, and the reaction proceeds at relatively low temperatures.

The main target of quantum-chemical investigations given in the paper is to reveal the reaction mechanism, stimulators role and to make the comparative analysis of the influence of different metals and their salts on the process. Cu, Ni, Co, Fe, Mn, and Cr are considered as stimulators.

The calculations are done in the limits of *ab initio* Hartree–Fock method [2] in TZV basis [3, 4, and 5] using GAMESS program complex [6]. Additional calculations in enhanced TZV bases have not revealed qualitative changes in the results. As *ab initio* calculations require considerable calculation resources, the investigations are carried out on the simplest model systems including one benzene ring and transition metal ion.

The parameters of metal/benzene ring complexes optimized by energy are given in the tables. The lengths of bonds of optimized structures for benzene and its compounds with the metals of the row considered are shown in Table 2.1 in which, the fraction bar divides non equivalent bond lengths formed while decreasing benzene ring symmetry. In the frameworks of calculations done, it is found that ion Cu(+) does not distort the benzene ring geometrical structure (see Figure 2.1b), but there is considerable distortion of benzene structure followed by the decrease of its symmetry under the influence of Co, Ni, Mn ions. Thus, complexes with Co and Ni have symmetry C_{2v}, at the same time two opposite carbon atoms move outside the ring plane; as a result the ring acquires "bath" conformation (see Figure 2.1c). In manganese complex symmetry—C_{3v}, the ring plane distortion is not observed, long and short C–C bonds alternate in the structure obtained.

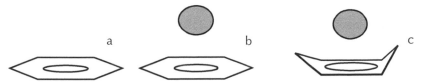

FIGURE 2.1 The images of metal–benzene interaction models. a Ring structure of benzene, b complex metal/benzene ring with Cu, c complex metal or benzene ring with Co or Ni or Mn.

TABLE 2.1 Bonds lengths in metal–organic complexes (in Angstroms).

	C–C	C–H	Me–C
C_6H_6	1.387	1.072	
$(CuC_6H_6)^{+1}$	1.396	1.071	2.579
$(CoC_6H_6)^{+1}$	1.400/1.396	1.071/1.070	2.521/2,494
$(MnC_6H_6)^{+1}$	1.385/1.410	1.07	2.521
$(NiC_6H_6)^{+2}$	1.414/1.396	1.072/1.073	2.370/2.298

The influence of metal ions on the strength of C–H bond in benzene and the interaction energy of metal ion with benzene ring are revealed. Besides, the exponents and lengths of chemical bonds in the compounds considered, and atomic charges are calculated.

C–H bond energy is calculated as the difference of complex energy in an equilibrium configuration and its energy without a proton (the proton energy is assumed to be zero):

$$E_{C-H} = E\,(C_6H_6Me)^{+n} - E\,(C_6H_6Me)^{+n} \tag{2.1}$$

To estimate the interaction energy of metal cation with aromatic ring the following formula is applied:

$$E_{Me-ring} = [E(C_6H_6) + E(Me^{+n})] - E(C_6H_6Me)^{+n} \tag{2.2}$$

The calculation results are given in Table 2.2. In accordance with the table data it is possible to conclude that metal ion influence on chemical bond in benzene ring increases in the row Cu, Co, Mn, Cr, Fe, Ni. In general, metal ions considerably facilitate the proton breaking-off from benzene ring accelerating dehydropolycondensation process. In parallel the calculations of the same complexes have been done in the frameworks of INTAS project by chemists team under Professor Molina's supervision (Spain) [7]. The calculations have been done using program product GAUSSIAN-98. They have also used ab initio Hartree–Fock method, but basis set 6-311+g*. Their results are mainly in accordance with the results presented in this paper.

TABLE 2.2 Interaction energies of metal ions and aromatic rings. The comparison of calculations (in Hartee) in TZV bases and enhanced TZV.

Structures of benzene-metal complexes	$Me+C_6H_6$		Proton breaking-off	
	TZV	**TZV***	**TZV**	**TZV***
C_6H_6			0.674	
$CuC_6H_6^+$	-0.052	-0.054	0.501	0.497
$NiC_6H_6^{++}$	-0.222	-0.231	0.305	0.297
$CoC_6H_6^+$	-0.056	-0.060	0.498	0.492
$FeC_6H_6^{++}$	-0.696		0.310	
$MnC_6H_6^+$	-0.054	-0.058	0.498	0.493
$CrC_6H_6^{++}$	-0.032		0.380	

TABLE 2.3 The orders of non equivalent bonds in different metals and aromatic ring systems.

	C_6H_6	$CuC_6H_6^+$	$CoC_6H_6^+$	$MnC_6H_6^+$
C–C	1.525	1.436	1.40/1.430	1.505/1.327
C–H	0.896	0.861	0.861	0.861
Me–C		0.127	0.136/0.170	0.166

TABLE 2.4 Charges and free valences of non equivalent atoms in different metals and aromatic ring systems.

	C_6H_6	$CuC_6H_6^+$	$CoC_6H_6^+$	$MnC_6H_6^+$
		Charge according to Mulliken		
C	-0.21	-0.209	-0.168/-0.214	-0.198/-0.202
H	0.21	0.279	0.277	0.276
Me		0.58	0.531	0.538
		Charge according to Lowdin		
C	-0.135	-0.09	-0.057/-0.063	-0.058/-0.059
H	0.135	0.165	0.168	0.169
Me		0.548976	0.355756	0.339225
		Bond valence		
C	4.046	3.962	3.962	3.95
H	0.914	0.88	0.88	0.885
Me		0.757	0.956	1.018

Also, the general mechanism of dehydropolycondensation reaction is investigated.

The graphs for potential energies of interaction between benzene ring and copper clusters (white squares in Figure 2.2) and copper chloride (black squares) are obtained from results of semi-empirical model MNDO [8]. Copper is chosen as an interaction ion, for CuC_6H_6 interaction is the weakest. While considering the interaction of copper chloride and benzene ring, chloride is located along the ring axis and copper ion—closer to the ring. While preparing the graphs, the distance between the complexes changes but complexes' geometry does not distort, copper and chlorine ions are located on the ring axis.

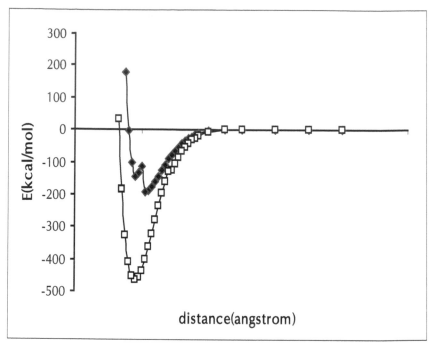

FIGURE 2.2 Interaction of benzene ring with copper atom (white squares) and copper ion (black squares). Axis Y—energy (kcal/mol), axis X—distance from cluster to ring plane.

Besides, the dependence graphs of proton interaction energy with complex MeC_6H_5 on the distance between the complex and proton are drawn. In this case, Cu and Co are considered. Both graphs are drawn in relative axes, that is, the energy of optimized complex metal-benzene ring as zero energy is chosen. Also we supposed that the ring geometry is not changed and metal ion is always on the ring axis. When the distances from the proton to the complex are different, the energy is not minimized additionally. The curves obtained are given in Figure 2.3. The graph for the complex with Cu is shown by the solid line, the graph for Co by triangles. The comparative analysis of the graphs is carried out to estimate the entropy contribution to proton-complex interaction at non-zero temperature. It is seen from the figure that in the region considered, the curves are identical to the shift.

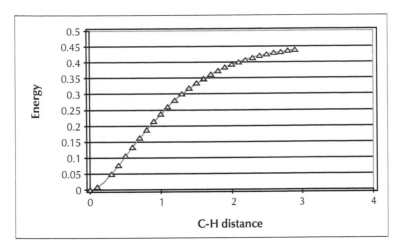

FIGURE 2.3 Proton interaction energy with complex MeC_6H_5 as the function of the proton distance away from its balanced position in benzene ring. CuC_6H_6 (*solid line*) and CoC_6H_6 (*triangles*).

It is seen from the tables that metal weakens C–H bond in benzene molecule and facilitates hydrogen atom removal from hydrocarbon substance. In order to investigate the following polymerization of aromatic hydrocarbons, the computer modeling of anthracene molecule interactions and several benzene molecules is done. The part of hydrogen atoms of the molecules mentioned is removed. The result of energy optimization of the complex mentioned within the frameworks of semi-empirical method MNDO is shown in Figure 2.4.

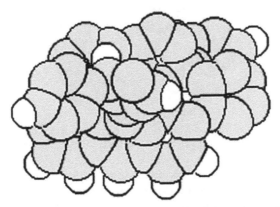

FIGURE 2.4 Complex optimized by energy containing phenanthrene molecule and six benzene molecules. Hydrogen atoms are partly removed.

It is seen from the figure that a highly defective branched structure is formed. Though the primary hybridization degree of carbon atoms is sp², atoms with sp³- and sp-hybridization are presented in the complex. Not all stopped bonds of carbon atoms close on other stopped bonds. In many cases the system stabilizes due to the increase of C–C bond order.

During the formation of carbon shells the metal clusters and their salts can stimulate the orientation. This interaction is shown in Figure 2.3 and Table 2.1. Some energy must be expended for carbon net bending to form a carbon tube out of a graphite sheet. The calculation results show that for tubes of different diameters obtained experimentally, the interaction energy of metal ion with benzene ring exceeds the bending energy of carbon net even with copper.

In the frameworks of semi-empirical tight-binding model in Gudwin parameterization [9] the modeling of polyaromatic carbon strip with metal particle is carried out. The metal particle is imitated by cylindrical field, in which carbon atoms are present. The field potential is estimated out of the curves shown in Figure 2.3. At the start time, the strip is given the rate directed to the cylinder axis; initially the band is located at an angle to the axis. The investigations are carried out using own program complex, which allows making quantum-chemical molecular-dynamic investigations of carbon systems. As a result of molecular-dynamic experiment the strip "winds around" the cylinder, its ends close forming covalent bond between carbon atoms, as it is shown in Figure 2.5.

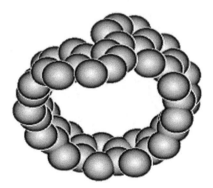

FIGURE 2.5　The formation of carbon shell on the metal particle surface.

The following conclusions can be made from the results obtained. A metal weakens C–H bond in benzene ring stimulating proton breaking-off and chemical bond formation between different rings, in other words facilitating dehydropolycondensation process. Besides, the role of metal salts consists in absorption of free protons by chlorine anions. In turn, metal particles attract hydrocarbon atoms, playing an orienting part in during their dehydropolycondensation.

Quantum-chemical investigations of metal-organic complexes and model systems point out the formation possibility of defective carbon shells with metals in the frameworks of the reaction considered. This fact is also confirmed by experimental investigations, in particular, microphotographs and diffractograms of the samples obtained, which demonstrate the presence of cylindrical graphite-like shells with metal inside.

2.1.2 Theoretical Fundamentals of Tubulene Formation Processes in Active Media.

Stimulated Dehydropoly condensation

The hypothesis of carbon and metal-carbon tubulene formation in active lamellar media is based on literary and own experimental data.

While combining aromatic hydrocarbons with mineral lamellar systems, solid introduction solutions are formed when aromatic compounds are placed between the layers. This variant is possible if combining fine powders of organic and mineral phases with the following remelting of the systems being formed. The above systems are complexes consisting of a certain number of aromatic rings and a quite mobile metal ion (atom) which interacts with a π-electron system of the rings facilitating the ring hydrogen breaking off and its transport into the medium due to hydrogen binding with complex anions in the matrix. Finally, this leads to dehydropolycondensation and the formation of a polymer aromatic compound situated in interface regions of the mineral medium, being saturated by hydrogen. Therefore, "medium capacity by hydrogen" is very important. The estimation of components ratio in reactive mass is made using hydrogen atom number in an aromatic compound and a coordination number (complex anion charge). For instance, molar ratio of copper chloride and anthracene is 10:1, anthracene—10H, $[CuCl_2]^-$—1H; if the metal oxidation state does not change during the process. The increase of the complex anion negative charge leads to much greater hydrogen absorption and, consequently, to the decrease of the medium and aromatic compound ratio. The mechanism of carbon structures formation can be shown as a series of the following consecutive parallel reactions:

Stage one: Dehydropolycondensation reaction (on the example of anthracene). Intermediate complex with charge transfer between aromatic hydrocarbon and aluminium chloride has a bright yellow (with anthracene), or dark red (phenanthrene) color.

$$(2.3)$$

Second stage: Fischer–Hafner reaction [10] of transition metal, metalarene complexes obtaining (on the example of cobalt). Metalarene complexes being formed, as well as initial hydrocarbons, easily participate in dehydropolycondensation reaction due to the increase of nucleophile properties of an aromatic ring during the aromatic ring activation by transition metal atom.

$$(2.4)$$

It should be expected that the separation of protons from the aromatic ring will be easier depending on the metal connected with the aromatic ring; in particular, the proton evolution from metal–organic complexes will be easier in Cr-Mn-Co-Ni row (this is confirmed by the experiment data).

Carbon content in the products obtained increases during consecutive dehydropolycondensation reactions between hydrocarbons and metalarene complex compounds. This leads to the formation of graphite planes containing transition metals connected with them.

By analogy, the aromatic hydrocarbons dehydropolycondensation mechanism through the complexes with charge transfer for polyphosphoric acid can be shown:

(2.5)

Here, the formation of intermediate complexes is possible

$$nH^+(Ar)_m[OP(O)OH]^-_n$$

This interaction picture is confirmed by the color change in the same way as during the formation of a complex with charge transfer, by electric resistance change and EPR-spectra respectively.

As the stability of the polymer chain being formed in the interface region of an active medium increases due to the distortion of complanarity, the conditions for the formation of a spiral structure with metal introduction into spirals being formed, appear. The dehydropolycondensation process transfers into stimulated carbonization one, when the interaction with protons evolution into the environment takes place between the spiral coils. Thus, it is assumed that the process of tubulene formation takes place on metal ions in the adsorbed layer "melt salt particles". There is a more preferable variant, when a lamellar polyaromatic structure with the interacting metal atoms rolls up into a spiral "a scroll". The preferability of this variant is caused by the fact that it is difficult to divide the aggregates of anthracene, phenanthrene and other aromatic compound molecules in a mineral medium and to place them evenly on salt and metal particles.

The interaction of aromatic hydrocarbons with the particles of active mineral medium is investigated by quantum-chemical calculations. Here, the interactions of such ions as Cu^+, Co^+, Mn^+, Ni^{2+} with an aromatic ring in $(C_6H_6Me)^{n+}$ complexes are compared. Also, to determine the possibility of dehydropolycondensation, it is necessary to carry out a comparative analysis of metal ion influence on the decomposition easiness of C–H bond and on the possibility of the complanarity disturbance of aromatic rings structure with the consequent polyaromatic strip bending and spiral formation.

The calculations are done by restricted Hartree–Fock method in TZV basis set [11, 12] using GAMESS program product [6] for model systems, which are the simplest metal-organic complexes $(C_6H_6Me)^{+n}$. C–H bond energy is calculated as the difference of complex energy in an equilibrium configuration and its energy without a proton (the proton energy is assumed to be zero):

$$E_{C-H} = E(C_6H_6Me)^{+n} - E(C_6H_5Me)^{+(n-1)} \qquad (2.6)$$

To estimate the interaction energy of metal cation with an aromatic ring the following formula is applied:

$$E_{Me-ring} = E(C_6H_6Me)^{+n} - [E(C_6H_6) + E(Me^{+n})] \qquad (2.7)$$

It follows from the calculation results that the presence of a transition metal makes the proton breaking-off from an aromatic ring easier. Thus, a metal can stimulate dehydropolycondensation process of aromatic hydrocarbons.

The investigation of geometric parameters of considered complexes, optimized by energy, shows that the considerable distortion of a benzene ring structure, followed by its symmetry lowering, takes place in the presence of Co, Ni, Mn ions. So, complexes with Co and Ni have C_{2v} symmetry, two opposite carbon atoms leave the ring plane, and as a result, the ring achieves "bath" conformation. The complex with manganese has C_{3v} symmetry. Here, the ring plane distortion is not observed, whereas the localization of C–C bonds takes place, long and short C–C bonds alternate in the structure obtained.

The estimation of the polyaromatic strip turning possibility is made by comparing the bond energies of anthracene and phenanthrene. It shows that the bending and branching of polyaromatic strip are advantageous from energetic point of view. In particular, the structure of phenanthrene is energetically more favorable (by 0.01 Hartee) than the one of anthracene.

In accordance with the results of quantum-chemical calculations, the possibilities of carbon tubular structures formation in the examined active media can be concluded. The interaction between the metal ion and the carbon ring increases in Cu-Mn-Co-Ni row. The proton breaking-off from the ring becomes easier in this row too. Copper influence on the ring structure is weak. Cobalt and manganese interact with the ring more intensively. The interaction intensity increases considerably in the complex with nickel. This can be explained by the increased charge of nickel complex. Calculations of the same complexes, but using another basis set 6-311G+* and the quantum chemistry program Gaussian 98, were made in the group of prof. J. Molina [7]. Mainly their results are in accordance with ours. The only considerable difference was found in the case of manganese complex. They found that the metal ion-ring interaction and C–H bond weakening in $(C_6H_6Mn)^+$ complex were sufficiently stronger than in the complexes with cobalt and copper.

Finally, we compare the interaction energy between Cu ion and the aromatic ring (for different distances between the ion and the ring plane) and the energy of graphene plane bending to form nanotubes [13]. It is found that even in the case of Cu the energy of ion-ring interaction is enough to provide bending of polyaromatic structure on a metal cluster for the diameter as small as 1nm. Thus, the tubulenes radii which can be found in the experiment are limited only by metal particle size.

The process of stimulated carbonization in the presence of polyphosphoric acid can be presented through polyene chain formation due to dehydrogenization, the chain rolling up into a spiral and the following cross-linking between the spiral branches—those are stimulated carbonization stages. The dehydrogenization of conjugated double bonds apparently follows the scheme for the description of dehydropolycondensation mechanism of aromatic hydrocarbons in polyphosphoric acid melt.

It is natural that the use of other polymers, for example, polyvinyl alcohol is determined by the metal nature in a complex anion and instability constant of the anion itself when H_2O is a ligand. The molecular mass of a polymer, its crystal structure, and the dispersity of its particles influence tubulenes formation. The formation of defect tubulenes with accrete tubes in different directions is quite possible. For instance, tubulenes formation from polyvinyl alcohol dissolved in water and combined with polyvanadium acid (HVO_3) or the acid intercalated by Cr and Mo ions takes place under the temperature action up to 363K or the electric field action (voltage ≤ 3V). Here, water evolution and the proceeding of stimulated carbonization with the inclusion of transition metals into tubulene's structure is also assumed, as with polyphosphoric acid or its metal containing derivatives. It should be noted that the dehydration process can proceed in several stages: first, hydrogen breaking off with the formation of ionized oxygen which in turn (second stage) forms the bonds with matrix metal atoms, and then proceeds destruction with water evolution. As the coordination number of the transition metal and its charge are usually greater in the matrix, it can be expected that it will take part in the interaction with conjugated double bonds in polyene chain and will be included into the shells of tubulenes being formed.

2.1.3 Quantum–Chemical Calculating Experiments for Prognosis of Redox Synthesis in Polymeric Matrixes with Salt Solution Applications or Metal Oxide Using

Based on Prigozhin's theory, usually the probability of self-organization processes increases at the directed action or in the particle flow [14]. Consequently these processes have to take place:

- At the directed interaction between chemical particles and active centers in micropores;
- In surface and interface layers of lamellar (polymeric organic and inorganic) systems;
- On membranes or in intracellular space of biological objects.

The driving force of self-organization processes (formation of nanostructures with definite shapes) is the difference of potentials between the interacting particles and walls of the object that stimulates these interactions. In turn, the directedness of the process between the particles is determined by the interactions between charges and dipoles. At the same time, the charge is shifted by external electric and electromagnetic fields of a certain intensity or energy field of the particles being formed that changes at the interaction between the nanoreactor walls, and flow of chemical particles being taken inside if the temperature on concentration gradient is available along the nanoreactor; for instance, during the adsorption of chemical particles with their following transformation into nanostructures on the corresponding metal–substrate template. Here the flow of electrons excited by the heat energy moves upwards. The fluctuations of heat energy on the template determine the level of transformations of initial compounds into nanostructures on separate sectors of the template [15]. At the electrochemical action onto carbon electrodes [16], in the ion flow carbon particles transform into nanostructures of various shapes similar to those of bodies of revolution.

The formation of nanostructures in xerogels and lamellar substances with energy saturated channels is slightly different. In such cases the flow of ionized particles is formed due to the difference in the state of charge of the channel walls (nanoreactors) and chemical particles contained in the solution or melt flowing into the channels. When the matrix is placed between electrodes at a small potential and when the solvent is evaporated under vacuum or temperature action, the channels and ions flow in them are oriented, and chemical particles of different nature (organic and inorganic) are transformed into nanostructures.

Much more processes have to take place in cathode and anode zones, as well as in inert electrode area of nanoreactors of active lamellar media. In such case, apart from redox processes, it is necessary to take into consideration the exchange processes and interaction reactions of reaction byproducts (sometimes low-molecular) with the walls of nanoreactors. Therefore, it is interesting to investigate the composition, structure, and potential being created on the walls of nanoreactors. The potential jump on the boundary "nanoreactor wall–reacting particles" is defined by the charge of the wall surface and the size of the reacting layer that, in turn, depends upon the external energy fed and the surface energy (energy stored during the nanoreactor formation). Presumably the energy required to obtain carbon tubules or fullerenes from their "embryos" is less by an order than the formation energy of "embryos" themselves. If redox process is considered the main process preceding the formation of tubules and fullerenes, the work for carrying the charge corresponds to the formation energy of corresponding nanoparticles in the reacting layer. Then the equation of energy conservation for a nanoreactor during the formation of a mole of nanoparticles will be as follows:

$$nF\Delta\varphi = RT\ln N_p/N_r \qquad (2.8),$$

where n—number reflecting the charge of chemical particles flowing into the nanoreactor; F—Faraday number; $\Delta\varphi$—difference of potentials between the nanoreactor walls and flow of chemical particles; R—gas constant; T—process temperature; N_p—molar share of nanoparticles obtained; N_r—molar share of initial reagents from which the nanoparticles are obtained.

It can be presumed that nanoreactor walls are not inert, but they are participating in the process. In this case the directedness of chemical processes is determined by the peculiarities of adsorption of chemical particles and possibility to form nanophases containing the particles and fragments or separate atoms of nanoreactor walls, as well as the possibility to transport ionized particles and electrons along the nanoreactor. At the same time the thickness of nanoreactor walls can reach the values several times exceeding the distance between walls.

Using the aforesaid equation, we can define the values of equilibrium constants when reaching the definite output of nanoparticles, sizes of nanoparticles and with the modification of the equation—shapes of nanoparticles being formed. The sizes of internal cavity or reaction zone of the nanoreactor and its geometry significantly influence the sizes and shape of nanoparticles being formed.

The succession of the ongoing processes is conditioned by the composition and parameters (energy and geometry) of nanoreactors. If nanoreactors represent nanopores or cavities in polymeric gels being formed during the removal of solvents out of gels and their transformation into xerogels or during the formation of crazes in the process of mechanical-chemical treatment of polymers and inorganic phase in the presence of an active medium, the essence of processes in nanoreactors is as follows:

First, a nanoreactor is formed in the polymeric matrix in which geometrical and energy parameters corresponds to the transition state of reagents participating in the reaction; afterwards the nanoreactor is filled with the reaction mass containing reagents and a solvent. The latter is removed and only the reagents oriented in a definite way are left in the nanoreactor. If a sufficient energy impulse is available, for instance, energy isolated during the formation of coordinating bonds between the fragments of reagents and functional groups located in nanoreactor walls, they interact with the formation of the required products. The initially produced coordinating bonds are destroyed simultaneously with a nanoproduct formation.

Thus, the main role of nanoreactors is to contribute to the formation of activated complex and is to decrease the activation energy of the main reaction between reagents.

To perform such processes, it is necessary to preliminary select the polymeric matrix containing nanoreactors (in the form of nanopores or crazes) suitable to the process. Such selection can be carried out with the help of computer chemistry. Further the calculation experiment is made with reagents placed into the nanoreactor with corresponding geometrical and energy parameters. The examples of such calculations in gels of polyvanadium acid and its derivatives, as well as in gels of polyvinyl alcohol are discussed [17, 18]. Experimental confirmation of the correctness of the selection of polymeric matrixes and nanoreactors to obtain carbon–metal-containing nanostructures is reported [19, 20]. Quantum-chemical investigation was carried out with the extraction of the fragments of polymeric chains with the introduction of corresponding metals into the model of ions or oxides. When several types of polymeric and oligomeric substances are used in the model for the formation of the matrix and nanoreactors, the fragments of the corresponding types of substances are applied. The definite result of the calculation experiment is obtained gradually. Any transformations at the initial stages are taken into account at the further ones. Figure 2.6a, b, c demonstrates the pictures of the initial calculation stage and the next stage of the interaction between nickel chloride and fragments imitating polyvinyl alcohol and polyethylenepolyamine.

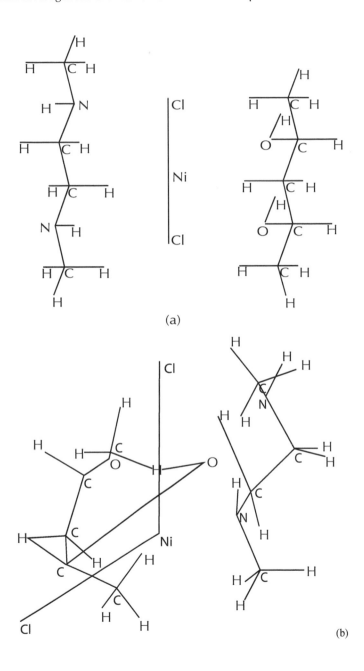

(a)

(b)

FIGURE 2.6 *(Continued)*

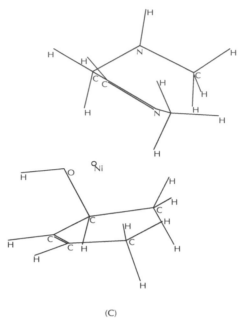

(C)

FIGURE 2.6 Stages of the calculation experiment of the interactions between nickel chloride and fragments of PVA-PEPA.

Sizes and shapes of nanostructures being formed are determined with the help of methods of molecular mechanics with a large array of atoms. However, macromolecular systems that are presently accepted as nanostructures were initially registered.

The possibilities of the calculation experiment prognosis:

• determination of the possibility to form nanostructures of a definite shape when applying the corresponding metal-containing and polymeric substances;

• revelation of optimal ratios of metal-containing and polymeric components to obtain the required nanoproduct.

For example, the probable processes at the interaction of PVA with the oxides of 3d-metals are considered in the frameworks of quantum-chemical approximation ZINDO/1, realized in the program product HyperChem v.6.03. The evaluation of the interaction possibility of molecule fragments was carried out by the bond length change in them as a result of the geometry optimization, which shows the system interaction and stability. The computational experiment was carried out at the ratios NiO:OH = 1:2; 1:4. Figure 2.7 shows the experimental results of the process, second stage after the initial processes of dehydration and dehydrogenation. The component ratio NiO:OH = 1: 4.

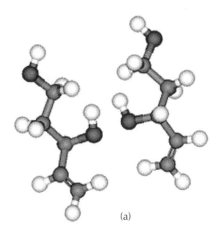

(a)

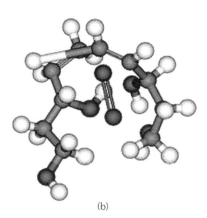

(b)

FIGURE 2.7 Two molecules of 3,5-dihydroxypentene-1 before the geometry optimization (a), after the geometry optimization with nickel oxide (b).

Thus the energetically more favorable states of the systems are revealed. The scrolling of the macromolecule fragments relatively to the nickel oxide in the shape of hemisphere with the formation of carbon shell is observed in the system. The Table 2.5 shows the bond lengths in the molecules before and after the geometry optimization for the systems "3,5dihydroxypentene-1–nickel oxide".

TABLE 2.5 Bond lengths in the molecules before and after the geometry optimization for the system "3,5dihydroxypentene-1–nickel oxide".

System	Bond	Bond length before optimization (Å)	Bond length after optimization (Å)
Two molecules 3,5dihydroxypentene-1 and nickel oxide	Ni–O	1.547	2.04565
	C–H	1.09	2.9268

When analyzing the interaction of the system "3, 5dihydroxypentene-1–nickel oxide", it can be concluded that the metal is reduced by polyene, which is formed during the dehydrogenation of PVA as a result of breaking C–H bond and formation of hydrogen radical. Based on the data of theoretical models, the ratios of components are selected according to coordination interactions. For nickel, cobalt, and copper compounds, the molar ratio of the components is 1:4 (metal: number of PVA functional groups). Based on the modeling results obtained, it can be expected that the availability of metals of different nature than nickel in the system will result in the distortion of nanostructure shapes obtained, thus influencing the second thermal-chemical stage of nanostructure obtaining.

2.2 SPATIAL ENERGY INVESTIGATION OF NANOSTRUCTURES FORMATION AND INTERACTIONS PROCESSES

2.2.1 Spatial Energy Parameter in Comparison with Lagrange and Hamilton Functions

It is demonstrated that for two-particle interactions the principle of adding reciprocals of energy characteristics of subsystems is fulfilled for the processes flowing along the potential gradient, and the principle of their algebraic addition—for the processes against the potential gradient.

The analysis of the kinetics of various physical and chemical processes shows that in many cases the reciprocals of velocities, kinetic or energy characteristics of the corresponding interactions are added.

Some examples: Ambipolar diffusion, resulting velocity of topochemical reaction, change in the light velocity during the transition from vacuum into the given medium and effective permeability of biomembranes.

In particular, such supposition is confirmed by the formula of electron transport possibility $\left(W_\infty\right)$ due to the overlapping of wave functions 1 and 2 (in steady state) during electron-conformation interactions:

$$W_\infty = \frac{1}{2}\frac{W_1 W_2}{W_1 + W_2} \qquad (2.9)$$

Equation (1) is used when evaluating the characteristics of diffusion processes followed by non radiating transport of electrons in proteins [21].

And also from classical mechanics, it is known that the relative motion of two particles with the interaction energy U(r) takes place as the motion of material point with the reduced mass μ:

$$\frac{1}{\mu} = \frac{1}{m_1} + \frac{1}{m_2} \qquad (2.10)$$

In the field of central force U(r), and general translational motion—as a free motion of material point with the mass:

$$m = m_1 + m_2 \qquad (2.11)$$

Such things take place in quantum mechanics as well [22].

The task of two-particle interactions taking place along the bond line was solved in the times of Newton and Lagrange:

$$\overset{\circ}{A} = \frac{m_1 v_1^2}{2} + \frac{m_2 v_2^2}{2} + U\left(\overline{r_2} - \overline{r_1}\right) \qquad (2.12),$$

where E is the total energy of the system, first and second elements—kinetic energies of the particles, third element—potential energy between particles 1 and 2, vectors $\overline{r_2}$ and $\overline{r_1}$—characterize the distance between the particles in final and initial states.

For moving thermodynamic systems the first commencement of thermodynamics is as follows [23]:

$$\delta E = d\left(U + \frac{mv^2}{2}\right) \pm \delta A, \qquad (2.13)$$

where, δE is the amount of energy transferred to the system;

element $d\left(U + \dfrac{mv^2}{2}\right)$ characterize the changes in internal and kinetic energies of the system;

$+ \delta A$—work performed by the system; $- \delta A$—work performed with the system.

As the work value numerically equals the change in the potential energy, then:

$$+ \delta A = \Delta U \text{ or } - \delta A = + \Delta U \qquad (2.14)$$

It is probable that not only in thermodynamics but in many other processes in the dynamics of moving particle's interaction not only the value of potential energy is critical, but its change as well. Therefore, similar to the equation (2.12), the following should be fulfilled for two-particle interactions:

$$\delta E = d\left(\frac{m_1 v_1^2}{2} + \frac{m_2 v_2^2}{2}\right) \pm \Delta U \qquad (2.15)$$

Here, $\Delta U = U_2 - U_1$ (2.16)

where U_2 and U_1—potential energies of the system in final and initial states.

At the same time, the total energy (E) and kinetic energy $\left(\dfrac{mv^2}{2}\right)$ can be calculated from their zero value, and then only the last element is modified in the equation (2.12).

The character of the change in the potential energy value $\left(\Delta U\right)$ was analyzed by its sign for various potential fields and the results are given in Table 2.6.

From the table, it is seen that the values— ΔU and accordingly $+ \delta A$ (positive work) correspond to the interactions taking place along the potential gradient, and ΔU and $-\delta A$ (negative work) occur during the interactions against the potential gradient.

The solution of two-particle task of the interaction of two material points with masses m_1 and m_2 obtained under the condition of the absence of external forces, corresponds to the interactions flowing along the gradient, the positive work is performed by the system (similar to the attraction process in the gravitation field).

The solution of this equation via the reduced mass (μ) is [24] the Lagrange equation for the relative motion of the isolated system of two interacting material points with masses m_1 and m_2, which in coordinate x is as follows:

$$\mu \cdot x'' = -\frac{\partial U}{\partial x}; \frac{1}{\mu} = \frac{1}{m_1} + \frac{1}{m_2}$$

Here U—mutual potential energy of material points; μ—reduced mass. At the same time, $x'' = a$ (feature of the system acceleration). For elementary portions of the interactions, Δx can be taken as follows:

$$\frac{\partial U}{\partial x} \approx \frac{\Delta U}{\Delta x}.$$

That is, $\mu a \Delta x = -\Delta U$. Then,

$$\frac{1}{1/(a\Delta x)}\frac{1}{(1/m_1 + 1/m_2)} \approx \Delta U;$$

$$\frac{1}{1/(m_1 a\Delta x) + 1(m_2 a\Delta x)} \approx -\Delta U;$$

Or
$$\frac{1}{\Delta U} \approx \frac{1}{\Delta U_1} + \frac{1}{\Delta U_2} \qquad (2.17),$$

where ΔU_1 and ΔU_2—potential energies of material points on the elementary portion of interactions, ΔU—resulting (mutual) potential energy of this interactions.

TABLE 2.6 Directedness of the interaction processes

No.	Systems	Type of potential field	Process	U	r_2/r_1 (x_2/x_1)	U_2/U_1	Sign ΔU	Sign δA	Process directedness in potential field
1	Opposite electrical charges	Electrostatic	Attraction	$-k\dfrac{q_1 q_2}{r}$	$r_2 < r_1$	$U_2 > U_1$	-	+	Along the gradient
			Repulsion	$-k\dfrac{q_1 q_2}{r}$	$r_2 > r_1$	$U_2 < U_1$	+	-	Against the gradient
2	Similar electrical charges	Electrostatic	Attraction	$k\dfrac{q_1 q_2}{r}$	$r_2 < r_1$	$U_2 > U_1$	+	-	Against the gradient
			Repulsion	$k\dfrac{q_1 q_2}{r}$	$r_2 > r_1$	$U_2 < U_1$	-	+	Along the gradient

TABLE 2.6 *(Continued)*

No.	Name		Type	Formula					Direction
3	Elementary masses m_1 and m_2	Gravitational	Attraction	$-\gamma\dfrac{m_1 m_2}{r}$	$r_2 < r_1$	-	$U_2 > U_1$	+	Along the gradient
			Repulsion	$-\gamma\dfrac{m_1 m_2}{r}$	$r_2 > r_1$	+	$U_2 < U_1$	-	Against the gradient
4	Spring deformation	Field of elastic forces	Compression	$k\dfrac{\Delta x^2}{2}$	$x_2 < x_1$	+	$U_2 > U_1$	-	Against the gradient
			Extension	$k\dfrac{\Delta x^2}{2}$	$x_2 > x_1$	+	$U_2 > U_1$	-	Against the gradient
5	Photoeffect	Electrostatic	Repulsion	$k\dfrac{q_1 q_2}{r}$	$r_2 > r_1$	-	$U_2 < U_1$	+	Along the gradient
6	Transfer systems	Field transfer	Passive	$I = -\sigma\dfrac{\Delta\varphi}{\Delta x}$ and other	$x_2 > x_1$	-	$U_2 > U_1$	+	Along the gradient
			Active		$x_2 < x_1$	+	$U_2 < U_1$	-	Against the gradient

Thus:

- In the systems in which the interactions proceed along the potential gradient (positive performance), the Lagrangian is performed and the resulting potential energy is found based on the principle of adding reciprocals of the corresponding energies of subsystems [25]. Similarly, the reduced mass for the relative motion of two-particle system is calculated.
- In the systems in which the interactions proceed against the potential gradient (negative performance), the algebraic addition of their masses as well as the corresponding energies of subsystems is performed (by the analogy with Hamiltonian).

From the equation (2.17), it is seen that the resulting energy characteristic of the system of two material points interaction is found based on the principle of adding reciprocals of initial energies of interacting subsystems.

Electron with the mass m moving near the proton with the mass M is equivalent to the particle with the mass:

$$\mu = \frac{mM}{m + M} \qquad [2.16].$$

In this system, the energy characteristics of subsystems are: electron orbital energy (w_i) and nucleus effective energy taking screening effects into account (by Clementi).

Therefore when modifying the equation (2.17), we can assume that the energy of atom valence orbitals (responsible for interatomic interactions) can be calculated [25] by the principle of adding reciprocals of some initial energy components based on the following equations:

$$\frac{1}{q^2/r_i} + \frac{1}{W_i n_i} = \frac{1}{P_E} \ or \ \frac{1}{P_0} = \frac{1}{q^2} + \frac{1}{(Wrn)_i}; P_E = P_0/r_i$$

(2.18)

Here, W_i—electron orbital energy [27]; r_i—orbital radius of i orbital [28]; $q = Z^*/n^*$ [29, 30], n_i—number of electrons of the given orbital, Z^* and n^*—nucleus effective charge and effective main quantum number, r—bond dimensional characteristics.

P_0 was called a spatial-energy parameter (SEP), and P_E—effective P-parameter (effective SEP). Effective SEP has a physical sense of some averaged energy of valence electrons in the atom and is measured in energy units, for example, electron-volts (eV).

The values of P_0-parameter are tabulated constants for the electrons of the given atom orbital.

For dimensionality SEP can be written down as follows:

$$[P_0] = [q^2] = [E] \cdot [r] = [h] \cdot [v] = \frac{kg \cdot m^3}{s^2} = J \cdot m,$$

where, $[E]$, $[h]$ and $[v]$—dimensions of energy, Planck constant and velocity.

The introduction of P-parameter should be considered as further development of quasi-classical notions using quantum-mechanical data on atom structure to obtain the criteria of energy conditions of phase-formation. At the same time, for the systems of similarly charged (e.g., orbitals in the given atom), homogeneous systems, the principle of algebraic addition of such parameters are preserved:

$$\Sigma P_{\ni} = \Sigma \left(P_0 / r_i \right);$$

$$\Sigma P_E = \frac{\Sigma P_0}{r} \qquad (2.19)$$

or $\quad \Sigma P_0 = P_0{}' + P_0{}'' + P_0{}''' + \cdots;$

$$r \Sigma P_{\ni} = \Sigma P_0 \qquad (2.20)$$

Here P-parameters are summed on all atom valence orbitals.

In accordance with the conclusions obtained, in these systems the interactions take place against the potential gradient.

The principle of adding the reciprocals of energy characteristics of heterogeneous, oppositely charged systems—by the equation (2.18) is performed for the interactions proceeding along the potential gradient.

To calculate the values of P_E-parameter at the given distance from the nucleus depending on the bond type either atom radius (R) or ion radius (r_I) can be used instead of r.

The reliability of such approach is briefly discussed in [25, 31]. The calculations demonstrated that the value of P_E-parameters numerically equals (within

2%) total energy of valence electrons (U) by the atom statistic model. Using the known correlation between the electron density (β) and interatomic potential by the atom statistic model, we can obtain the direct dependence of P_E-parameter upon the electron density at the distance r_i from the nucleus:

$$\beta_i^{2/3} = \frac{AP_0}{r_i} = AP_E ,$$

where A—constant.

The rationality of such approach is confirmed by the calculation of electron density using wave functions of Clementi and its comparison with the value of electron density calculated via the value of P_E-parameter.

The correlations of the modules of maximum values of radial part of Ψ-function with the value of P_0-parameter are carried out and the linear dependence between these values is found. Using some properties of wave function with regard to P-parameter, we obtain the wave equation of P-parameter formally analogous to the equation of Ψ-function.

Lagrange function is the difference between kinetic (T) and potential (U) energies of the system:

$$L = T - U \cdot$$

Hamilton function can be considered as the sum of potential and kinetic energies, that is, as the total mechanic energy of the system:

$$H = T + U \cdot$$

From this equation and in accordance with energy conservation law we can see that:

$$H + L = 2T, \tag{2.21}$$

$$H - L = 2U \tag{2.22}.$$

Let us try to assess the movement of the isolated system of a free atom as a relative movement of its two subsystems: nucleus and orbital.

The atom structure is formed of oppositely charged masses of nucleus and electrons. In this system, the energy characteristics of subsystems are the orbital energy of electrons (W_i) and effective energy of atom nucleus taking screening effects into account.

In a free atom, its electrons move in Coulomb field of nucleus charge. The affective nucleus charge characterizing the potential energy of such subsystem taking screening effects into account equals q^2/r_i, where,

$$q = Z^*/n^*.$$

Here, Z^* and n^*—effective nucleus charge and effective main quantum number [29, 30]; r_i—orbital radius [31].

It can be presumed that orbital energy of electrons during their motion in Coulomb field of atom nucleus is mainly defined by the value of kinetic energy of such motion.

Thus, it is assumed that:

$$T \sim W \text{ and } U \sim q^2/r_i$$

In such approach, the total of the values W and q^2/r_i is analogous to Hamilton function (H):

$$W + q^2/r_i \sim H \qquad (2.23).$$

The analogous comparison of P-parameter with Lagrange function can be carried out when investigating Lagrange equation for relative motion of isolated system of two interacting material points with masses m_1 and m_2 in coordinate x, equations (2.18).

Thus, it is presumed that: $P_E \sim L$.

Then equation (2.24) is as follows:

$$\left(W + \frac{q^2}{r_i} \right) + P_E \approx 2W \qquad (2.24)$$

Using the values of electron bond energy as the orbital electron energy [27], the values of P_E-parameters of free atoms (Table 2.7) are calculated by equations (2.18). When calculating the values of effective P_E-parameter, mainly the atom radius values by Belov–Bokiy or covalent radii (for non-metals) were applied as dimensional characteristics of atom (R).

Mutual overlapping of orbitals of valence electrons in the atom in a first approximation can be considered via the averaging of their total energy dividing it by a number of valence electrons considered (N):

$$\left(\frac{q^2}{r_i} + W \right) \frac{1}{N} + P_E \approx 2W \qquad (2.25)$$

That is, the expression

$$\left(\frac{q^2}{r_i} + W \right) \frac{1}{N}$$

characterizes the average value of total energy of i-orbital in terms of one valence electron, and in this approach it is the analog of Hamilton function—H.

In free atoms of 1a group of Periodic system, s-orbital is the only valence orbital whose electron cloud deformations for three elements of long periods (K, Rb, Cs) were considered via the introduction of the coefficient $K = n/n^*$, where n—main quantum number, n^*—effective main quantum number:

$$\left(\frac{q^2}{r_i} + W \right) \frac{1}{KN} + P_E \approx 2W \qquad (2.26)$$

This means that $K = \dfrac{4}{3.7}; \dfrac{5}{4}; \dfrac{6}{4.2}$ for K, Rb, and Cs, respectively. For the others $K = 1$.

Besides, for the elements of 1a group the value $2r_i$ (i.e., the orbital radius of i-orbital) was used as a dimensional characteristic in the first component of equation (2.26). For hydrogen atom the ion radius was applied.

By the initial equations (2.22 and 2.23), the values

$$\left[\left(\frac{q^2}{r_i}+W\right)\frac{1}{N}+P_E\right]$$

and $2W$ were calculated and compared for the majority of elements. Their results are given in Table 2.7 Besides, the equation (2.25) can be transformed as:

$$\left(\frac{q^2}{r_i}+W\right)\frac{1}{N}-P_E \approx 2(W-P_E) \qquad (2.27)$$

and the correlation obtained is analogous to the equation (2.23). Actually, by the given analogy $W \sim T$ and $P_E \sim L$.

That is $(W-P_E) \sim (T-L)$, as $T-L=U$, then $(W-P_E) \sim U$ and equation (2.28) in general is analogous to the equation (2.23).

The analysis of the data given in Table 2.7 reveals that the proximity of the values investigated is mostly within 5% and only sometimes exceeds 5%, but never exceeds 10 %.

All this indicates that there is a certain analogy of equations (2.22) and (2.26), (2.23) and (2.27), and in a first approximation the value of P_E-parameter can be

considered the analog of Lagrange function and value $\left(\frac{q^2}{r_i}+W\right)\frac{1}{N}$ —analog of Hamilton function [31].

Thus:

- Two main types of adding energy characteristics of subsystems are defined by the directions of their structural interactions along the potential gradient or against it.
- Spatial-energy parameter obtained based on the principle of adding reciprocals of energy characteristics of subsystems can be assessed as Lagrangian analog.

TABLE 2.7 Comparison of energy atomic characteristics

Element	Valence electrons	W (eV)	r_i (Å)	q^2 (eVÅ)	P_0 (eVÅ)	R (Å)	$P_E = P_0/R$ (eV)	N	$\left(\frac{q^2}{r_i}+W\right)\frac{1}{N}+D_E$	2W (eV)
1	2	3	4	5	6	7	8	9	10	11
H	$1S^1$	13.595	$r_{\dot{E}}=$ 1.36	14.394	8.0933	1.362	2.9755	1	27.155	27.190
Li	$2S^1$	5.3416	1.586·2	5.8892	3.475	1.55	2.2419	1	9.440	10.683
Be	$2S^2$	8.4157	1.040	13.159	7.512	1.13	6.6478	2	17.182	16.831
B	$2P^1$	8.4315	0.776	21.105	4.9945	0.91	5.4885	3	17.365	16.863
	$2S^2$	13.462	0.769	23.890	11.092	0.91	12.189	3	27.032	26.924
C	$2P^2$	11.792	0.596	35.395	10.061	0.77	13.066	6	24.923	23.584
	$2S^2$	19.201	0.620	37.240	14.524	0.77	18.862	4	38.824	38.402
N	$2P^1$	15.445	0.4875	52.912	6.5916	0.55	11.985	7	29.562	30.890
	$2S^2$	25.724	0.521	53.283	17.833	0.71	25.117	5	50.716	51.448
O	$2P^1$	17.195	0.4135	71.383	6.4663	0.59	10.960	8	34.089	34.390
	$2S^2$	33.859	0.450	72.620	21.466	0.59	36.383	6	68.923	67.718
F	$2P^1$	19.864	0.3595	93.625	6.6350	0.64	10.367	9	41.511	39.728
Na	$3S^1$	4.9552	1.713·2	10.058	4.6034	1.89	2.4357	1	10.327	9.9104

TABLE 2.7 *(Continued)*

		1	2	3	4	5	6	7	8	9
Mg	$3S^1$	6.8859	1.279	17.501	5.8588	1.60	3.6618	2	13.946	13.772
Al	$3P^1$	5.713	1.312	26.443	5.840	1.43	4.084	3	12.707	11.426
	$3S^2$	10.706	1.044	27.119	12.253	1.43	8.5685	3	20.796	21.412
Si	$3P^2$	8.0848	1.068	29.377	10.876	1.34	8.1169	4	17.014	16.170
	$3S^2$	14.690	0.904	38.462	15.711	1.17	13.428	4	17.737	29.380
	$3P^2+3S^2$								44.75	45.50
P	$3P^3$	10.659	0.9175	38.199	16.594	1.10	15.085	7	21.551	21.318
	$3S^1$	18.951	0.803	50.922	11.716	1.10	10.651	3	38.106	37.902
S	$3P^2$	11.901	0.808	48.108	13.740	1.04	13.215	6	25.122	23.802
	$3S^2$	23.933	0.723	64.852	22.565	1.04	21.697	4	50.105	47.866
Cl	$3P^1$	13.780	0.7235	59.844	8.5461	1.00	8.5461	5	27.845	27.560
1	2	3	4	5	6	7	8	9	10	11
K	$4S^1$	$4.0130\,\frac{3.7}{4}$	2.6122	$10.993\,\frac{3.7}{4}$	4.8490	2.36	2.0547	1	7.7137	8.026
Ca	$4S^1$	5.3212	1.690	17.406	5.929	1.97	3.0096	2	10.820	10.642
Sc	$4S^2$	5.7174	1.570	19.311	9.3035	1.64	5.6729	3	11.679	11.435

TABLE 2.7 (Continued)

Ti	$4S^2$	6.0082	1.477	20.879	9.5934	1.46	6.5708	4	11.230	12.016
V	$4S^2$	6.2755	1.401	22.328	9.8361	1.34	7.3404	4	12.894	12.549
	$4S^2$	6.2755	1.401	22.328	9.8361	1.34	7.3404	5	11.783	12.549
Cr	$4S^1$	6.5238	1.453	23.712	6.7720	1.27	5.3323	3	12.947	13.048
	$4S^2$	6.5238	1.453	23.712	10.535	1.27	8.2953	5	12.864	13.048
Mn	$4S^2$	6.7451	1.278	25.118	10.223	1.30	7.8638	5	13.144	13.490
Fe	$4S^1$	7.0256	1.227	26.572	6.5089	1.26	5.1658	3	14.716	14.051
	$4S^2$	7.0256	1.227	26.572	10.456	1.26	8.2984	5	14.035	14.051
Co	$4S^2$	7.2770	1.181	27.973	10.648	1.25	8.5184	5	14.705	14.544
Ni	$4S^2$	7.5176	1.139	29.248	10.815	1.24	8.7218	5	15.361	15.035
Cu	$4S^2$	7.7485	1.191	30.117	11.444	1.28	8.9406	5	15.548	15.497
Zn	$4S^2$	7.9594	1.065	32.021	11.085	1.37	8.0912	5	15.580	15.919
Ga	$4P^1$	5.6736	1.254	34.833	5.9081	1.39	4.2504	5	10.941	11.347
	$4S^2$	11.544	0.960	44.940	14.852	1.39	10.685	5	22.356	23.086
Ge	$4P^2$	7.8190	1.090	41.372	12.072	1.39	8.6649	6	16.294	15.636
	$4S^2$	15.059	0.886	58.223	18.298	1.39	13.164	5	29.319	30.116
	$4P^2+4S^2$								45.613	45.752

TABLE 2.7 *(Continued)*

1	2	3	4	5	6	7	8	9	10	11
As	4P¹	10.054	0.992	49.936	8.275	1.17	7.0726	5	19.152	20.108
	4S²	18.664	0.826	71.987	21.587	1.40	15.419	5	36.582	37.328
Se	4P¹	10.963	0.909	61.803	8.5811	1.14	7.5272	6	20.656	21.924
	4S²	22.787	0.775	85.678	25.010	1.17	21.376	6	43.599	45.474
Br	4P¹	12.438	0.8425	73.346	9.1690	1.96	4.6781	5	24.577	24.876
Rb	5S¹	$3.7511\,\frac{4}{5}$	$2.287\cdot2$	$14.309\,\frac{4}{5}$	5.3630	2.48	2.1625	1	7.666	7.5022
Sr	5S¹	4.8559	1.836	21.224	6.2790	2.15.2	1.4663	2	9.6713	9.7118
y	5S²	6.3376	1.693	22.540	10.030	1.81	5.5417	3	12.092	12.675
Zr	5S²	5.6414	1.593	23.926	10.263	1.60	6.4146	4	11.580	11.283
Nb (5S²4d³)	5S²	5.8947	1.589	25.822	10.857	1.45	7.4876	5	11.917	11.789
Mo (5S²4d⁴)	5S²	6.1140	1.520	28.027	11.175	1.39	8.0396	6	12.132	12.223
Tc	5S²	6.2942	1.391	30.076	11.067	1.36	8.1376	7	12.126	12.588

TABLE 2.7 *(Continued)*

Ru $(5S^2 4d^6)$	$5S^2$	6.5294	1.410	31.986	11.686	1.34	8.7208	8	12.373	13.059
Rh $(5S^2 4d^7)$	$5S^2$	6.7240	1.364	33.643	11.871	1.34	8.8588	8	12.782	13.448
Pd $(5S^2 4d^8)$	$5S^2$	6.9026	35.377	1.325	12.057	1.37	8.8007	8	13.00	13.805
Ag $(5S^2 4d^{10})$	$5S^1$	7.0655	1.286	26.283	6.752	1.44	4.6889	3	13.867	14.131
Cd	$5S^1$	7.2070	1.184	38.649	6.9898	1.56	4.4806	4	14.443	14.414
Jn	$5P^1$	5.3684	1.382	41.318	6.2896	1.66	3.7869	5	10.842	10.737
	$5S^2$	10.141	1.093	52.103	15.551	1.66	9.3681	5	20.930	20.282
Sn	$5P^1$	7.2124	1.240	47.714	7.5313	1.42	5.3037	5	14.442	14.424
	$5S^2$	12.965	1.027	65.062	18.896	1.42	13.307	6	26.027	25.930
Sb	$5P^1$	9.1072	1.1665	57.530	8.9676	1.39	6.5519	5	18.237	18.214
	$5S^2$	15.833	0.969	77.644	21.993	1.39	15.822	5	31.816	31.666
Te	$5P^1$	9.7907	1.087	67.285	9.1894	1.37	6.7076	6	18.656	19.581
	$5S^2$	19.064	0.920	90.537	25.283	1.37	18.454	6	38.033	38.128

TABLE 2.7   (Continued)

1	2	3	4	5	6	7	8	9	10	11
J ($5S^2 5p^5$)	$5P^1$	10.971	1.0215	77.651	9.7936	1.35	7.2545	5 / 7	(21.753)	21.942
Cs	$6S^1$	$3.3647\ \frac{6}{4.2}$	2.518·2	$16.193\ \frac{6}{4.2}$	5.5628	2.68	2.0757	1	6.6914	6.7294
Ba	$6S^2$	4.2872	2.060	22.950	9.9812	2.21	4.5164	4	8.3734	8.5744
La ($6S^2 5p^1 4f^1$)	$6S^1$	4.3528	1.915	34.681	6.7203	1.87	3.5937	4	9.2094	8.706
Hf	$6S^1$	5.6863	1.476	33.590	6.7151	1.59	4.2233	4	11.334	11.373
Ta	$6S^1$	5.9192	1.413	36.285	6.7971	1.46	4.6555	5	10.975	11.838
W(V)	$6S^1$	6.1184	1.360	38.838	6.8528	1.40	4.8949	5	11.830	12.237
Re(V)	$6S^1$	6.2783	1.310	40.928	6.8483	1.37	4.9988	5	12.503	12.557
Os(VI)	$6S^1$	6.4995	1.266	42.620	7.7344	1.35	5.7292	6	12.423	12.999
Ir(VI)	$6S^1$	6.6788	1.227	44.655	7.7691	1.35	5.7549	6	12.934	13.358
Pt(VI)	$6S^1$	6.8377	1.221	46.231	7.0718	1.38	5.1245	6	12.575	13.675

TABLE 2.7 *(Continued)*

Au	6S¹	6.9820	1.187	47.849	7.0641	1.44	4.9056	5	14.364	13.964
	6S²	6.9820	1.187	47.849	12.311	1.44	8.5491	8	14.461	13.964
Hg	6S¹	7.1037	1.126	49.432	6.8849	1.60	4.3031	5	14.504	14.207
	6S²	7.1037	1.126	49.432	12.086	1.60	7.5538	8	13.929	14.207
Tl	6P¹	5.2354	1.319	60.054	6.1933	1.71	3.6218	5	19.567	19.654
	6S¹	9.8268	1.060	65.728	8.9912	1.71	5.1995	8	9.9675	10.421
Pb	6P²	7.0913	1.215	61.417	13.460	1.75	7.6914	8	14.897	14.183
	6S²	12.416	1.010	79.515	19.066	1.75	12.711	8	24.104	29.832
Bi	6P¹	8.7076	1.2125	71.171	9.0406	1.82	4.9674	5	18.448	17.45
	6S¹	15.012	0.963	92.892	12.50	1.51	8.2841	5	30.578	30.024
	6S²	15.012	0.963	92.892	22.050	1.51	14.603	8	28.537	30.024
P_0	6P¹	9.2887	1.1385	80.881	9.3523	1.50	6.2349	6		18.577
								8	(17.950)	
	6S¹	17.909	0.923	106.65	14.312	1.50	9.5413	5	36.233	35.818
	6S²	17.909	0.923	106.65	25.237	1.50	16.825	8	33.507	35.818
At	6P¹	10.337	1.078	91.958	9.9074	1.39	7.1276	8	19.083	20.674
	6S²	20.828	0.885	119.70	28.185	1.39	20.277	8	39.787	41.756

2.2.2   Dependence of Constants of Electromagnetic Interactions upon Electron Spatial-Energy Characteristic

By the analogy with basic equation of ESCA method, the simple dependence of magnetic and electric constants upon electron spatial-energy parameter is obtained with semi-empirical method. The relative calculation error is under 0.1%.

The comparison of multiple regularities of physical and chemical processes allows assuming that in such and similar cases the principle of adding reciprocals of volume energies or kinetic parameters of interacting structures are realized.

Thus the Lagrangian equation for relative motion of the system with two inter-acting material points with masses m_1 and m_2 in coordinate x with reduced mass m_r, can be written down as follows:

$$\frac{1}{\Delta U} \approx \frac{1}{\Delta U_1} + \frac{1}{\Delta U_2},$$

where, ΔU_1 and ΔU_2 —potential energies of material points, ΔU —resultant potential energy of the system.

The electron with mass m, moving near the proton with mass M, is equivalent

to the particle with mass $\quad m_r = \dfrac{mM}{m+M}$.

Therefore, assuming that the energy of atom valence orbitals (responsible for interatomic interactions) can be calculated following the principle of adding the reciprocals of some initial energy components, the introduction of P–parameter as an averaged energy characteristic of valence orbitals is assigned [25] based on the following equations:

$$\frac{1}{q^2/r_i} + \frac{1}{W_i n_i} = \frac{1}{P_E}$$

$$\text{or} \quad \frac{1}{P_0} = \frac{1}{q^2} + \frac{1}{(Wrn)_i};$$

$$P_E = P_0/r \qquad\qquad (2.28)$$

Where, W_i—orbital energy of electrons [27]; r_i—orbital radius of i-orbital [28]; $q = Z^*/n^*$, n_i—number of electrons of the given orbital, Z^* and n^*—nucleus effective charge and effective main quantum number [29, 30]; r—bond magnitude.

In equations (2.28), the parameters q^2 and Wr can be considered as initial (primary) values of P_0-parameter that are tabulated constants for the electrons of the given atom orbital. For its dimensionality it can be written down as follows:

$$[P_0] = [q^2] = [E] \cdot [r] = [h] \cdot [\upsilon] = \frac{kg \cdot m^3}{s^2} = $$

where, [E], [h] and [υ]—dimensionality of energy, Plank constant and velocity.

At the same time, for like-charged systems (for example, orbitals in the given atom), homogeneous systems, the principle of algebraic addition of such parameters is preserved:

$$\Sigma PE = \Sigma (P0/ri);$$

$$\Sigma PE = \frac{\Sigma P_0}{r} \qquad (2.29).$$

Applying the equation to hydrogen atom for initial values of P-parameters we can obtain the following:

$$K \left(\frac{e}{n} \right)_1^2 = K \left(\frac{e}{n} \right)_2^2 + mc^2 \lambda \qquad (2.30)$$

where, e—elementary charge, n_1 and n_2—main quantum numbers, m—electron mass, c—electromagnetic wave velocity, λ—wave length, K—constant.

Using the known correlations $V = C/\lambda$ and $\lambda = h/mc$ (where, h—Plank constant, v—wave frequency) from the formula (5), we can obtain the equation of spectral regularities in hydrogen atom, in which $2\pi^2 e^2/hc = K$.

Taking into account the main quantum characteristics of atom sublevels, and based on the equation we have the following:

$$\Delta P_E \approx \frac{\Delta P_0}{\Delta x} \text{ or}$$

$$\partial P_E = \frac{\partial P_0}{\partial x},$$

where, the value ΔP_0 equals the difference between P_0-parameter of i-orbital and P_{cn}—parameter of counting (parameter of the main state at the given set of quantum numbers).

According to the rule of adding P-parameters of like-charged or homogeneous systems for two orbitals in the given atom with different quantum characteristics and in accordance with energy conservation law we have:

$$\Delta P_E'' - \Delta P_E' = P_{E,\lambda},$$

where, $P_{E,\lambda}$—spatial energy parameter of quantum transition.

Taking the interaction $\Delta\lambda = \Delta x$ as a magnitude we have:

$$\frac{\Delta_0''}{\Delta\lambda} - \frac{\Delta P_0'}{\Delta\lambda} = \frac{P_0}{\Delta\lambda} \text{ or } \frac{\Delta_0'}{\Delta\lambda} - \frac{\Delta P_0''}{\Delta\lambda} = -\frac{P_0}{\Delta\lambda}$$

Let us divide again by term by $\Delta\lambda = \dfrac{\left(\dfrac{\Delta_0'}{\Delta\lambda} - \dfrac{\Delta P_0''}{\Delta\lambda}\right)}{\Delta\lambda} = -\dfrac{P_0}{\Delta\lambda^2}$

where,

$$\frac{\left(\dfrac{\Delta_0'}{\Delta\lambda} - \dfrac{\Delta P_0''}{\Delta\lambda}\right)}{\Delta\lambda} \approx \frac{d^2 P_0}{\Delta\lambda^2}, \text{ that is,}$$

$$\frac{d^2 P_0}{d\lambda^2} + \frac{P_0}{\Delta\lambda^2} \approx 0.$$

Taking into account only those interactions when $2\pi\Delta x = \Delta\lambda$ (closed oscillator), we get the equation similar to Schroedinger equation for stationary state in coordinate x:

$$\frac{d^2 P_0}{dx^2} + 4\pi^2 \frac{P_0}{\Delta\lambda^2} \approx 0.$$

Since $\Delta\lambda = \dfrac{h}{mv}$, then:

$$\frac{d^2 P_0}{dx^2} + 4\pi^2 \frac{P_0}{h^2} m^2 v^2 \approx 0$$

or $\dfrac{d^2 P_0}{dx^2} + \dfrac{8\pi^2 m}{h^2} P_0 E_k = 0$

where, $E_k = \dfrac{mv^2}{2}$ is electron kinetic energy.

Besides, we obtained the correlation [25] between the values of P_E-parameter and energy of valence electrons in atom statistic model, which allowed assuming that P-parameter is the direct characteristic of electron charge density. Therefore, the exchange spatial-energy interactions based on the equalization of electron densities apparently have broad manifestation in physical–chemical processes [31–34].

The value of the relative difference of P-parameters of atom's components was used as a complex characteristic of structural interactions:

$$\alpha = \frac{P_1 - P_2}{(P_1 + P_2)/2} \ or \ \Delta P = \frac{\alpha}{2}\Sigma P \tag{2.31}$$

The equalization of electron densities of interacting components brings their unbalanced system into the equalized state. Similarly with diffusion processes of transfer. The general equation of mass transfer (m) in x direction (diffusion) is as follows:

$$11 \ m = -D\frac{\Delta\rho}{\Delta x} St,$$

where, D—diffusion coefficient, S—transfer square, t—transfer time, $\dfrac{\Delta\rho}{\Delta x}$ —density gradient. Or

$$m\Delta x = -D\Delta\rho St.$$

As applicable to elementary particles, we can consider that $\Delta\rho$ —value characterizing the energy mass density change, where $mc^2 = E$ (MeV).

As $E_r = P_e$—electron parameter, then there must be a functional bond between P_e-parameter and constants of electromagnetic interactions. The electron classical radius:

$$r = \frac{e^2}{mc^2}, P_e \sim e^2$$

The value of initial P_e-parameter for the electron:

$$P_e = 0.5110 \cdot 2.8179 = 1.43995 \, MeVfm_{.}$$

In experimental and theoretical investigations in electron spectroscopy of chemical analysis (ESCA) the following equation is used as the basic one [35]:

$$\Delta W_i = e \sum \frac{1-K}{d_i} q_i \qquad (2.32)$$

where, ΔW_i—"shift" in the bond energy (change in the bond energy); d_i—internuclear distances; K—constant characterizing the element (approximately equal to the electrostatic interaction between the skeleton and valence electrons in a free atom); q_i—effective transferred charge. As applicable to the binary bond, we have:

$$\Delta W_i d_i = \Delta P = (1 - K)q^2 \qquad (2.33)$$

Equation (8) is similar to equation (6), but the calculations in it are made for initial (primary) P-parameters q^2 and Wd.

Considering the electric charge transfer as the electric current generating the magnetic field under vacuum with magnetic constant μ_0, we can assume, by analogy with equation (2.33), that in more elementary processes:

$$(1-k)q^2 \sim (k\mu_0)^2$$
$$\text{Then, } P_e \sim (k\mu_0)^2,$$

where, k—proportionality coefficient.

The calculations revealed that in these cases the simple correlation for the magnetic constant is fulfilled:

$$(2\pi\mu_0)^2 = 3 \ _sA^2 \rightarrow 4\pi^2\mu_0^2 = 3 \ _sA^2 \rightarrow k\mu_0 = \tfrac{1}{2}A \qquad (2.34),$$

where, $r = \dfrac{e^2}{mc^2}$; number 3 is apparently determined by the number of elec

trons interacting with three quarks or three protons. Transfer coefficient from $MeVfm$ into Jm: $10^6 \times 1.602 \times 10^{-19} \times 10^{-15} = 1.602 \times 10^{-28} \ Jm$.

Then, $P_e = 1.43995 \times 1.602 \times 10^{-28} = 2.3068 \times 10^{-28}$ Jm.

Calculation of μ_0 by equation (9) gives

$$\mu_0 = 1.2554 \times 10^{-6} \frac{Gn}{m}.$$

The relative error in comparison with the actual value of μ_0 is below 0.1%.

Value k^2 gets the dimensionality Jm^3/Gn^2, where the numerator characterizes the volume values of P-parameter.

As $P_e \sim e^2$, then the dimensionality $[P_e^{1/2}c]$ gives the physical sense of magnetic constant as the current element for the elementary charge: [Am].

The known correlation between three fundamental constants of electromagnetic interactions can be represented as follows:

$$\mu_0 c = 1 / \left(\varepsilon_0 c\right)$$

where, ε_0—electric constant.

Using this dependence together with equation (2.34), and after the relevant calculations we can obtain the equation P-parameter bond with constants of electromagnetic interactions:

$$k\mu_0 c = \frac{k}{\varepsilon_0 c} = P_e^{\frac{1}{2}} c^2 \approx 1366 (AM) \frac{M}{c} \qquad (2.35)$$

The simple dependence of electromagnetic constants upon initial electron spatial-energy characteristics is obtained with semi-empirical method.

2.3 THE APPLICATION OF AVRAMI EQUATIONS FOR CALCULATING EXPERIMENTS OF METAL OR CARBON NANOCOMPOSITES SYNTHESIS IN POLYMERIC MATRIXES

Different shapes of nanostructures appear during the formation of lamellar or linear substances. For instance, graphite, clay, mica, asbestos, and many silicates have lamellar structure, and consequently, they contain nanostructures with the shape of rotation bodies. It can be explained by the facts that bands of lamellar structure are rolled into clews and spirals with further sewing between them and formation of spheres, cylinders, ellipsoids, cones, and so on. Nanostructures formed can be classified based on complexity. The formation of film structures resembling petals that can be put together into segments and semi-spheres afterwards, is possible for carbon-containing structures that are formed mainly from hexagons with a certain number of pentagons and insignificant number of heptagons. Under certain energy actions, the distortion of coplanar (flat) aromatic rings are known, when π-electrons are shifted and charges are separated on the ring, the ring polarity goes up. For polymeric chains and bands formed from macromolecules the possibilities of formation of permolecular structures, corresponding to the possibility of taking a definite shape, can be predicted with the help of Avrami's equation [36] for one-, two-, and three-dimensional crystallization of macromolecules. Here, energy exchange during the formation of ordered nanostructures and without energy exchange with the surroundings, that is, under athermal and thermal embryoformation is considered.

The analysis of changes in the values of power n in Avrami's equation is appropriate:

$$1-v^c = \exp\{-kt^n\} \qquad (2.36),$$

where, $1-v^c$—corresponding degree of order during the formation a definite shape, v^c—volumetric share of crystallized substance ($v^c = 1-\exp(-4\pi v^3 t^3/3)$ —for athermal embryo-formation, $v^c = 1-\exp(-\pi I^* v^3 t^4/3)$—for thermal embryo-formation, v—linear rate of crystal growth (ordered nanostructure), t—nanostructure growth time from the beginning of embryo growth, I^*—rate of embryo formation.

Then k in the first case corresponds to $4\pi v^3/3$, in the second—$\pi I^* v^3/3$.

Let us show the types of embryo growth depending upon n from Avrami's equation and embryo-formation type (Table 2.8).

TABLE 2.8   Types of embryo growth by Avrami's equation [36].

Growth type	Embryo-formation type	n from the equation	Equation
One-dimensional growth			
Line (chain formation)	Athermal	1	$1-v^c=\exp\{-\pi N v^2 \tau^2\}$, where N – number of embryos on the surface unit
Same	Thermal	2	$1-v^c=\exp\{-\pi I^* v^2 \tau^3/3\}$
Two-dimensional growth			
Band	Athermal	≤1	$1-v^c=(-N d^2 v\tau/2)$, where d—diameter of fibrils*, which is unchanged
Band	Thermal	≤2	$1-v^c=\exp\{-\pi I^* d^2 v\tau^2/4\}$
Circle	Athermal	2	$1-v^c=\exp\{-\pi N v^2 \tau^2\}$
Circle	Thermal	3	$1-v^c=\exp\{-\pi I^* v^2 \tau^3/3\}$
Three-dimensional growth			
Fibril*	Athermal	≤1	$1-v^c = \exp\{-N d^2 v\tau/2\}$
Fibril*	Thermal	≤2	$1-v^c = \exp\{-\pi I^* d^2 v\tau^2/4\}$

TABLE 2.8   *(Continued)*

Lamellar circle	Athermal	≤ 2	$1-v^c = \exp(-\pi N dv^2\tau^2)$, where d—fold thickness
Lamellar circle	Thermal	≤ 3	$1-v^c = \exp(-\pi I^* dv^2\tau^3/3)$
Sphere	Athermal	3	$1-v^c = \exp\{-4\pi v^3\tau^3/3\}$
Sphere	Thermal	4	$1-v^c = \exp\{-\pi I^* v^3\tau^4/3\}$
Solid beams	Athermal	≥ 5	$1-v^c = \exp[-k\tau^5]$
Solid beams	Thermal	≥ 6	$1-v^c = \exp[-k\tau^6]$
Truncated sphere	Athermal	2–3	$1-v^c = \exp[-\pi dv^2\tau^2]$ $1-v^c = \exp[-\pi I^* dv^2\tau^3]$
Truncated sphere	Thermal	3–4	The same equations

According to the data, per molecular structures of polymers, that are nano-sized, are formed practically analogously to those obtained from inorganic, band or film nanostructures.

Linear structures can be the basis for obtaining clews, fiber, and rope nano-structures.

When being rolled band nanostructures can form lamellar (folded) nanostructures, fibrillar (rod) nanostructures, spiral, and bamboo-like nanostructures. When several bands fold (Figure 2.8a) [37] rod extensive nanostructures, called fibrils in polymer science, are formed. If one band folds several times in space, it folds periodically with identical steps and formation of a fold d of definite thickness (Figure 2.8b) [37], lamellas able to roll into the circle are obtained.

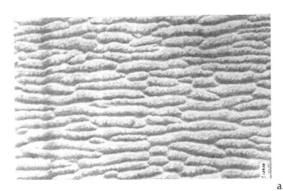

a.

FIGURE 2.8   *(Continued)*

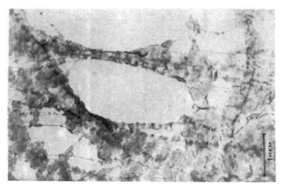

b.

FIGURE 2.8   Nanostructures formed from bands: a fibrils, b lamellas.

As mentioned before, at two-dimensional growth circles that form petals and further more complex nanostructures resembling flowers can be formed [38]. When being rolled, the circles formed can produce distorted semi-spheres or can serve as a basis for the transformation into "a beady", it was proposed to use the thermodynamics of small systems and Avrami's equations to describe the formation processes of carbon nanostructures during re-crystallization (graphitization) [39, 40]. These equations are successfully applied [36] to forecast per molecular structures and prognosticate the conditions on the level of parameters resulting in the obtaining of nanostructures of definite size and shape. The equation was also used to forecast the formation of fibers [41]. The application of Avrami's equations in the processes of nanostructure formation: a) embryo formation and crystal growth in polymers [36]—$(1-v) = \exp[-k\tau^n]$, where v—crystallinity degree, τ—duration, k—value corresponding to the process specific rate, n—number of the degrees of freedom changing from 1 to 6; b) graphitization process with the formation of carbon nanostructures [39, 40]—$v = 1-\exp[-B\tau^n]$, where v—volume share that was changed, τ—duration, B—index connected with the process rate, n—value determining the process directedness; c) process of fiber formation [41]—$\omega = 1-\exp[-z\tau^n]$, where ω—share of the fiber formed, τ— process duration, z— statistic sum connected with the process rate constant, n—number of the degrees of freedom (for the fiber n equals 1). Instead of k, B or z based on the previous considerations, the activity of nanoreactors can be used (a). Then the equation for defining the share (W) of nanostructures formed can be written down as follows by analogy with the aforesaid equations:

$$1-W = \exp[-\beta a \tau^n] \tag{2.37}$$

where, β—coefficient taking into account the changes in the activity in the process of nanoproduct formation.

Substituting the value of a, we get the dependence of the share of nanostructures formed upon the ratio of the energy of nanoreactor internal surface to its volume energy:

$$W = 1—\exp[-\beta(\varepsilon_s/\varepsilon_v)\tau^n] = 1—\exp[-\beta(\varepsilon_s^0/\varepsilon_v^0 \, S/V)\tau^n] \qquad (2.38)$$

If the nanoreactor internal walls become the shells of nanostructures during the process, the nanostructures so obtained are the nanoreactor mirror reflections.

The formation processes of metal-containing nanostructures in carbon or carbon-polymeric shells in nanoreactors can be related to one type of reaction series using the terminology of the theory of linear dependencies of free energies (LFE) [42]. Then, it is useful to introduce definite critical values for the volume, surface energy of nanoreactor internal walls, as well as the temperature critical value. When the ration $\lg k/k_c$ is proportional $-\Delta\Delta F/RT$, the ratio W/W_c can be transformed into the following expression:

$$W/W_c = b\cdot\exp\{-(k/k_c)\cdot(\tau/\tau_c)^n\} = b\cdot\exp\{-(\tau/\tau_c)^n \exp(-\Delta\Delta F/RT)\} = b\cdot\exp\{(\tau/\tau_c)^n[\exp k_T k_{vs}(\varepsilon_v/\varepsilon_{Vk} - \varepsilon_s/\varepsilon_{Sk})\theta/T]\} \qquad (2.39)$$

where b—proportionality coefficient considering the temperature factor, k_{vs}—coefficient considering correlations $\varepsilon_v/\varepsilon_{Vk}$ and $\varepsilon_s/\varepsilon_{Sk,}$ ε_v and ε_{Vk}—volume energies of nanoreactor and "equilibrium" nanoreactor calculated via the ratios of their volumes; ε_s and ε_{Sk}—surface energy and its equilibrium value, T and θ—temperature of the process and temperature of the equilibrium process; τ—time required to develop the process of nanostructure formation; n—index of the process directedness to the formation of nanostructures of definite shapes. The values of volume and surface energies are given after the transformation of $\Delta\Delta F$ in accordance with [43], in which the physical sense of Taft constants is substantiated using the indicated energies.

At the same time, the share of nanostructures (W) during the redox process can depend upon the potential of nanoreactor wall interaction with reagents, as well as the number of electrons participating in the process. The metal ions in nanoreactor are reduced and its internal walls are partially oxidized (transformation of hydrocarbon fragments into carbon ones). Then the Helmholtz thermodynamic potential (ΔF) is proportional to the product of $zF\Delta\varphi$ and Avrami equations will be expressed by the following formulae in accordance with the above models, one of them can be as follows:

$$W = 1 - k \exp[-\tau^n \exp(zF\Delta\varphi/RT)] \qquad (2.40),$$

where k—proportionality coefficient considering the temperature factor, n—index of the process directedness to the formation of nanostructures of definite shapes; z—number of electrons participating in the process; $\Delta\varphi$—difference of potentials on the border "nanoreactor wall—reaction mixture"; F—Faraday number; R—universal gas constant. When n equals 1, one-dimensional nanostructures are obtained (linear nanosystems and narrow bands). If n equals 2 or changes from 1 to 2, narrow flat nanostructures are formed (nanofilms, circles, petals, broad nanobands). If n changes from 2 to 3 and over, spatial nanostructures are formed, since n also means the number of degrees of freedom. If in this equation we take k as 1 and consider the process in which copper is reduced with simultaneous formation of nanostructures of a definite shape, the share of such formations or transformation degree can be connected with the process duration.

The experimental modeling of obtaining nanofilms after the alignment of copper compounds with polyvinyl alcohol at 200°C revealed that optimal duration when the share of nanofilms approaches 100% equals 2.5 hrs. This corresponds to the calculated value based on the aforesaid Avrami's equation. The calculations are made supposing the formation of copper nanocrystals on the nanofilms. It is pointed out that copper ions are predominantly reduced to metal. Therefore, it was accepted for the calculations that n equals 2 (two-dimensional growth), potential of redox process during the ion reduction to metal ($\Delta\varphi$) equals 0.34V, temperature (T) equals 473K, Faraday number (F) corresponds to 26.81 (A·hr/mol), gas constant R equals 2.31 (W·hr/mol·degree). The analysis

of the dimensionality shows the zero dimension of the ratio, $\dfrac{zF\Delta\phi}{RT}$. The calculations are made when changing

the process duration with a half-hour increment:

Duration (hrs)	0.5	1.0	1.5	2.0	2.5
Content of nanofilms (%)	22.5	63.8	89.4	98.3	99.8

If nanofillms are scrolled together with copper nanowires, β is taken as equaled to 3, the temperature increases up to 400°C, the optimal time when the transformation degree reaches 99.97%, corresponds to the duration of 2hrs, thus also coinciding with the experiment. According to the calculation results if following the definite conditions of the system exposure, the duration of the exposure has the greatest influence on the value of nanostructure share. The selection of the corresponding equation form depends upon the shape and sizes of nanoreactor (nanostructure) and defines the nanostructure growth in nanoreactor or the influence distribution of the nanostructure on the structurally changing medium. With one-dimensional growth and when the activation zero is nearly zero, the equation for the specific rate of the influence distribution via the oscillations of one bond can be written down as follows:

$$W = 1—\exp[-\beta\, v\tau^n]$$
(2.41),

where v—oscillation frequency of the bond through which the nanostructure influences upon the medium, β—coefficient considering the changes in the bond oscillation frequency in the process. In the case discussed the parameter βv can be represented as the ratio of frequencies of bond oscillations v_{is}/v_{fs}, which are changing during the process. At the same time v_{is} corresponds to the frequency of skeleton oscillations of C–C bond at $1,100 cm^{-1}$, v_{fs}—symmetrical skeleton oscillations of C=C bond at $1,050 cm^{-1}$. In this case the equation looks as follows:

$$W = 1 - \exp\left\{-\tau^n \cdot \frac{v_{is}}{v_{fs}}\right\}$$
(2.42)

For the example discussed the content of nanofilms in % will be changing together with the changes in the duration as follows:

Duration (hrs)	0.5	1.0	1.5	2.0	2.5
Content of nanofilms (%)	23.0	64.9	90.5	98.5	99.9

By the analogy with the above calculations the parameters a in the equation (5) should be considered as a value that reflects the transition from the initial to final state of the system and represents the ratios of activities of system states. Under the aforesaid conditions the linear sizes of copper (from ion radius to atom radius) and carbon-carbon bond (from C–C to C=C) are changing during the process. Apparently the structure of copper ion and electron interacts with electrons of the corresponding bonds forming the layer with linear sizes $r_i + l_{C-C}$ in the initial condition and the layer with the size $r_a + l_{C=C}$ in the final condition. Then the equation for the content of nanofilms can be written down as follows:

$$W = 1 - \exp\left\{ -\tau^n \cdot \frac{r_\alpha + l_{C=C}}{r_i + l_{C-C}} \right\} \qquad (2.43)$$

At the same time r_i for Cu^{2+} equals 0.082 nm, r_a for four-coordinated copper atom corresponds to 0.113 nm, bond energy C–C equals 0.154 nm, and C=C bond—0.142 nm. Representing the ratio of activities as the ratio of corresponding linear sizes and taking the value n as equaled to 2, at the same time changing τ in the same intervals as before, we get the following change in the transformation degree based on the process duration:

Duration, hours	0.5	1.0	1.5	2.0	2.5
Content of nano-films, %	23.7	66.0	91.2	98.7	99.9

Thus, with the help of Avrami's equations or their modified analogs we can determine the optimal duration of the process to obtain the required result. It opens up the possibility of defining other parameters of the process and characteristics of nanostructures obtained (by shape and sizes).

The influence of nanostructures on the media and compositions can be assessed with the help of quantum-chemical experiment and Avrami's equations.

At the same time, the metal orientation proceeds in interface regions and nano-pores of polymeric phase which conditions further direction of the process to the formation of metal/carbon nanocomposite. In other words, the birth and growth of nanosize structures occur during the process in the same way as known from the macromolecule physics [8], in which Avrami's equations are successfully used. The application of Avrami's equations to the processes of nanostructure formation was previously discussed in the papers dedicated to the formation of ordered shapes of macromolecules [36], formation of carbon nanostructures by electric arc method [39], obtaining of fiber materials [41].

As follows from Avrami's equation:

$$1 - \upsilon = \exp[-k\tau^n], \qquad (2.44)$$

where v—crystallinity degree, τ—duration, k—value corresponding to specific process rate, n—number of degrees of freedom changing from 1 to 6, the factor under the exponential is connected with the process rate with the duration (time) of the process. Under the conditions of the isothermal growth of the ordered system "embryo", it can be accepted that the nanoreactor activity will be proportional to the process rate in relation to the flowing process. Then the share of the product being formed (W) in nanoreactor will be expressed by the following equation:

$$W = 1—\exp(-a\tau^n) = 1—\exp[-(\varepsilon_s/\varepsilon_V)\tau^n] = 1—\exp\{-[(\varepsilon^0_s d/\varepsilon^0_V)S/V]\tau^n\}, \quad (2.45)$$

where, a—nanoreactor activity, ε_s—surface energy reflecting the energy of interaction of reagents with nanoreactor walls, ε_V—nanoreactor volume energy, $\varepsilon^0_s d$—multiplication of surface layer energy by its thickness, ε^0_V—energy of nanoreactor volume unit, S—surface of nanoreactor walls, V—nanoreactor volume.

When the metal ion moves inside the nanoreactor with redox interaction of ion (mol) with nanoreactor walls, the balance setting in the pair "metal containing—polymeric phase" can apparently be described with the following equation:

$$zF\Delta\varphi = RT\ln K = RT\ln(N_p/N_r) = RT\ln(1—W), \quad (2.46)$$

where z—number of electrons participating in the process; $\Delta\varphi$—difference of potentials at the boundary "nanoreactor wall—reactive mixture"; F—Faraday number; R—universal gas constant; T—process temperature; K—process balance constant; N_p—number of moles of the product produced in nanoreactor; N_r—number of moles of reagents or atoms (ions) participating in the process which filled the nanoreactor; W—share of nanoproduct obtained in nanoreactor.

In turn, the share of the transformed components participating in phase interaction can be expressed with the equation which can be considered as a modified Avrami's equation:

$$W = 1—\exp[-\tau^n\exp(zF\Delta\varphi/RT)], \quad (2.47)$$

where τ—duration of the process in nanoreactor; n—number of degrees of freedom changing from 1 to 6.

During the redox process connected with the coordination process, the character of chemical bonds changes. Therefore, correlations of wave numbers of the changing chemical bonds can be applied as the characteristic of the nanostructure formation process in nanoreactor:

$$W = 1—\exp[-\tau^n(v_{нс}/v_{кс})], \quad (2.48)$$

where v_{HC} corresponds to wave numbers of initial state of chemical bonds, and v_{KC}—wave numbers of chemical bonds changing during the process.

Modified Avrami's equations were tested to prognosticate the duration of the processes of obtaining metal/carbon nanofilms in the system "Cu—PVA" at 200°C [44]. The calculated time (2.5hrs) correspond to the experimental duration of obtaining carbon nanofilms on copper clusters.

The influence of nanostructures on the media and compositions was discussed based on quantum-chemical modeling [11]. After comparing the energies of interaction of fullerene derivatives with water clusters, it was found that the increase in the interactions in water medium under the nanostructure influence is achieved only with the participation of hydroxyfullerene in the interaction. The energy changes reflect the oscillatory process with periodic boosts and attenuations of interactions. The modeling results can identify that the transfer of nanostructure influence onto the molecules in water medium is possible with the proximity or concordance of oscillations of chemical bonds in nanostructure and medium. The process of nanostructure influence onto media has an oscillatory character and is connected with a definite orientation of particles in the medium in the same way as reagents orientate in nanoreactors of polymeric matrixes. To describe this process, it is advisable to introduce such critical parameters as critical content of nanoparticles, critical time and critical temperature [46]. The growth of the number of nanoparticles (n) usually leads to the increase in the number of interaction (N). Also such situation is possible when with the increase of n critical value; N value gets much greater than the number of active nanoparticles. If the temperature exceeds the critical value, this results in the distortion of self-organization processes in the composition being modified and decrease in nanostructure influence onto media.

KEYWORDS

- **Carbon nanocomposites**
- **Dehydropolycondensation**
- **Lagrangian equation**
- **Polymeric matrixes**
- **Quantum-chemical modeling**

REFERENCES

1. Kodolov, V. I. Shabanova, I. N., Kuznetsov, A. P., Nikolaeva, O. A. et al. (1999). *Analytical Control, 4*.
2. Stepanov, N. F., & Pupyshev, V. I. (1991). *Quantum mechanics of molecules and quantum chemistry* (p. 384). Moscow: MGU.
3. Dunning, T. H. (1971). Journal of Chemical Physics, *55*, 716–723.
4. McLean, A. D., Chandler, G. S. (1980). *Journal of Chemical Physics, 72*, 5639–5648.
5. Rappe, A. K., Smedley, T. A., & Goddard III, W. A. (1981). *Journal of Physical Chemistry, 85*, 2607–2611.
6. Schmidt, M.W. et al. (1993). *Journal of Computational Chemistry, 14*, 1347–1363.
7. Molina, J. M., & Melchor, S. F. *Private Communication*.
8. Sadlej, J. (1985). Semi-empirical methods of quantum chemistry. *International Journal of Quantum Chemistry, 30*(3), 437.
9. Goodwin, L. (1991). *Journal of Physics: Condensed. Matter, 3*, 3869.
10. Fischer, E. O., & Hafner, W. Z. (1955). *Naturforsch, 10b*, 665.
11. Wachters, A .J. H. (1970). *The Journal of Chemical Physics, 52*, 1033.
12. Dupuiis, M. et al. (1989). *ComputerPhysics Communication, 52*, 415.
13. Bethune, D. S., Klang, C. H., de Vries, M. S. et al. (1993). *Nature, 363*, 605.
14. Prigozhin, I., & Defay, R. (1966). *Chemical thermodynamics* (Russian translation) (p. 510). Novosibirsk: Nauka.
15. Berezkin, V. I. (2000). Fullerenes as embryos of soot particles. *Physics of Solids, 42*, 567–572.
16. Khokhriakov, N. V., & Kodolov, V. I. (2005). Quantum-chemical modeling of nanostructure formation. *Nanotechnics, 2*, 108–112.
17. Kodolov, V. I., Khokhriakov, N. V., Nikolaeva, O. A., & Volkov, V. L. (2001). Quantum-chemical investigation of alcohols dehydration and dehydrogenization possibility in interface layers of vanadium oxide systems. *Chemical Physics and Mesoscopy, 3*(1), 53–65.
18. Kodolov, V. I., Didik, A. A., Volkov, A. Yu., & Volkova, E. G. (2004). Low-temperature synthesis of copper nanoparticles in carbon shell. Bulletin of HEIs. *Chemistry and Chemical Engineering, 47*(1), 27–30.
19. Lipanov, A. M., Kodolov, V. I., Khokhriakov, N. V. et al. (2005). Challenges in the production of nanoreactors for the synthesis of metallic nanoparticles in carbon shells. *Alternative Energetic and Ecology (ISJAEE), 22*(2), 58–63.
20. Nikolaeva, O. A., Kodolov, V. I., Zakharova, G. S. et al. (2004). *Method of obtaining carbon-metal-containing nanostructures*. Patent of the RF 2225835, registered on 20.03.2004.
21. Rubin, A. B. (1987). *Biofizika (Biophysics)* (Vol. 1, p. 319). Moscow: Vysshaya Shkola.
22. Blokhintsev, D. I. (1961). *Basics of quantum mechanics* (p. 512). Moscow: Vysshaya Shkola.
23. Yavorsky, B. M., Detlaf, A. A. (1968). *Directory of physics* (p. 939). Moscow: Nauka.
24. Christy, P., Pitti, A. (1969). *Substance structure: Introduction to modern physics* (p. 596, Translated from English). Moscow: Nauka.

25. Korablev, G. A. (2005). *Spatial-energy principles of complex structures formation* (p. 426). Brill Academic Publishers and VSP: Netherlands.
26. Airing, G., Walter, J., & Kimbal, J. (1948). *Quantum chemistry* (p. 528). Illinois: University of Illinois.
27. Fischer, C. F. (1972). Average-energy of configuration Hartree–Fock results for the atoms helium to radon. *Atomic Data and Nuclear Data Tables, 4,* 301–399.
28. Waber, J. T., & Cromer, D. T. (1965). Orbital radii of atoms and ions. *Journal of Physical Chemistry, 42*(12), 4116–4123.
29. Clementi, E., & Raimondi, D. L. (1963). Atomic screening constants from S.C.F. functions. *Journal of Physical Chemistry, 38*(11), 2686–2689.
30. Clementi, E., Raimondi, D.L. (1967). Atomic Screening constants from S.C.F. Functions, 1 *Journal of Physical Chemistry, 47*(14), 1300–1307.
31. Korablev, G. A., & Zaikov, G. E. (2006). Energy of chemical bond and spatial-energy principles of hybridization of atom orbitals. *Journal of Applied Polymer Science, 101*(3), 2101–2107.
32. Korablev, G. A., & Zaikov, G. E. (2008). Spatial-energy interactions of free radicals. *Successes in Gerontology, 21*(4), 535.
33. Korablev, G. A., & Zaikov, G. E (2009). Spatial-energy parameter as a materialised analog of wave function. *Progress on Chemistry and Biochemistry* (pp. 355-376). New York: Nova Science Publishers Inc.
34. Korablev, G. A., & Zaikov, G. E. (2006). Chemical bond energy and spatial-energy principles of atom orbital hybridization. *Chemical Physics RAS. M., 25*(7), 24.
35. Zigban, K., Norling, K., Valman, A. et al. (1971). In (ed.), *Electron Spectroscopy* (p 493).Moscow: Mir.
36. Vunderlikh, B. (1979). In M. Mir (ed.), *Physics of macromolecules, 2,* 574.
37. Kargin, V.A., & Slonimsky, G. L. (1960). Short *Essays in Physical Chemistry of Polymers* (p. 13–14). Moscow: Akademiya Moscow. (In Russian)
38. Proceedings of Conference NATO–ASI. (2002*). Synthesis, functional properties and applications*, July–August, 2002.
39. Fedorov, V. B., Khakimova, D. K., Shipkov, N. N., & Avdeenko, M. A. (1974). *Thermodynamics of carbon materials* (Vol. 219(3), pp. 596–599). Academy of Sciences of the USSR Doklady: USSR.
40. Fedorov, V. B., Khakimova, D. K., Shorshorov, M. H. et al. (1975). *Kinetics of Graphitation* (Vol. *222*(2), pp. 399–402). Academy of Sciences of the USSR Doklady: USSR.
41. Serkov, A. T. (Ed.) (1975). *Theory of Chemical Fiber Formation* (p. 548). Moscow: Himiya.
42. Palm, V. A. (1967). In L. Khimiya (ed.), *Fundamentals of the Quantitative Theory of Organic Reactions* [in Russian] (p. 356).
43. Kodolov, V. I. (1965). On modeling possibility in organic chemistry. *Organic Reactivity*. Tartu: TSU, *2*(4), 11–18.
44. Kodolov, V. I., Khokhriakov, N. V., Trineeva, V. V., & Blagodatskikh, I. I. (2008). Activity of nanostructures and its display in nanoreactors of polymeric matrixes and in active media. *Chemical Physics and Mesoscopy, 10*(4), 448–460.
45. Kodolov, V. I. (2009).*Chemical Physics and Mesoscopy, 11*(1), 134–136. (Commentary to the paper (2008) by V. I. Kodolov et al.)

46. Kodolov, V. I., Trineeva, V. V. (2012). Perspectives of idea development about nano-systems self-organization in polymeric matrixes. *The Problems of Nanochemistry for the Creation of New Materials* (pp. 75–100). Torun: IEPMD.

CHAPTER 3

THE NANOSTRUCTURES OBTAINING AND THE SYNTHESIS OF METAL OR CARBON NANOCOMPOSITES IN NANOREACTORS

N. V. KHOKHRIAKOV, V. I. KODOLOV, G. A. KORABLEV,
V. V. TRINEEVA, and G. E. ZAIKOV

CONTENTS

3.1 BASIC METHODS OF OBTAINING AND PRODUCTION OF NANOSTRUCTURES, INCLUDING METAL OR CARBON NANOCOMPOSITES

3.1.1 The Methods of Nanostructures Obtaining

Nanostructures of various shapes can be obtained either by the dispersion of a substance in a definite medium, or condensation or synthesis from low-molecular chemical particles (Figure 3.1). Thus, the techniques for obtaining nanostructures can be classified by the mechanism of their formation. Other features, by which the methods for nanostructure production can be classified, comprise the variants of nanoproduct formation process by the change of energy consumption. A temperature or an energy factor is usually evident here. Besides, a so-called apparatus factor plays an important role together with the aforesaid features. At present, nanostructural "formations" of various shapes and compositions are obtained in a rather wide region of actions upon the chemical particles and substances.

Due to the overlapping of different classification features, it is appropriate to present a set of diagrams by main features [1–3]. For instance, the methods for nanoparticle formation by substance dispersion and chemical particle condensation can be identified with physical and chemical methods, though such decision is incorrect, since substance destruction methods can contain both chemical and physical impacts. In turn, when complex nanostructures are formed from simple ones, both purely physical and chemical factors are possible. However, in the process of substance dispersion high-energy sources, such as electric arc, laser pyrolysis, plasma sources, mechanical crushing or grinding should be applied.

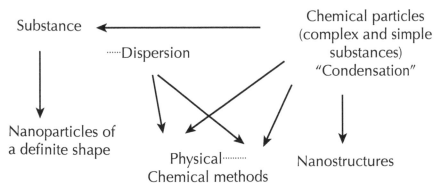

FIGURE 3.1 Classification diagram of nanostructure formation techniques by the features "dispersion" and "condensation".

At the same time, the conceptions "dispersion" and "condensation" are conditional and can be explained in various ways (Figure 3.2 and 3.3). Among the physical methods of "dispersion" high-temperature methods and methods with relatively low temperatures or high-energy and low-energy ones are distinguished.

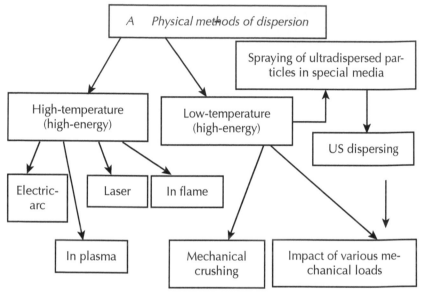

FIGURE 3.2   Classification diagram of physical methods of dispersion.

Considerably fewer chemical methods of dispersion are known, though, in the physical methods listed chemical processes are surely present, since it is difficult to imagine spraying and grinding of a substance without chemical reactions of destruction. Therefore, it is more appropriate to speak about physical methods of impact upon the substance that lead to their dispersion (decomposition, destruction at high temperatures, radiations or mechanical loads). Then, chemical methods are identified with methods of impact upon the substances of chemical particles and media.

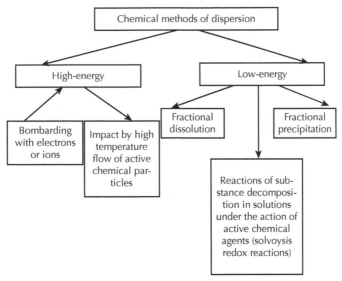

FIGURE 3.3   Classification diagram of dispersion chemical methods.

Basically nanostructures are obtained from pico-sized chemical particles by means of chemical or physical-chemical techniques, in which the activity of a chemical particle but not its activation during "condensation" is determinant (Figure 3.4). Therefore, the classification diagram of chemical techniques for nano-structure formation under the action of chemical media can be given as follows:

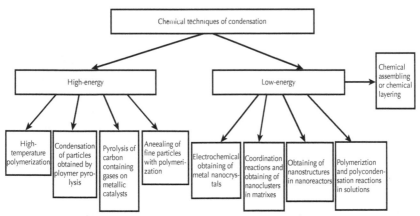

FIGURE 3.4   Classification diagram of chemical techniques of "condensation".

Actually, such diagrams are approximate in the same way as conditional are separate points of separation and difference between them. Several techniques can be referred to combined or physical methods of impact upon active chemical particles. For instance, CVD method comprises high-energy technique that leads to the formation of gaseous phase that can be attributed to chemical methods of dispersion, and then active chemical particles formed during the pyrolysis "transform" into nanostructures. The polymerization processes of gaseous phase particles are carried out by means of probe technological stations, and this can be referred to the physical techniques of "condensation". So, under the physical methods of formation of various nanostructures, including nanocrystals, nanoclusters, fullerenes, nanotubes, and so on, it means the techniques in which physical impact results in the formation of nanostructures from active pico-sized chemical particles (Figure 3.5).

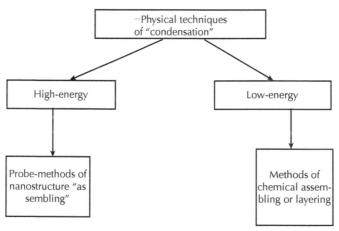

FIGURE 3.5 Classification diagram of physical methods of "condensation".

The proposed classification does not reflect multi-vicissitude of methods for nanostructure formation. Usually nanostructure production comprises the following: 1) preparation of "embryos" or precursors of nanostructures; 2) production of nanostructures; 3) isolation and refining of nanostructures of a definite shape. The production of nanostructures by various methods, including mechanical ones, for instance, extrusion, grinding and similar operations can proceed in several stages.

Mechanical crushing, combined methods for grinding in media and influence of action power and medium, where the substance is crushed and sprayed, upon the size and shape of nanostructures formed will be discussed in this chapter. At the same time, multi-stepped combined methods for obtaining nanostructures with different shapes and sizes have been applied in recent years. Ways of substance dispersion by mechanical methods till nanoproducts are obtained (fine powders or ultradispersed particles) can be given in the following diagram (Figure 3.6):

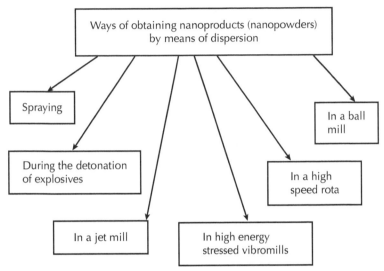

FIGURE 3.6 Diagram of classification of ways for obtaining powder-like nanoproducts by means of dispersion.

One of the most widely applied methods used for grinding different substances is the crushing and grinding in mills: ball, rod, colloid combined with the action of shock force, abrasion force, centrifugal force and vibration upon the materials. Since, initially, carbon nanostructures, fullerenes and nanotubes were extracted from the carbon dust obtained in electrical charge, the possibility of the formation of corresponding nanostructures when using conventional mechanical methods of action upon substances seems doubtful. However, the investigation of the products obtained after graphite crushing in ball mills allows making the conclusion that the method proposed can be quite competitive high-temperature way for forming nanostructures when the mechanical power provides practically the same conditions as the heat. In this case, it is difficult to imagine a directed action of the combined forces upon the substance. However, at the set speed of the mill drum the directed action of the "stream" of steel balls upon the material can be predicted. As an example, let us give the description of one of such processes during graphite grinding. Carbon nanostructures resembling "onions" by their shape are obtained when grinding graphite in ball mills [1]. The product is ground in planetary ball mill in the atmosphere of pure argon. Mass ratio of steel balls to the powder of pure graphite equals 40. The rotation speed of the drum is 270rpm. The grinding time changed from 150 to 250hrs. It was observed that after 150hrs of grinding the nanoproduct obtained resembles by characteristics and appearance the nanoparticles obtained in electric-arc method.

Iron-containing nanostructures were also obtained in planetary ball mills after milling the iron powders in heptane adding oleic acid [1].

3.1.2 The Methods of Metal Containing Carbon Nanostructures Obtaining. Methods of Redox Synthesis

The tubulenes formation is carried out using three methods: 1) from aromatic hydrocarbons on the boundaries of aromatic hydrocarbon melts and mixture of salt and metal particles; 2)from aromatic hydrocarbons in eutectic melts of salts with participating aluminum chloride; 3) from polyvinyl alcohol in the melts of polyphosphoric acid or in the solution of polyvanadium acid including Cr or Mo.

In the first variant antracene or phenanthrene as fine powders are mixed with fine powders of the following metal chlorides: copper (I), cobalt (II), nickel (II), manganese (II) (in brackets metal oxidation state) in molar ratio (salt: aromatic compound) 10 or 5, depending on metal oxidation state and coordination number. Fine particles of metal (which is in the salt composition) are introduced as a stimulating additive (0.1 part from the salt content). Fine particles are $\sim10^{-7}$m, except copper powder which comprises two particle fractions $\sim10^{-6}$–10^{-5}m. The process is carried out during the heating up to 623 K, until the stable black color of the reactive mass is reached.

After that the reactive mass is washed up in large volumes of distilled water and hydrochloric acid (sometimes nitric acid) to remove salt and the products of their hydrolysis. The residue is washed by water up to pH=7 and treated by organic solvents and their mixtures, including benzene and carbon tetrachloride, to remove non-reacted hydrocarbons and polymer-olygomer reaction products. The product, dried after the extraction from the mass, is dispersed in alcohol or acetone with the help of ultrasonic field to decrease the possibility of particles sticking. The product yield in the form of fine powders varies from 50 up to 80%, depending on the mixture composition. The powders obtained are mainly of black color, insoluble in water and resistant to high temperatures (up to 800 K).

The second method suggests that fine powders of antracene and phenantrene are mixed with the powders of transition metal salts (1:1 in moles) and introduced into $AlCl_3$–NaCl melt (1:1) (T_{mlt}=453 K), using the active medium excess, 2–10 moles of $AlCl_3$ per hydrocarbon mol. The mixture obtained is stood at 573K. In the same way hydrocarbon melt in polyphosphoric acid excess is obtained in the following ratio: active medium:hydrocarbon 10:1. The mixture color turns black during melting. The product yields deviate from 87 up to 98%, depending on reactive mass composition. The products obtained are fine powders of dark brown to black color, insoluble in water and organic solvents; they do not react with acid and base solutions and are heat-resistant. During the oxidation by concentrated boiling nitric acid during 15min, the mass loss is not more than 7% depending on the sample.

- In the third case polyphosphoric acid, manganese chloride or polyvana-dium acid and its derivatives containing Cr or Mo are used as mineral media, melts and solutions of which include "implantation metals". Molar ratio of medium mass and PVA unit is not less than 2.
- When polyphosphoric acid and its mixtures are used with metal salts, the process is carried out in the melt at 453–463 K. If polyvanadic acid is used as a medium, the process is carried out in water solution.
- During the mixing and consequent action of heat and electric fields the medium color becomes black. The whole mass becomes black during the heat action; but if it is treated by electric field, the mass turns black near the anode. Then carbon-metal containing tubulenes may be formed. The reactive mass is washed by water at 363–373K to remove medium particles and dried. Sometimes acids (vanadium acid solutions and its derivatives) are not removed. Then carbon-metaloxide cylindrical systems can be formed. The products obtained are investigated by X-ray photoelectron spectroscopy (XPES), transmission electron microscopy and electron microdiffraction methods [4–6].

3.1.3 The Carbon-Metal Containing Nanostructures Obtaining in Nano-reactors of Polymeric Matrixes

The perspective area of nanochemistry is the chemistry in nanoreactors. Nanoreactors can be compared with specific nanostructures representing limited space regions in which chemical particles orientate creating "transition state" prior to the formation of the desired nanoproduct. Nanorectors have a definite activity which predetermines the creation of the corresponding product. When nanosized particles are formed in nanoreactors, their shape and dimensions can be the reflection of shape and dimensions of the nanoreactor.

The proposed method of corresponding nanostructures synthesis consists in the conducting of redox processes which proceed in nanoreactors of polymeric matrixes and is accompanied by the reduction of metal ions included into the cavities of organic polymer gels. At the same time, hydrocarbon shells are simultaneously oxidized to carbon.

The process starts with the preparation of polymer solutions, for instance, polyvinyl alcohol (PVA), and metal compounds, for instance, 3d-metal chlorides. Afterwards, the solutions with a certain concentration are mixed in the ratio "PVA-metal chloride" equals to 20:1–1:5. Then the prepared solutions are dried till they obtain gel-like colored films with further temperature elevation up to 100°C. The color changes depending upon the metal, the films obtained are controlled spectrophotometrically, and also with help of transmission optical microscopy and atomic force microscopy. When the film color changes to black, the films are heated in the furnace according to the following program: 100-200-300-400°C. As a result, the dark porous semi-product with many microcracks is formed, that is milled in spherical or jet mill. The nanopowder obtained is steamed and dispersed in hot water. After filtration, the powder is dried and tested with the help of Raman spectroscopy, X-ray photoelectron spectroscopy, transmission electron microscopy and electron micro diffraction.

As a result, carbon-metal-containing multiwall nanotubes and astralenes are obtained. To depend on conditions nanoproducts can be metallic nanoparticles and nanowires in carbon shells. The yield of nanoproducts, calculated on Carbon of PVA, is near to 80–90%.

For the synthesis of nanoparticles and nanowires from the mixture of metal salts and polyvinyl alcohol (PVA), the aqueous solutions of salts were mixed in a certain ratio with the aqueous solution of PVA. The mean molar ratio of PVA in the mixture was 5. The experiments were carried out on the glass substrates; after the obtained mixtures had been dried, they formed colored transparent films. On some samples, the films were broken due to a large surface tension. The films were heated at t = 250°C until their color, composition and morphology changed. For the control over the process, a complex of methods was used, that is, photocolorimetry, optical microscopy, X-ray photoelectron spectroscopy and atomic power microscopy.

When PVA was added to the powders of metal chlorides, the color of the mixture changed: the mixture of copper chloride became yellow-green, the cobalt chloride mixture—blue, and nickel chloride—pale-green (Figure 3.7).

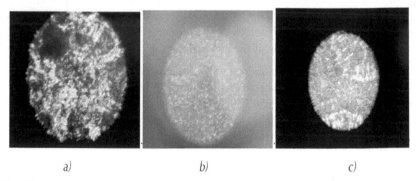

a)   b)   c)

FIGURE 3.7 The photographs of the samples containing PVA and copper chloride (a), cobalt chloride (b), and nickel chloride (c).

Observing the color changes, one can draw a conclusion that when PVA interacts with metal chlorides, the formation of complex compounds takes place.

Among the above metals, iron is most active. Brown-red inclusions on the photograph evidence the formation of the complex iron compounds. In addition, on all the photographs depicting the mixtures containing metal chlorides, one can see a net of weaves, which are most likely the reflections of nanostructures.

In order to compare these structures, the investigations of the morphology of the films changing over a certain range of temperatures were carried out with the help of atomic power microscopy (Figure 3.8).

When the nanoproduct pictures obtained by atomic power microscopy and optical microscopy are compared with the TEM micrograph of the nanoproduct treated thermally and with aqueous solution for the matrix removal, one can notice some correspondence between them. The nanoproduct represents interweaving tubulens containing Cu (I), Cu(II). In Figure 3.7, there are also optical effects indicating light polarization at light transmission through the films owing to the defects appearing during the formation of the complex compounds at the initial stage of the process.

Due to the fact, that metal ions are active, in the polymer medium they immediately appear in the environment of the PVA molecules and form bonds with the hydroxyl groups of this polymer. Polyvinyl alcohol replicates the structure of the particle that it surrounds; however, due to the tendency of the molecules of the metal salts or other metal compounds to combine, PVA as if envelops the powder particles, and therefore the forms of the obtained nanostructures can be different. The optical microscopy method allows determining the structure of the nanostructures at the early stage.

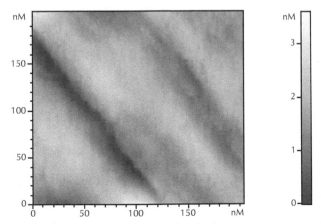

FIGURE 3.8 The Micrographs of the surface geometry of the PVA film with nickel chloride.

When the samples are heated, dehydration occurs, and as a result, metal-containing nanotubes form. These processes are thoroughly described in works [7, 8]. Dehydration leads to the darkening of the film. After the samples have been heated, on the photograph the remaining net of weaves can be seen, that is, the structure morphology has remained. To some extent, this fact indicates that the initially formed structure of matrices is inherited. The methods of optical spectroscopy and X-ray photoelectron spectroscopy allow determining the energy of the interaction of the chemical particles in the nanoreactors with the active centers of the nanoreactor walls, which stimulate reduction-oxidation processes (Figure 3.9).

FIGURE 3.9   Micrographs of tubulenes.

Depending on the nature of the metal salt and the electrochemical potential of the metal, different metal reduction nanoproducts in the carbon shells differing in shape are formed. Based on this result we may speak about a new scientific branch—nanometallurgy. The stages of nanostructures synthesis may be represented by the following scheme:

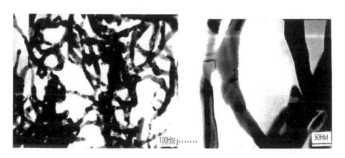

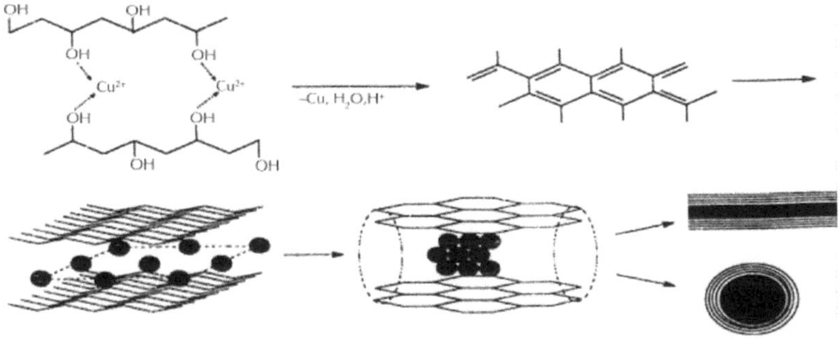

FIGURE 3.10   Scheme of copper-carbon nanostructures obtaining from copper ions and polyvinyl alcohol.

The possible ways for obtaining metallic nanostructures in carbon shells have been determined. The investigation results allow speaking about the possibility of the isolation of metallic and metal-containing nano-particles in the carbon shells differing in shape and structure. However, there are still problems related to the calculation and experiment because using the existing investigation methods it is difficult unambiguously to estimate the geometry and energy parameters of nanoreactors under the condition of 'erosion' of their walls during the formation of metallic nanostructures in them.

3.1.4 Methods of Metal or carbon Nanocomposites Production in Nanoreactors of Polymeric Matrixes Filled by Metal Containing Phases

The metal or carbon nanocomposites production proceeds in two stages. At the first stage, the nanoreactor filled by the corresponding metal containing phase is formed in the determined polymeric matrix (polyvinyl alcohol, polyvinyl chloride, polyvinyl acetate). At the second stage, metal or carbon nanocomposite is formed with the metal containing phase reduction and simultaneously the polymeric phase hydrocarbon part oxidation. Because of the association of polymeric phase with metal containing phase the walls of nanoreactor participate in redox process with cover formation for metal containing clusters obtained [1, 9–11]. Metal or carbon nanocomposites will be more active than carbon or silicon nanostructures because their masses are bigger at identical sizes and shapes. Therefore the vibration energy transmitted to the medium is also high.

Metal or carbon nanocomposite (Me/C) represents metal nanoparticles stabilized in carbon nanofilm structures [12–14]. In turn, nanofilm structures are formed with carbon amorphous nanofibers associated with metal containing phase. As a result of stabilization and association of metal nanoparticles with carbon phase, the metal chemically active particles are stable in the air and during heating as the strong complex of metal nanoparticles with carbon material matrix is formed.

The test results of nanocomposites obtained are given in Table 3.1.

TABLE 3.1 Characteristic of metal or carbon nanocomposites (Met/C HK)

Type of Met/C NC	Cu/C	Ni/C	Co/C	Fe/C
Composition, Metal or carbon [%]	50/50	60/40	65/35	70/30
Density, [g/cm³]	1.71	2.17	1.61	2.1
Average dimension, [nm]	20(25)	11	15	17

TABLE 3.1   *(Continued)*

Specific surface, [m²/g]	160 (average)	251	209	168
Metal nanoparticle shape	Close to spherical, there are dodecahedrons	There are spheres and rods	Nanocrystals	Close to spherical
Caron phase shape (shell)	Nanofibers associated with metal phase forming nanocoatings	Nanofilms scrolled in nanotubes	Nanofilms associated with nanocrystals of metal containing phase	Nanofilms forming nanobeads with metal containing phase
Atomic magnetic moment [8], [µB]	0.0	0.6	1.7	2.2
Atomic magnetic moment (nanocomposite), [µB]	0.6	1.8	2.5	2.5

To investigate the processes, optical transmission microscopies, spectral photometry, IR spectroscopy, atomic force microscopy (AFM) are applied at the first stage. For the corresponding correlations "polymer—metal containing phase" the dimensions, shape and energy characteristics of nanoreactors are found with the help of AFM [9, 15]. Depending on a metal participating in coordination, the structure and relief of xerogel surface change.

To investigate the processes at the second stage of obtaining metal or carbon nanocomposites X-ray photoelectron spectroscopy, transmission electron microscopy and IR spectroscopy are applied.

In turn, the nanopowders obtained were tested with the help of high-resolution transmission electron microscopy, electron microdiffraction, laser analyzer, X-ray photoelectron spectroscopy and IR spectroscopy.

The distinctive feature of the considered technique for producing metal or carbon nanocomposites is a wide application of independent, modern, experimental and theoretical analysis methods to substantiate the proposed technique and investigation of the composites obtained (quantum-chemical calculations, methods of transmission electron microscopy and electron diffraction, method of X-ray photoelectron spectroscopy, X-ray phase analysis, etc.). The technique developed allows synthesizing a wide range of metal or carbon nanocomposites by composition, size and morphology depending on the process conditions. In its application it is possible to use secondary metallurgical and polymer raw materials. Thus, the nanocomposite structure can be adjusted to extend the function of its application without pre-functionalization. Controlling the sizes and shapes of nanostructures by changing the metal-containing phase, to some extent, apply completely new, practicable properties to the materials which sufficiently differ from conventional materials.

The essence of the method [11] consists in coordination interaction of functional groups of polymer and compounds of 3d-metals as a result of grinding of metal-containing and polymer phases (Figure 3.11). Further, the composition obtained undergoes thermolysis following the temperature mode set with the help of thermogravimetric and differential thermal analyses. At the same time, one observes the polymer carbonization, partial or complete reduction of metal compounds and structuring of carbon material in the form of nanostructures with different shapes and sizes.

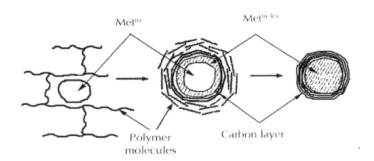

FIGURE 3.11   Mechanism of nanostructure formation in nanoreactors of polymer matrixes.

Metal or carbon nanocomposite (Me/C) represents metal nanoparticles stabilized in carbon nanofilm structures [12, 13]. In turn, nanofilm structures are formed with carbon amorphous nanofibers associated with metal containing phase. As a result of stabilization and association of metal nanoparticles with carbon phase, the metal chemically active particles are stable in the air and during heating as the strong complex of metal nanoparticles with carbon material matrix is formed.

Figure 3.12 demonstrates the microphotographs of transmission electron microscopy specific for different types of metal or carbon nanocomposites.

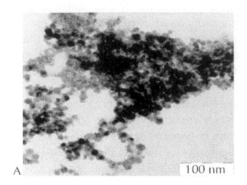

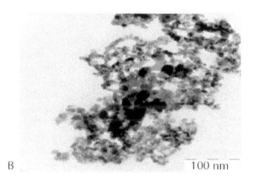

FIGURE 3.12   *(Continued)*

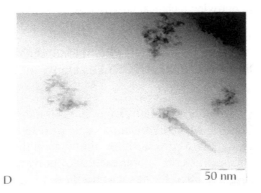

D   50 nm

FIGURE 3.12   Microphotographs of metal or carbon nanocomposites: A–Cu/C; B–Ni/C; C–Co/C; D–Fe/C.

One of the main properties of metal or carbon nanocomposites obtained is the ability to form fine suspensions [14] in various media (organic solvents, water, solutions of surface-active substances). The average size of nanoparticles in fine suspensions is 10–25 nm depending on the type of metal or carbon nanocomposite (Figure 3.13).

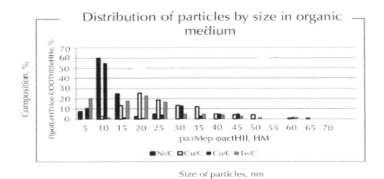

FIGURE 3.13 Distribution of particles by size in organic medium depending on the type of metal or carbon nanocomposite.

The short information about nanostructures formation mechanism in polymeric matrix nanoreactors as well as about the methods of synthesis and control during metal or carbon nanocomposites production represents. The main attention is given for the ability of nanocomposites obtained to form the fine dispersed suspensions in different media and for the distribution of nanoparticles in media.

The examples of improving technical characteristics of foam concretes and glue compositions are given.

3.2 NANOREACTORS DESIGNING PROBLEMS AND THE METHODS OF FORMATION OF NANOREACTORS FILLED BY METAL CONTAINING PHASE

The synthesis of Carbon or Metal or carbon nanostructures usually proceeds with redox reactions, in which hydrocarbon part of reactive mass is oxidized and metal containing phase part partly or almost completely is reduced [1]. The synthesis of nanostructures in nonreactors of polymeric matrixes is represented the perspective trend for nanochemistry development. Nanoreactors can be compared with specific nanostructures representing limited space regions in which chemical particles orientate creating "transition state" prior to the formation of the desired nanoproduct. Nonreactors have a definite activity which predetermines the creation of the corresponding product. When nanosized particles are formed in nanoreactors, their shape and dimensions can be the reflection of shape and dimensions of the nanoreactor [9].

The formation of nanostructures or metal or carbon nanocomposites in polymeric matrixes depends on the nature of metal containing phase and the nature of polymeric matrix, and also the conditions of formation of nanoreactors filled by secondary metal containing phase, as well as the conditions of redox synthesis.

In the case, when the interaction between metal salt solution and polymer solution takes place, metal ion, according to scheme (Figure 3.10), interacts with the functional groups of macromolecules in the interlayer space or with the functional groups of individual macromolecule. In this time the embryos of metal or metal containing clusters are formed. The sizes and forms of cluster embryos are determined by the metal nature. The clusters obtained associate with the macromolecules oriented inside nanoreactor walls. The creation of different nanosized nanoreactors and nanostructures is possible in dependence on the concentration of polymer solution.

In the case, when the interaction of solid metal containing phase (metal oxides) with polymeric phase in active medium at the intensive grinding occurs, the metal containing clusters formed get in active zones (pores, interlayer space) of polymeric phase and interact with the functional groups of polymeric matrixes. Sizes and forms of nanoreactors created depend on the sizes of active zones in polymeric matrixes. For instance, the nanoreactors obtained inside polyvinyl alcohol matrix with large pores distinguish the great distribution on sizes and forms. Metal or carbon nanocomposites obtained in these nanoreactors have middle activity.

The investigation of redox synthesis of metal or carbon nanocomposites in nanoreactors of polymeric matrixes is realized in three stages:

1. The computational designing of nanoreactors filled by metal containing phase and quantum chemical modeling of processes within nanoreactors.

2. The experimental designing and nanorectors filling by metal containing phase with using two methods—2.1. The mixing of salt solution with the solution of functional polymer (polyvinyl alcohol, polyvinyl chloride, polyvinyl acetate). 2.2. The common degenaration of polymeric phase with metal containing phase.

3. The properly redox synthesis of Metal or carbon nanocomposites in nanoreactors of polymeric matrixes at narrow temperature intervals.

The first and second stages concern to preperatiory stages. On the second stage the functional groups in nanoreactor walls participate in coordination reactions between metal ions (2.1 method) or clusters of metal containing phase (2.2 method).

The computational experiment was carried out with software products GAMESS and HyperChem, with visualization. The definite result of the computational experiment is obtained stage by stage. Any transformations at the initial stages are taken into account at further stages. The prognostic possibilities of the computational experiment consist in defining the probability of the formation of nanostructures of definite shapes when using the corresponding metal containing and polymeric substances, when studying the character of interaction of metal ion, atom, cluster or its compound, fragment of metal containing phase with fragments of nanoreactor walls. The optimal dimensions and shape of internal cavity of nanoreactors, optimal correlation between metal containing and polymeric components for obtaining the necessary nanoproducts are found with the help of quantum-chemical modeling. The availability of the metal in polymeric matrix results, in accordance with modeling results, in its regular distribution in the matrix and self-organization of the matrix.

For the corresponding correlations "polymer—metal containing phase" the dimensions, shape and energy characteristics of nanoreactors are found with the help of AFM [9, 15]. Depending on a metal participating in coordination, the structure and relief of xerogel surface change. The comparison of phase contrast pictures on the corresponding films indicates greater concentration of the extended polar structures in the films containing copper, in comparison with the films containing nickel and cobalt (Figure 3.14).

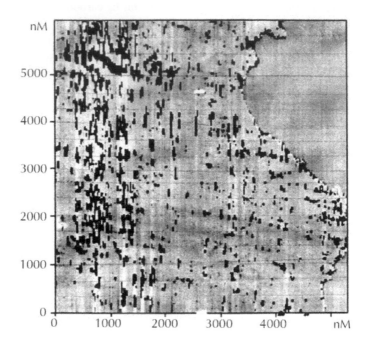

a

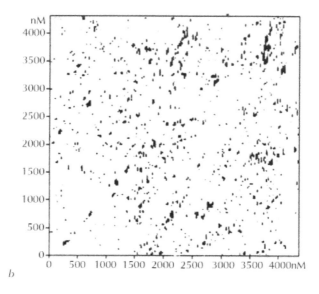

b

FIGURE 3.14   *(Continued)*

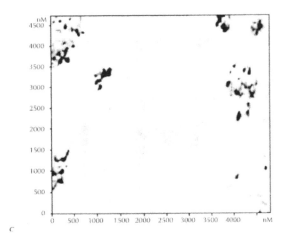

c

FIGURE 3.14   Pictures of phase contrast of PVA surfaces containing copper (a), nickel (b) and cobalt (c).

The processing of the pictures of phase contrast to reveal the regions of energy interaction of cantilever with the surface in comparison with the background produces practically similar result with optical transmission microscopy. Corresponding to data of AFM the sizes of nanoreactors obtained from solutions of metal chlorides and the mixture of polyvinyl alcohol (PVA) with polyethylene polyamine (PEPA) are determined (Table 3.2).

Table 3.2   Sizes of nanoreactors found with the help of atomic force microscopy

Composition (PVA:PEPA)	Sizes of AFM formations				
	Length	Width	Height	Area	Density
$CoCl_2 = 2:1:1$	400–800	150–400	30–40	60–350	5.5
$NiCl_2 = 2:1:1$	80–100	80–100	25–35	6–12	120
$CuCl_2 = 2:1:1$	80–100	80–100	20–30	6–20	20
$CoCl_2 = 2:2:1$	600–900	300–600	100–120	180–500	3.0
$NiCl_2 = 2:2:1$	40–80	40–60	10–30	2–4	350

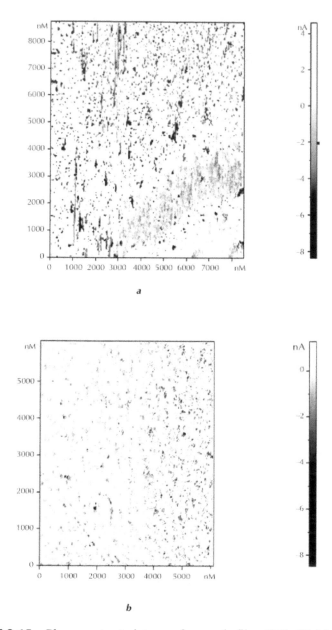

FIGURE 3.15 Phase contrast pictures of xerogels films PVA–Ni (a) and PVA–Cu (b). Below the fields of energetic interaction of cantilever with surface of xerogels PVA–Ni (a) and PVA–Cu (b) are given.

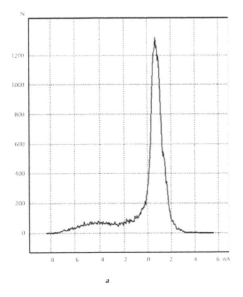

a

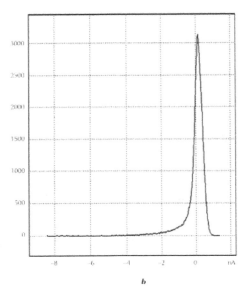

b

FIGURE 3.16   The pictures of energetic interaction of cantilever with surface of xerogels PVA–Ni (a) and PVA–Cu (b).

According to AFM results investigation the addition of Ni/C nanocomposite in PVA leads to more strong coordination in comparison with analogous addition Cu/C nanocomposite.

The mechanism of formation of nanoreactors filled with metals was found with the help of IR spectroscopy.

Thus, at the second stage the coordination of metal containing phase and corresponding orientation in nanoreactor take place.

The proposed scheme of obtaining carbon or metal containing nanostructures in nanoreactors of polymeric matrixes includes the selection of polymeric matrixes containing functional groups. 3d metals (iron, cobalt, nickel, and copper) are selected as the elements coordinating functional groups. The elements indicated easily coordinate with functional groups containing oxygen, nitrogen, halogens. Depending on metal coordinating ability and conditions for nanostructure obtaining (in liquid or solid medium with minimal content of liquid) one obtains "embryos" of future nanostructures of different shapes, dimensions and composition. It is advisable to model coordination processes and further redox processes with the help of quantum chemistry apparatus.

The computational experiment was carried out with software products GAMESS and HyperChem, with visualization. The definite result of the computational experiment is obtained stage by stage. Any transformations at the initial stages are taken into account at further stages. The prognostic possibilities of the computational experiment consist in defining the probability of the formation of nanostructures of definite shapes when using the corresponding metal containing and polymeric substances, when studying the character of interaction of metal ion, atom, cluster or its compound, fragment of metal containing phase with fragments of nanoreactor walls.

The optimal dimensions and shape of internal cavity of nanoreactors, optimal correlation between metal containing and polymeric components for obtaining the necessary nanoproducts are found with the help of quantum-chemical modeling. The availability of d metal in polymeric matrix results, in accordance with modeling results, in its regular distribution in the matrix and self-organization of the matrix, for example, silver cluster atom distribution in the fragment imitating the hardened epoxy resin (Figure 3.17).

a)

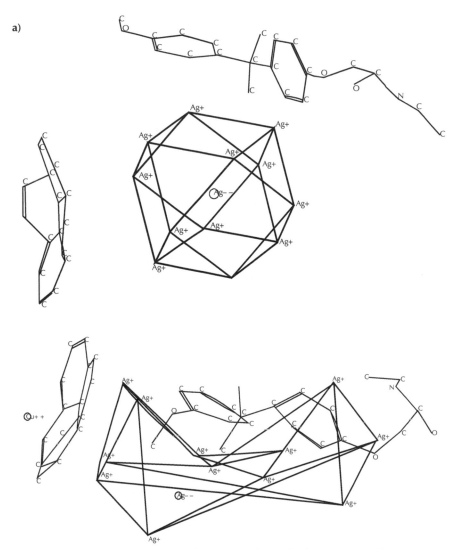

FIGURE 3.17 "Stretching" of silver nanocluster in the presence of nanostructure fragment and fragment of hardened epoxy resin molecule ED-20: a initial state; b result of system geometry calculation.

From the given model it is seen that the nanocluster "stretches" along the fragment of hardened molecule ED-20 with changing the nanostructure position in such a way that the curved graphene plane consisting of 16 carbon atoms is placed between the oriented silver atoms and copper ion. The resulting picture can indicate the silver atom orientation in epoxy polymer fragment. At the same time, the metal orientation proceeds in interface regions and nanopores of polymeric phase which conditions further direction of the process to the formation of metal or carbon nanocomposite. In other words, the birth and growth of nanosize structures occur during the process in the same way as known from the macromolecule physics [16].

3.3 THE CONDITIONS OF REDOX SYNTHESIS OF METAL OR CARBON NANOCOMPOSITES AND THE NANOCOMPOSITES CHARACTERIZATION

At the third stage it is required to give the corresponding energy impulse to transfer the "transition state" formed into carbon/metal nanocomposite of definite size and shape. To define the temperature ranges in which the structuring takes place, DTA-TG investigation is applied.

It is known that small changes of weight loss (TG curve) at invariable exothermal effect (DTA curve) testify to the self-organization (structural formation) in system. Below typical curves of DTA-TG investigation lead for prognosis of redox synthesis of metal or carbon nanocomposites (Figure 3.18).

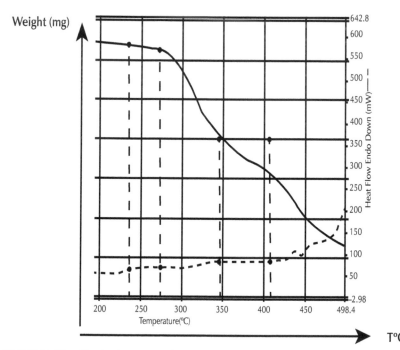

FIGURE 3.18 Typical curves of DTA-TG investigations of xerogels films containing metal and polymeric phases.

According to data of DTA-TG investigation (Figure 3.18) optimal temperature field for film nanostructure obtaining is 230–270°C, and for spatial nanostructure obtaining—325–410°C.

It is found that in the temperature range under 200°C nanofilms, from carbon fibers associated with metal phase as well (Figure 3.19), are formed on metal or metal oxide clusters. When the temperature elevates up to 400°C, 3D nanostructures are formed with different shapes depending on coordinating ability of the metal.

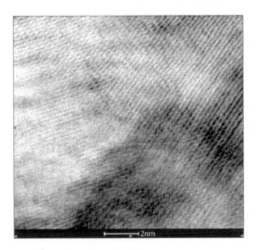

FIGURE 3.19 Microphotographs obtained with the help of transmission electron microscopy. Cu/C nanocomposite.

To investigate the processes at the second stage of obtaining metal or carbon nanocomposites X-ray photoelectron spectroscopy, transmission electron microscopy and IR spectroscopy are applied. The sample for IR spectroscopy was prepared when mixing metal or carbon nanocomposite powder with one drop of vaseline oil in agate mortar to obtain a homogeneous paste with further investigation of the paste obtained on the appropriate instrument. As the vaseline oil was applied when the spectra were taken, one can expect strong bands in the range $2,750–2,950 cm^{-1}$. Two types of nanocomposites rather widely applied during the modification of various polymeric materials were investigated. These were: copper or carbon nanocomposite and nickel/carbon nanocomposite specified below. In turn, the nanopowders obtained were tested with the help of high-resolution transmission electron microscopy, electron microdiffraction, laser analyzer, X-ray photoelectron spectroscopy and IR spectroscopy.

The method of metal or carbon nanocomposite synthesis applied has the following advantages:

- Originality of stage-by-stage obtaining of metal or carbon nanocomposites with intermediary evaluation of the influence of initial mixture composition on their properties.
- Wide application of independent modern experimental and theoretical analysis methods to control the technological process.
- Technology developed allows synthesizing a wide range of metal or carbon nanocomposites depending on the process conditions.
- Process does not require the use of inert or reduction atmospheres and specially prepared catalysts.
- Method of obtaining metal or carbon nanocomposites allows applying secondary raw materials.

Characteristics of Nanocomposites Obtained

Under metal or carbon nanocomposite one understands the nanostructure containing metal clusters stabilized in carbon nanofilm structures. The carbon phase can be in the form of film structures or fibers. The metal particles are associated with carbon phase. The metal nanoparticles in the composite basically have the shapes close to spherical or cylindrical ones. Due to the stabilization and association of metal nanoparticles with carbon phase, chemically active metal particles are stable in air and during heating as the strong complex of metal nanoparticles with the matrix of carbon material are formed. In this chapter the IR spectra of Cu/C and Ni/C nanocomposites are discussed (Figure 3.20, 3.21), which find a wider application as the material modifiers.

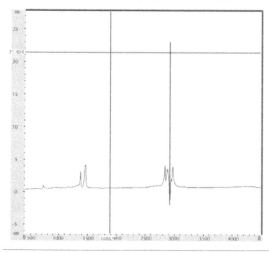

FIGURE 3.20   The IR spectra of copper or carbon nanocomposite powder.

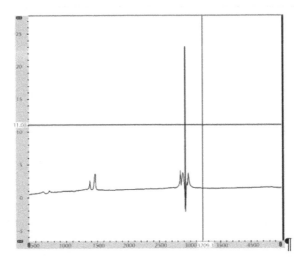

FIGURE 3.21   The IR spectra of nickel/carbon nanocomposite powder.

On IR spectra (Figure 3.20, 3.21) of two nanocomposites the common regions of IR radiation absorption are registered. Further, the bands appearing in the spectra and having the largest relative area were evaluated. One can see the difference in the intensity and number of absorption bands in the range 1,300–1,460cm^{-1}, which confirms the different structures of composites. In the range 600–800cm^{-1} the bands with a very weak intensity are seen, which can be referred to the oscillations of double bonds (π-electrons) coordinated with metals. In case of Cu/C nanocomposite a weak absorption is found at 720cm^{-1}. In case of Ni/C nanocomposite, except for this absorption, the absorption at 620cm^{-1} is also observed.

In IR spectrum of copper or carbon nanocomposite two bands with a high relative area are found:

- at 1,323cm^{-1} (relative area—9.28)
- at 1,406cm^{-1} (relative area—25.18).

These bands can be referred to skeleton oscillations of polyarylene rings. In IR spectrum of nickel/carbon nanocomposite the band mostly appears at 1,406cm^{-1} (relative area—14.47).

According to the investigations with transmission electron microscopy the formation of carbon nanofilm structures consisting of carbon threads is characteristic for copper or carbon nanocomposite. In contrast, carbon fiber structures, including nanotubes, are formed in nickel/carbon nanocomposite. There are several absorption bands in the range 2,800–3,050cm^{-1}, which are attributed to valence oscillations of C–H bonds in aromatic and aliphatic compounds. These absorption bonds are connected with the presence of vaseline oil in the sample. It is difficult to find the presence of metal in the composite as the metal is stabilized in carbon nanostructure. At the same time, it should be pointed out that apparently nano-composites influence the structure of vaseline oil in different ways. The intensities and number of bands for Cu/C and Ni/C nanocomposites are different:

- For copper or carbon nanocomposite in the indicated range—5 bands, and total intensity corresponds by the relative area—64.63
- For nickel/carbon nanocomposite in the same range—4 bands with total intensity (relative area)—85.6.

The distribution of nanoparticles in water, alcohol and water-alcohol suspensions prepared based on the above technique are determined with the help of laser analyzer. In Figure 3.22 and 3.23 you can see distributions of copper or carbon nanocomposite in the media different polarity and dielectric penetration. When comparing the figures one can see that ultrasound dispergation of one and the same nanocomposite in media different by polarity results in the changes of distribution of its particles. In water solution the average size of Cu/C nanocomposite equals 20nm, and in alcohol medium—greater by 5nm.

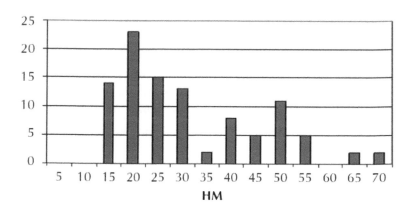

FIGURE 3.22　Distribution of copper or carbon nanocomposites in alcohol.

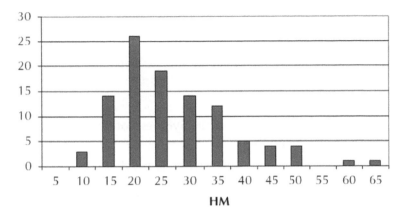

FIGURE 3.23 Distribution of copper or carbon nanocomposites in water.

Assuming that the nanocomposites obtained can be considered as oscillators transferring their oscillations onto the medium molecules, one can determine to what extent the IR spectrum of liquid medium will change.

In the chapter the possibilities of developing new ideas about self-organization processes during redox synthesis within nanoreactors of polymeric matrixes as well as about nanostructures and nanosystems are presented on the example of metal or carbon nanocomposites. It is proposed to consider the obtaining of metal or carbon nanocomposites in nanoreactors of polymeric matrixes as self-organization process similar to the formation of ordered phases.

The perspectives of this investigation are looked through in an opportunity of thin regulation of processes and the entering of corrective amendments during processes.

3.4 THE ESTIMATION OF SYNTHESIS RESULTS AND HYPOTHESIS OF SYNTHESIS MECHANISM IN NANPREACTORS OF POLYMERIC MATRIXES

Among the many various methods of obtaining metal clusters and carbon/metal containing nanostructures the synthesis in nanoreactors is considered the most perspective [1, 17]. The main advantage of the above synthetic method in comparison with the others is the possibility to predict the result obtained depending on the matrix and nanoreactor selected, as well as the basic metal. However this method has not been sufficiently developed. For its development theoretical and experimental results are required. This chapter discusses the ways for the production and investigation of nanoreactors in polymeric matrixes and synthesis of nanostructures with corresponding activity in them.

Nanocomposites obtained in nanoreactors can be applied as modifiers of different compositions and materials. At the same time, the mechanism of the influence of super small quantities of metal or carbon nanocomposites on the structure and properties of materials modified is still not completely clear. Therefore the corresponding explanations and assumptions are confirmed, to some extent, by the authors' results on the example of inorganic and organic systems given in the chapter.

3.4.1 Theoretical Substantiation of the Synthesis of Nanostructures and Their Influence on Compositions

Previously [7, 8] the models of nanoreactors in which carbon/metal containing nanostructures can be obtained were considered. The synthesis takes place due to the redox reaction between metal containing substances and nanoreactor walls which are the macromolecules of polymer matrix with functional groups participating in the interaction with metal containing compound or metal ion. If nanoreactors are nanopores or cavities in polymer gels which are formed in the process of solvent removal from gels and their transformation into xerogels or during the formation of crazes in the process of mechanical-chemical processing of polymers and inorganic phase in the presence of active medium, the process essence is as follows:

First, the nanoreactor is formed in the polymeric matrix which, by geometry and energy parameters, corresponds to the transition state of the reagents participating in the reaction, and then the nanoreactor is filled with reactive mass comprising reagents and solvent. The latter is removed and only the reagents oriented in a particular way stay in the nanoreactor and, if the sufficient energy impulse is available, for instance, energy isolated at the formation of coordination bonds between the fragments of reagents and functional groups in nanoreactor walls interact with the formation of necessary products. The initially appeared coordination bonds destruct together with the nanoproduct formation.

The difference of potentials between the interacting particles and object walls stimulating these interactions is the driving force of self-organization processes (formation of nanoparticles with definite shapes). The potential jump at the boundary "nanoreactor wall—reacting particles" is defined by the wall surface charge and reacting layer size. If the redox process is considered as the main process preceding the nanostructure formation, the work for charge transport corresponds to the energy of nanoparticle formation process in the reacting layer. Then the equation of energy conservation for nanoreactor during the formation of nanoparticle mol will be as follows:

$$nF\,\Delta\varphi = RTln\,N_p\,/\,N_r, \tag{1}$$

where n—number reflecting the charge of chemical particles moving inside "the nanoreactor"; F—Faraday number; Δ—difference of potentials between the nanoreactor walls and flow of chemical particles; R—gas constant; T—process temperature; N_p—mol share of nanoparticles obtained; N_r—mol share of initial reagents from which the nanoparticles are obtained.

Using the above equation one can determine the values of equilibrium constants when reaching the certain output of nanoparticles, sizes of nanoparticles and shapes of nanoparticles formed with the appropriate equation modification. The internal cavity sizes or nanoreactor reaction zone and its geometry significantly influence the sizes and shapes of nanostructures.

The sequence of the processes is conditioned by the composition and parameters (energy and geometry) of nanoreactors. To accomplish such processes it is advisable to preliminarily select the polymeric matrix containing the nanoreactors in the form of nanopores or crazes as process appropriate. Such selection can be realized with the help of computer chemistry. Further the computational experiment is carried out with the reagents placed in the nanoreactor with the corresponding geometry and energy parameters. Examples of such computations were given in [18]. The experimental confirmation of polymer matrix and nanoreactor selection to obtain carbon or metal containing nanostructures was given in [19, 20]. Avrami equations are widely used for such processes which usually reflect the share of a new phase produced. The share of nanostructures (W) during the redox process can depend on the potential of nanoreactor walls interaction with the reagents, as well as the number of electrons participating in the process. At the same time, metal ions in the nanoreactor are reduced and its internal walls are partially oxidized (transformation of hydrocarbon fragments into carbon ones). Then the isochoric-isothermal potential (ΔF) is proportional to the product $zF\Delta$ and Avrami equation will look as follows:

$$W = 1 - k_1 \exp\left[-\tau^n \exp\left(zF\,\Delta\varphi\,/\,RT\right)\right] \qquad (2)$$

where k_1—proportionality coefficient taking into account the temperature factor; n—index of the process directedness to the formation of nanostructures with certain shapes; z—number of electrons participating in the process; $\Delta\varphi$—difference of potentials at the boundary "nanoreactor wall—reactive mixture"; F Faraday number; R—universal gas constant. When "n" equals 1, one-dimensional nanostructures are obtained (linear nanostructures, nanofibers). If "n" equals 2 or changes from 1 to 2, flat nanostructures are formed (nanofilms, circles, petals, wide nanobands). If "n" changes from 2 to 3 and more, spatial nanostructures are formed as "n" also indicates the number of degrees of freedom. The selection of the corresponding equation recording form depends on the nanoreactor (nanostructure) shape and sizes and defines the nanostructure growth in the nanoreactor.

In case of nanostructure interaction with the medium molecules, the medium self-organization is effective with the proximity or concordance of the oscillations of separate fragments of nanostructures and chemical bonds of the medium molecules [21–24]. At the same time, it is possible to use quantum-chemical computational experiment with the known software products.

According to the chapter [25] the process vibration nature is discovered. This fact corroborates by IR spectroscopic investigations of nanostructures aqueous soles.

Self-organization processes in media and compositions can be compared with the processes of crystalline phase origin and growth. At the same time, the growth can be one-, two- and three-dimensional. For such processes Avrami's equations are widely used which usually reflect the share of the new phase appearing. In this case, the degree of nanostructure influence on active media and compositions is defined by the number of nanostructures, their activity in this composition and interaction duration. The temperature growth during the formation of new-phases in self-organizing medium prevents the process development. To determine the degree of nanostructure influence on the processes of composition self-organization, the following formula is proposed:

$$W = n / N \exp(an\tau^{\beta} / T), \tag{3}$$

where W—degree of nanostructure influence on structurally changing medium, n—number of nanoparticles in the medium, N—number of "branchings" conditioned by "cross" influences on the medium, a—nanoparticle activity, τ—duration of the medium modification process, T—structuring temperature, β degree—index of the process directedness to the formation of definite shapes of microstructures formed in nanocomposites. To describe this process, it is necessary to introduce such critical parameters as critical content of nanoparticles, critical time and critical temperature. The assumptions proposed were experimentally tested on the compositions for producing foam concretes, materials based on cold hardening epoxy resins or polyvinylchloride. The calculations revealed the following results: the content of nanoparticles for foam concretes can be in the range 0.001–0.003% (copper containing nanofilms) or 0.003–0.007% (nickel containing nanofilms), and for epoxy resins – 0.001–0.005% (copper containing nanofilms).

Based on the results of computational experiments the following polymeric matrixes were selected: polyvinyl alcohol (PVA), polyvinyl chloride (PVC), polyvinyl acetate (PVAc), mixture of PVA with polyethylene polyamine (PEPA), derivatives of polyvanadium acid and compounds of such metals as iron, cobalt, nickel and copper. At the same time, after the mixing and sometimes during mechanical and chemical processing the gels saturated with the corresponding metals were produced. During the stage-by-stage control of the processes in nanoreactors the following set of methods was applied: spectrophotometry; optical transmission microscopy; IR spectroscopy; X-ray photoelectron spectroscopy (XPES); atomic force microscopy (AFM); DTA-TG investigation.

At the same time, the directedness of the processes of nanostructure formation, shape and sizes of nanoreactors, as well as future nanostructures, temperature mode of thermal-chemical treatment of xerogels are found. The nanoreactors sizes are found by atomic force microscopy and depend on the nature of polymeric and metal containing phases [26].

Temperature intervals to obtain the nanostructures of different shapes are found by DTA-TG data: a) obtaining of nanofilms in the temperature range under 200°C; b) obtaining of globular or cylindrical nanostructures in the temperature range under 400°C. Temperature ranges found with DTA-TG investigations differ for different mixtures (metal compound—polymer). At 200°C lamellar nanofilms under 5 nm thick with metal containing nanocrystals between the layers are usually formed. The nanostructures obtained can be referred to as metal or carbon or, in case of synthesis in PVA-PEPA nanoreactors, metal or carbon polymeric nanocomposites.

A metal or carbon nanocomposite represents metal nanoparticles stabilized in carbon nanofilmed structures. In these samples the carbon phase can be in the form of film structures or fibers. The metal particles are associated with the carbon phase. Metal nanoparticles in the composite with the size range 7–25 nm (the size is determined by the metal nature) are mainly spherical.

3.5 THE EXAMPLE OF NANOSTRUCTURES OBTAINING WITH APPLICATION SECONDARY METALLURGICAL MATERIALS

It is possible that secondary metallurgical raw materials differs in higher reactivity due to the structure defectiveness, on the other hand, the use of industrial wastes sharply decreases the costs of the nanoproduct produced. At the same time low-temperature techniques of obtaining stabilized nanoparticles applying mechanical-chemical and thermal-chemical methods are attracting interest. However during such syntheses the processes of the formation of carbon metal-containing nanostructures are not sufficiently studied, as well as the factors defining their morphology. Besides the use of secondary resources (metallurgical dust) is important for the provision of the environment ecological safety. The aforementioned proves the topicality of the work in the direction of using metallurgical dusts for the synthesis of carbon metal-containing nanostructures, as well as the investigation of the formation mechanism and synthesis influence on the structure of nanoproducts obtained.

3.5.1 Substantiation of the Method for Obtaining and Selecting the Components for the Synthesis of Carbon Metal-Containing Nanostructures

The polymer gel contains nanopores in which chemical reactions resulting in the formation of nanoparticles can flow as in nanoreactors. The rate of such reactions increases in comparison with the reaction rate in macroscopic space, and the nanoparticle size is regulated by the pore sizes [27]. According to the obtaining technique suggested, the formation of metal nanoparticles takes place during the redox reaction both inside the polymer nanopore and on the surface. Interface interactions on the boundary "metal compound—polymer" result in the formation of nanoparticles of inorganic phase. In the process of grinding and chemical interaction of polyvinyl alcohol with metal compound the nanoparticles of metallic phase are localized in the pores by the method of countercurrent diffusion, as in "crazes" [28]. At the same time the following stages are implemented: a) penetration of metal ions into the polymeric matrix, b) diffusion of reagents into the matrix described by the first Fick law, c) chemical reaction itself. If the obtained size of metal or oxide nanoparticle corresponds to the polymer pore size, the particle or ion penetrates into the polymer nanopores. If the particle size exceeds the size of nanopores formed, the second stage of metal particle formation proceeds in the polymer—their penetration into the polymer surface cracks formed. As a result of agglomeration of the nanoparticles formed, nanoplates or nanowires are obtained both on the surface of carbon matrix produced and inside between the layers. The cluster maximum size in a nanopore is defined by the number of metal atoms penetrating into the nanopore, from which a nanoparticle is formed. In porous materials with pore sizes up to 10nm isolated nanoparticles with the sizes from one to several nanometers depending upon the nanopore size and initial concentration can be formed [27]. Nanoparticles of metals and metal oxides of such a small size possess specific atom dynamics, which can be seen as atom dynamics inside the cluster and motion of the cluster as a whole inside the pore.

The active gel of polyvinyl alcohol (PVA) is chosen as a polymeric substance for the experiment, since its gels and porosity have been studied rather well [16]. Polyvinyl alcohol is carbonating polymer and due to the availability of OH-groups it forms complex compounds during the interaction with metal compounds. The nanoparticles are obtained as result of redox processes with the participation of nanopores formed in PVA gel. The experiments were carried out using dusts of nonferrous metallurgy. The qualitative and quantitative composition of the nonferrous dust was obtained while investigating the sample of metallurgical dust by the method of X-ray phase analysis. The investigations were carried out on the diffractometer DRON-3. The diffractogram of metallurgical dust is given in Figure 3.24.

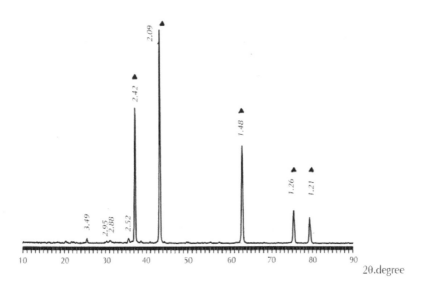

FIGURE 3.24   Diffractogram of metallurgical dust.

In accordance with the diffractogram the dusy has the following composition: 81.2% NiO, 8.1% NiS, 6% CuO, 2.5% CuS, 2% CoO.

The probable processes at the interaction of PVA with the oxides of 3d-metals are considered in the frameworks of quantum-chemical approximation ZINDO/1 realized in the program product HyperChem v.6.03. The evaluation of the interaction possibility of molecule fragments was carried out by the bond length change in them as a result of the geometry optimization, which shows the system interaction and stability. The computational experiment was carried out at the ratios NiO: OH = 1:2; 1:4. Figure 3.25 shows the experimental results of the process second stage after the initial processes of dehydration and dehydrogenation. The component ratio NiO: OH = 1: 4.

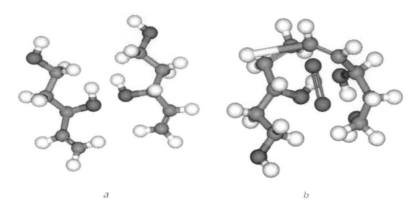

a   *b*

FIGURE 3.25   Two molecules of 3.5-dihydroxypentene-1 before the geometry optimization (a), after the geometry optimization with nickel oxide (b).

Thus the energetically more favorable states of the systems are revealed. The scrolling of the macromolecule fragments relatively to the nickel oxide in the shape of hemisphere with the formation of carbon shell is observed in the system. The Table 3.3 shows the bond lengths in the molecules before and after the geometry optimization for the systems "3.5-dihydroxypentene-1-nickel oxide".

TABLE 3.3   Bond lengths in the molecules before and after the geometry optimization for the system "3.5-dihydroxypentene-1-nickel oxide".

System	Bond	Bond length, before the optimization, Å	Bond length, after the optimization, Å
Two molecules 3.5-di-hydroxypentene-1-nickel oxide	Ni–O	1.547	2.04565
	C–H	1.09	2.9268

When analyzing the interaction of the system "3.5-dihydroxypentene-1-nickel oxide" it can be concluded that the metal is reduced by polyene, which is formed during the dehydrogenation of PVA as a result of breaking C–H bond and formation of hydrogen radical. Based on the data of theoretical models the ratios of components are selected according to coordination interactions. For nickel, cobalt, copper compounds the molar ratio of the components = 1:4 (metal: number of PVA functional groups). Based on the modeling results obtained it can be expected that the availability of metals of different nature than nickel in the system will result in the distortion of nanostructure shapes obtained, thus influencing the second thermal-chemical stage of nanostructure obtaining.

The temperature modes were defined by the character of PVA changes under the temperature action. When PVA is heated above 100°C, its' elasticity and solubility decreases. Chemical transformations of polyvinyl alcohol when heating up are very complicated and, in most cases, are the result of inner- and intermolecular dehydration. Inner-molecular dehydration initiates the formation of cyclic links in separate macromolecules, at the same time intermolecular dehydration results in the appearance of links constructed on the principle of vinyl ethers providing net structure for the polymer. Short-time thermal action (150–220°C) results in some rigidity elevation, elasticity decrease and complete loss of the solubility in water of the polymer, being the result of chain joining. At a higher temperature the polymer demonstrates the properties appropriate for polyenes with conjugated double bonds; consequently the water is also separated inside the macromolecule links [27]. Since in accordance with DTA-TG data the exothermic peak corresponding to the sharp decrease in the sample weight is observed above 400°C, this effect can be related to thermal-oxidation process due to the interaction with air. Therefore the synthesis maximum temperature of 400°C is selected.

The samples are obtained when mixing the components in molar ratios indicated with further drying of the gels obtained at 70°C. As a result, the xerogel samples are obtained with internally distributed inorganic fraction. The samples are investigated by the method of optical transmission microscopy. The results are given in the Figure 3.26a, b.

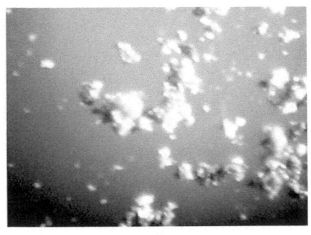

a

FIGURE 3.26 *(Continued)*

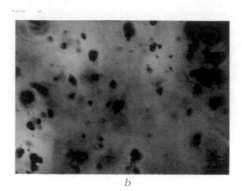

b

FIGURE 3.26   Microphotographs of the samples of metallurgical dust before (a) ×400 and after (b) ×1000 treatment with 10% PVA solution.

The particles of nonferrous metallurgical dust are inclined to agglomeration with the formation of agglomerates of definite shape and size. The size of agglomerates corresponds to 2.5–30 mcm (Figure 3.26a). When the PVA solution is added the particle size decreases. The agglomerates decompose; the maximum size of particles becomes 4mcm. By the way, according to the microphotograph the decomposition is not finished. The darker regions around the larger particles are observed, thus proving that the decomposition process is not finalized (Figure 3.26b). The color of the gel indicates that inorganic phase in the form of ions or fine particles are located in the PVA pores. Based on the results of this investigation one can speak of the decrease in metallurgical dust strength under the action of surface-active substance, metallurgical dust decomposition, Rehbinder effect.

Further the xerogels obtained are crushed and subjected to stage temperature treatment up to 400°C in closed crucibles in muffle furnace with the soaking time from 1 to 2hrs at 400°C. As a result ultra-fine black powders are obtained.

The samples obtained are investigated by the method of transmission electron microscopy (TEM) and electron diffraction to reveal the composition and structure. The investigation is carried out on the transmission electron microscope JEM-200CX with the accelerating voltage of 160 kV and attachment of electron diffraction.

According to the investigations by transmission electron microscopy and electron diffraction the samples are based on metal-containing nanostructures (spheres, rods, and plates) stabilized in carbon shell or matrix (Figure 3.27, 3.28). The particle sizes do not exceed 80nm. In all the samples carbon nanotubes are present. The composition of inorganic phase for the samples containing nonferrous metallurgical dust: nickel oxide, nickel, cobalt oxide, copper, and copper oxide (I).

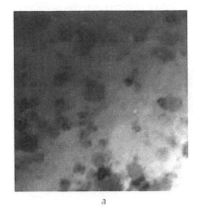

a

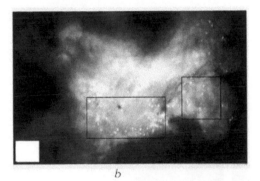

b

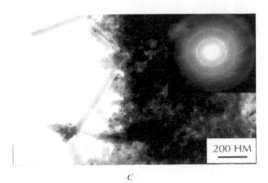

200 HM

c

FIGURE 3.27   The TEM microphotographs of the sample "PVA and non-ferrous metallurgical dust" (T = 400°C, t = 1hr): a metal and metal-oxide particles stabilized in carbon matrix with the particle size from 0.7 to 7 nm; b picture in the dark field mode, in nickel reflexes, the nickel particles with the sizes of 6nm glow; c growth of carbon nanotubes with the sizes up to 30 nm.

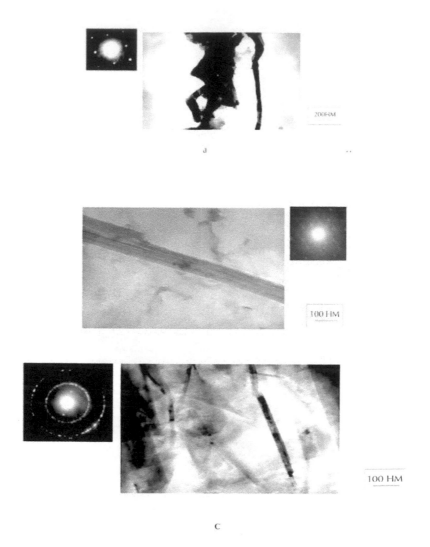

FIGURE 3.28   The TEM microphotographs of the sample "PVA and non-ferrous metallurgical dust" (T = 400°C, t = 2hrs): a nanorods from monocrystalline grains of nickel oxide with the diameters from 40 to 80 nm; b

carbon nanotubes with the diameters of 20nm; c nanowires and nanoplates of nickel oxide stabilized in carbon matrix, the sizes up to 20 nm.

Thus the metal is reduced from the compounds and different nanostructures are formed depending upon the synthesis duration. Wired structures are formed as a result of the agglomeration of metal nanoparticles. The investigations conform to the ones given in [29, 30]. In accordance with [31] nickel nanoparticles are agglomerated with the formation of wired or tubular structures on any conducting or non-conducting substrate in contrast to for example, copper nanoparticles that form "isolated islands". The possibility of obtaining carbon metal-containing nanostructures when utilizing metallurgical wastes is demonstrated. When obtaining metal and metal-oxide particles, nanorods and nanoplates with the method suggested, redox reactions flow with the formation of metal substances with lower metal oxidation degree. For temperature treatment within 1hr at 400°C globular nanoparticles are characteristic, carbon nanotubes are formed. When soaking for 2hrs at 400°C nanorods and nanoplates are formed, carbon nanotubes are present in the sample. Accordingly the duration of temperature treatment influences the shape, size and structure of nanoproducts obtained. In this method of obtaining metal and metal-oxide nanoparticles the change in the temperature soaking results in the considerable change in the nanoparticle structure and, consequently, the change in their properties.

KEYWORDS

- **Dispersion**
- **Laser pyrolysis**
- **Nanoreactors**
- **Polymeric matrixes**
- **Redox**

REFERENCES

1. Kodolov, V. I. & Khokhriakov, N. V. (2009). *Chemical physics of the processes of nanostructure and nanosystem formation and transformation* (1. p. 361, 2. p. 415). Izhevsk: Publishing House of Izhevsk State Agricultural Academy.

2. Kodolov, V. I. & Trineeva, V. V. (2012). Perspectives of idea development about nano-systems self- organization in polymeric matrixes. *The problems of nanochemistry for the creation of new materials* (pp. 75–100). Poland, Torun: IEPMD.

3. Shabanova, I. N., Kodolov, V. I., Terebova, N. S., & Trineeva, V. V. (2012). *X Ray Electron Spectroscopy in Investigation of Metal or carbon Nanosystems and Nanostructured Materials* (p. 250). Izhevsk: Udmurt State University Publication.

4. Kodolov, V. I., Kuznetsov, A. P., Nicolaeva, O. A., et al. (2001). *Surface and Interface Analysis, 32*, 10–14

5. Kodolov, V. I., Shabanova, I. N., Makarova, L. G., et al. (2001). *Journal of Structural Chemistry, 42*(2), 260–264 (in Russian)

6. Kodolov, V. I., Khokhriakov, N. V., Nikolaeva, O. A., & Volkov, V. L. (2001). *Chemical Physics and Mesoscopy, 3*(1), 53–65 (in Russian)

7. Volkova, E. G., Volkov, A. Yu., Murzakaev, A. M., et al. (2003). *The Physics of Metals and Metallography, 95*(4), 342–345

8. Didik, A. A., Kodolov, V. I., Volkov, A. Yu., et al. (2003). *Inorganic Material, 39*(6), 693–697

9. Kodolov, V. I., Khokhriakov, N. V., Trineeva, V. V., & Blagodatskikh, I. I. (2008). Activity of nanostructures and its presence in nanoreactors of polymeric matrixes and active media. *Chemical Physics & Mesoscopy, 10*(4), 448–460.

10. Kodolov, V. I. Commentary to the paper by Kodolov, V. I., et al. (2009). *Chemical Physics & Mesoscopy, 11*(1), 134–136.

11. Vasil'chenko, Yu. M., Akhmetshina, L. F., Shklyeva, D. A., et al. (2010). Synthesis of carbon metal containing nanostructures in PVC and PVAc gels with metallurgical dust. *Nanomaterials Yearbook* 2009. From nanostructures, nanomaterials and nano-technologies to nanoindustry (pp. 283–288). New York: Nova Science Publications Inc.

11. Kodolov, V. I., Kodolova, V. V., Semakina, N. V., et al. (2008). *Patent RU 2337062.*

12. Kodolov, V. I., Vasil'chenko, Yu. M., Shklyeva, D. A., et al. (2010). *Patent RU 2393110.*

13. Kodolov, V. I., Vasil'chenko, Yu.M., Akhmetshina, L. F., et al. (2010). *Patent RU 2423317.*

14. Trineeva, V. V., Lyakhovitch, A. M., & Kodolov, V. I. (2009). Prognosis of metal containing carbon nanostructures formation processes with the using of AFM method. *Nanotechnics,*5 (20), 87–90.

15. Vunderlikh, B. (1979). *Physics of macromolecules* (2, 574). Moscow: Mir.

16. Bronstein, L. M. & Shifrina, Z. B. (2009). Nanoparticles in dendrimers: from synthesis to application. *Russian nanotechnologies, 4*(9–10), 32–54.

17. Khokhriakov, N. V. & Kodolov, V. I. (2005). Quantum-chemical modeling of nano-structure formation. *Nanoengineering, 10*(2), 108–112.

18. Kodolov, V. I. Didik, A. A., Volkov, A. Yu., & Volkova, E. G. (2004). Low-temperature synthesis of copper nanoparticles in carbon shells. HEIs' news. *Chemistry and Chemical Engineering, 47*(1), 27–30.

19. Lipanov, A. M., Kodolov, V. I., Khokhriakov, N. V., et al. (2005). Challenges in creating nanoreactors for the synthesis of metal nanoparticles in carbon shells. *Alternative Energy and Ecology, 22*(2), 58–63.

20. Kodolov, V. I., Khokhriakov, N. V., Kuznetsov, A. P., et al. (2007). Perspectives of nanostructure and nanosystem application when creating composites with predicted behavior. *Space challenges in XXI century, vol. 3, Novel materials and technologies for space rockets and space development* (pp. 201–205). Moscow: Torus press.
21. Krutikov, V. A., Didik, A. A., Yakovlev, G. I., et al. (2005). Composite material with nanoreinforcement. *Alternative Energy and Ecology, 24*(4), 36–41.
22. Kodolov, V. I., Khokhriakov, N. V., & Kuznetsov, A. P. (2006). To the issue of the mechanism of nanostructure influence on structurally changing media during the formation of "intellectual" composites. *Nanotechnics, 3*(7), 27–35.
23. Khokhriakov, N. V., & Kodolov, V. I. (2005). Quantum-chemical modeling of nanostructure formation. *Nanotechnics*, (2), 108–112.
24. Khokhriakov, N. V., & Kodolov, V. I. (2011). Influence of hydroxyfullerene on the structure of water. *International Journal of Quantum Chemistry, 111*(11), 2620–2624.
25. Kodolov, V. I., Khokhriakov, N. V. Trineeva, V. V., & Blagodatskikh, I. I. (2010). Problems of Nanostructure Activity Estimation, Nanostructures Directed Production and Application. *Nanomaterials Yearbook. From Nanostructures, Nanomaterials and Nanotechnologies to Nanoindustry* (pp. 1–18). New York: Nova Science Publishers, Inc.
26. Suzdalev, I. P. (2006). *Nanotechnology: Physico-Chemistry of Nanoclusters, Nanostructures, Nanomaterials.* Moscow: KomKniga,
27. Volynsky, A. L., Yarysheva, L. M., & Bakeev, N. F. (2007). Universal method for obtaining nanocomposites on polymeric base. *Russian Nanotechnologies, 2*(3–4), 58–68.
28. Didik, A. A., Kodolov, V. I., Volkov, A. Yu., Volkova, E. G., & Khalmayer, K. H. (2009). Low-temperature technique for obtaining carbon nanotubes. *Inorganic Materials, 39*(6), 693–697.
29. Zezin, A. B., Rogacheva, V. B., Valueva, S. P., et al. (2006). From triple interpolyelectrolyte-metal complexes to nanocomposites polymer-metal. *Russian nanotechnologies, 1*(1–2), 191–200.
30. Korshak, V. V. (1970). *Chemical Composition and Temperature Characteristics of Polymers* (p. 420). Moscow: Nauka.

CHAPTER 4

THE FUNCTIONALIZATION OF METAL OR CARBON NANOCOMPOSITES OR THE INTRODUCTION OF FUNCTIONAL GROUPS IN METAL OR CARBON NANOCOMPOSITES

V. I. KODOLOV, L. F. AKHMETSHINA, M. A. CHASHKIN, M. A. VAKHRUSHINA, and V. V. TRINEEVA

CONTENTS

4.1 THE BASIC INFORMATION ABOUT THE ACTIVATION AND THE FUNCTIONALIZATION OF NANOSTRUCTURES

The activation of nanostructures with functional groups formation may be named as the Functionalization. The functionalization of metal/carbon nanocomposites is usually carried out for the improvement of nanostructures interaction with media and for the increasing of correspondent nanostructures fine dispersive suspensions stability. The functionalization of nanostructures is one from methods their activation to participate in the interaction with the definite media.

The functionalization is considered as the chemical activation and is used together with the physical activation. At present the information in scientific literature about the following physical methods of activation for the obtaining of the correspondent nanostructures activity takes place:

- Mechanical spraying (carbon nanostructures obtaining, metal/carbon nanocomposites obtaining);
- Ultrasound dispersion (work vibration—28kHz, power—2kW);
- Electromagnetic action (vibration of electromagnetic field—50Hz);
- Super high vibration action (field vibration—2.45GHz, power—0.8kW).

The estimation of results of nanostructure activation processes is carried out on the time of equal results reaching. In the investigation [1] it is registered that the best result for the nano porous catalyst production is achieved when ultrasound action is applied. The middle results take place when electromagnetic field or super high vibration action is used. The chemical functionalization of nanostructures ground on the grafting of the different functional groups or molecules to the surface of nanostructures formed [2–5]. This grafting changes the electron structure of graft nanocomposite and increases its interaction with medium molecules. In this case the possibility of nanoparticles coagulation is decreased.

Redox synthesis is one from widely applied methods of nanostructures functionalization [6–17]. The different oxidized media and clear Fluorine are used for processes realization [6–10, 14]. The processes of further interaction of functional groups obtained with other agents, for instance, with substances containing amino-groups or hydroxyl-groups are described in [12, 13, 15]. In these cases the electrochemical methods of grafting are possible [16, 17]. Below the examples of different groups (molecules) grafting is given. In [12] it is described the making of special cantilever for AFM by the further grafting of amino groups to carboxyl groups disposed on carbon nanotube surface (Figure 4.1).

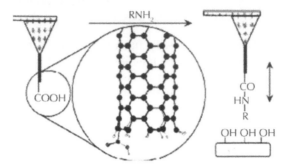

FIGURE 4.1   Scheme of special cantilever making.

For the increasing of nanotubes suspensions the functionalization of carbon nanotubes proceeds successfully [13] that represent by scheme (Figure 4.2):

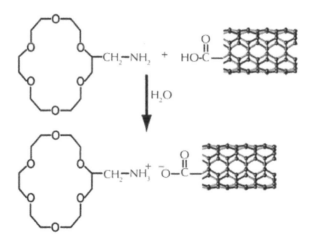

FIGURE 4.2   Scheme of nanotube functionalization.

In the chapter [15] the information is given about the alcohol suspension stability increasing by the fluorine functional groups disposed on the carbon nanotubes surface (Figure 4.3).

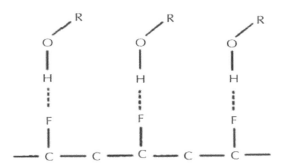

FIGURE 4.3 The picture of nanotube interaction with alcohol.

The mechanochemical method of nanocomposite functionalization will be great perspective method. The substitution of this method consists grinding of nanocomposites together with active media containing functional groups. The functional groups of correspondent media (phosphor and nitrogen containing groups) according to results of these processes investigations are grafting on the metal/carbon nanocomposite surfaces.

4.2 THE FUNCTIONALIZATION OF METAL/CARBON NANOCOMPOSITES BY PHOSPHORUS CONTAINING COMPOUNDS

At present, metal/carbon nanocomposites are such perspective modifiers. The introduction of nanocomposites into the coating can improve the material behavior during the combustion; retard the combustion process [18, 19].

In this regard, it is advisable to modify metal/carbon nanocomposites with ammonium phosphates to increase the compatibility with the corresponding compositions. The grafting of additional functional groups is appropriate in strengthening the additive interaction with the matrix and, thus, in improving the material properties. The grafting can contribute to improving the homogeneity of nanostructure distribution in the matrix and suspension stability. Due to partial ionization additional groups produce a small surface charge with the result of nanostructure repulsion from each other and stabilization of their dispersion. Phosphorylation leads to the increase in the influence of nanostructures on the medium and material being modified, as well as to the improvement of the quality of nanostructures due to metal reduction [20–22].

4.2.1   Materials

Metal/Carbon Nanocomposites

To modify the materials nickel/C, copper/C and iron/carbon nanocomposites were obtained with low-temperature synthesis in nanoreactors of polymeric matrixes of polyvinyl chloride [23] and polyvinyl alcohol [24]. Iron oxide (3), copper oxide and nickel oxide were selected as metal containing phase.

To define the sizes, shape and structure the initial nanostructures were investigated with transmission electron microscopy (TEM), electron diffraction (ED), X-ray photoelectron spectroscopy and IR spectroscopy. The detailed information about the sizes, shape and properties of Fe/C, Ni/C and Cu/C nanocomposites is represented in [25–26].

Ammonium Phosphates

Phosphates are the salts and ethers of phosphorous acids. From the salts, orthophosphates and polymeric (or condensed) phosphates are distinguished. The latter are divided into polyphosphates with linear structure of phosphate-anions, meta phosphates with ring (cyclic) phosphate-anion and ultra phosphates with net, branched structure of phosphate-anion. Among ammonium phosphates the most popular are: (a) 3-ammonium phosphate ((NH4)3PO4 (APh), stable only in water solution; (b) ammonium polyphosphate (NH4PO3) n (APPh). The latter is:

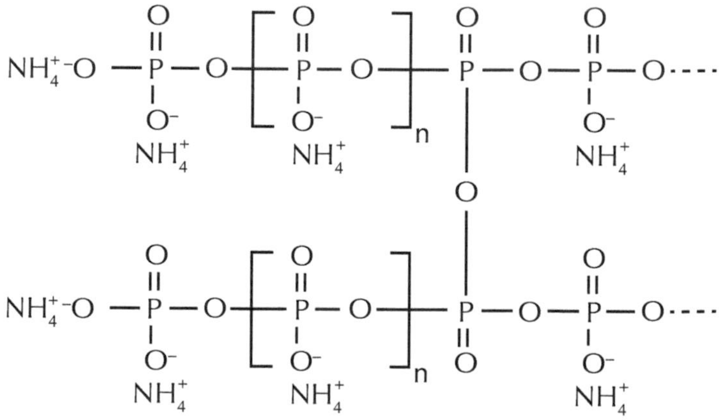

Ammonium phosphate and ammonium polyphosphate were used to graft additional functional groups to nanocomposites. Ammonium phosphate $(NH_4)_3PO_4$

is a colorless hard substances, it degrades at 30–40°C producing NH_3. Ammonium polyphosphate (APPh) is inorganic salt of polyphosphoric acid and ammonium. The chain length of such oligomer can exceed 1,000. Short and linear chains (n < 100) are more sensitive to water and are less heat-resistant. Longer chains (n > 1,000) have very low solubility in water (<0.1g/100ml). The product properties are: white crystalline powder, particle sizes—8–15mcm, phosphorus content—31–32%, nitrogen content—14–15%, water content (H_2O) <0.25%, density—1.7–1.9g/cm³, degrade temperature >275°C, pH (10% suspension)—6–8.

4.2.2 Quantum-Chemical Modeling

To define the possibility of the actual process of ammonium polyphosphate interaction with nanostructures, the models imitating the functionalization process of carbon nanostructures were built and optimized with the help of software HyperChem v. 6.03.

The functional groups were grafted by joint mechanical and chemical treatment of metal/carbon nanocomposite (metal/C NC) and APPh, therefore the interaction of metal/C NC fragment and APPh fragment are considered (Figure 4.4).

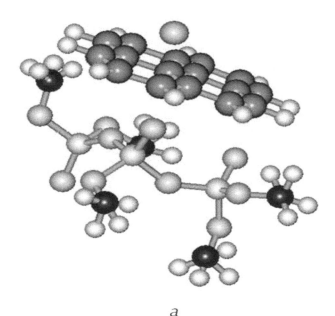

a

FIGURE 4.4 *(Continued)*

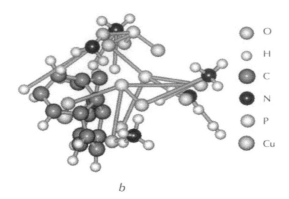

b

FIGURE 4.4 Modeling of functionalization process on the example of copper/carbon nanocomposite before geometry optimization (a), and after optimization (b).

The calculation demonstrates the change in the distance between oxygen atom and nitrogen atom (N^+-O^-) in APPh molecule; it varies in some cases from 1.36Å to 2.21Å, which indicates the breaking-off of bonds between oxygen and nitrogen with further isolation of NH_3 from the reaction mass. The distance between phosphorus atom and oxygen atom (P=O) increases, thus confirming the bond transition from (P=O) in ($P^+\rightarrow O^-$) with further orientation of hydrogen atoms being released from graphene plane surface.

Thus quantum-chemical modeling allows assuming the possibility of polyphosphate practical application for nanocomposite modification.

4.2.3 Equipment and Investigation Technique

When analyzing the structure of nanocomposites such physical and chemical investigation methods are used as X-ray photoelectron spectroscopy (XPES) and IR spectroscopy, the sorption ability of nanocomposites is found.

- *The IR spectroscopy method*: The IR Fourier spectrometer FSM 1201 was used to obtain IR spectra of functionalized nanocomposites and suspensions on alcohol basis. The pictures were taken in wave number range 399–4,500cm^{-1}.

- *Technique for determining the sorption ability of nanostructures:* The sorption ability of samples is determined as the content of the substance adsorbed in percent in relation to the initial adsorbent mass. To evaluate the sorption ability of carbon nanostructures the sorption with activated carbon which was used as the reference sample was carried out in parallel.

Alcohol was used as sorbate. The sorption efficiency was evaluated by extraction ratio:

$$\Gamma = \frac{\Delta m}{m_1 - m_0} \cdot 100\% \qquad (4.1),$$

where m_0—weighing bottle mass, m_1—weighing bottle and powder mass (initial), m_2—weighing bottle and powder mass (final), $\Delta m = m_2 - m_1$.

- *The XPES*: To examine the composition, structure and magnetic moment of functionalized nanocomposites the samples of nanopowders for XPES investigation were prepared. X-ray photoelectron spectra were obtained on electron spectrometer EMS-3.
- The length of the carbonized part of the material sample was measured after the flame source action within 1.5min on specially prepared plates placed horizontally during the flame action.

Samples Preparation for Flammability Determination

The samples to be tested are the plates with the dimensions $150 \times 15 \times 3$ (mm). The plates consist of foam polyethylene and paper glued together with phosphorus containing glue modified with metal/carbon nanocomposites with and without phosphorus. At the same time, check samples are prepared. These are the plates of foam polyethylene and paper glued together with the glue BF-19 filled with ammonium polyphosphate with phosphorus content in the glue 3, 4, 5% from its mass.

4.2.4   Interaction of Nanocomposites with Ammonium Phosphates

There are several techniques to grate functional groups. By the bond strength the processes of connecting to nanostructures are divided into two groups: with the formation of strong covalent bonds and without such bonds (due to hydrophobic interaction, formation of hydrogen bonds). The method of obtaining phosphorus containing metal/carbon nanocomposites consists in the interaction of preliminary obtained metal/carbon nanocomposites with ammonium phosphates, in particular APPh, in specifically calculated proportion. On the experiment results, the optimal proportion for all three types of nanostructures is the correlation of ammonium phosphate (APh) or APPh to NC = 1:1. The APPh interacts with NC in mechanical mortar following the experimentally found mode with water for better grinding. Afterwards the samples are dried in a dry heat oven and studied.

4.2.5   Investigations with X-Ray Photoelectron Spectroscopy Application

By the XPES data it was found that in some cases nanostructures before their modification with ammonium phosphates contained oxides of not completely reduced metals, the carbon structure was not completely formed (Figure 4.5). After functional groups were grafted to nanocomposites (Figure 4.6), a great number of bonds appropriate for nanostructures were formed, such as C–C, C–Me, the number of C–H bonds decreased. Below you can find the XPES spectrum and its analysis on the example of Fe/C nanocomposite.

Before the modification with APPh the process of iron reduction in nanocomposites is complicated and nanoproduct output is insignificant. The reduction process activity goes up after APPh is introduced. In this case, the metal is reduced and metal/carbon nanostructures are formed. After the modification C–Me bond is vivid on the spectra.

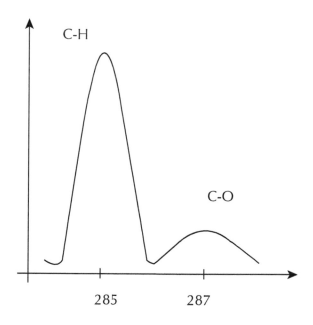

FIGURE 4.5   X-ray electron C1s spectrum of Fe/C nanocomposite before modification.

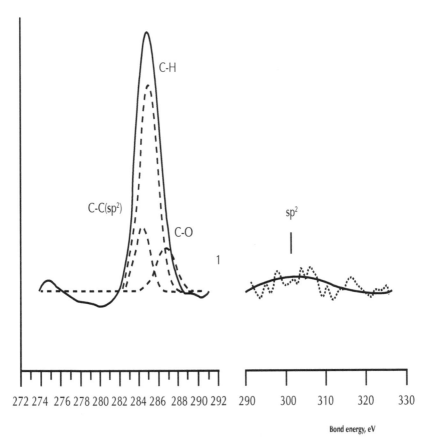

272 274 276 278 280 282 284 286 288 290 292   290   300   310   320   330

Bond energy, eV

FIGURE 4.6   X-ray electron C1s spectrum of Fe/C nanocomposite after modification.

The sample magnetic moment increases indicating the structure improvement as, for example, the magnetic moment of iron/carbon NC increased from 2.2 up to $2.5\mu_B$, and nickel/carbon NC—from 1.8 up to $3\mu_B$; the magnetic moment of nickel reference sample was 0.6. Thus the modification results in the improvement of nanocomposite structures, metal reduction, appearance of additional functional groups increasing the nanostructure activity in materials.

4.2.6   Investigations by IR Spectroscopy

In Figure 4.7, one can see the IR spectrum on the example of copper/carbon nanocomposite. The IR data demonstrate the appearance of the bands corresponding to phosphorus containing groups in the range 850–1,250cm^{-1}.

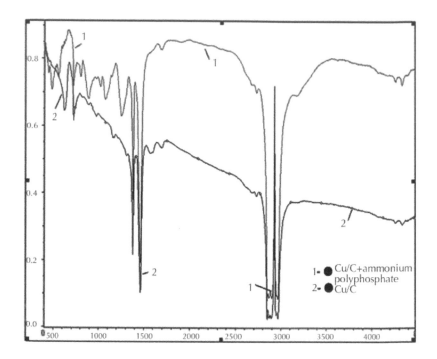

FIGURE 4.7   The IR spectrum of modified copper/carbon nanocomposites (1) in comparison with non modified nanostructures (2).

The IR spectrum indicates the peak of benzene ring flat deformation at wave number $1,500–1,600 cm^{-1}$, π-electrons shift relative to the plane and benzene ring curves to the side of phosphoryl group contained in ammonium polyphosphate.

When comparing IR spectra of the samples, it should be pointed out that the absorption intensity on the spectrum of phosphoryl (18) sample is much greater (nearly in two times) in comparison with the sample spectrum which does not contain phosphorus. This indicates the nanocomposite activity growth.

4.2.7   Sorption Ability to Alcohol

The sorption of samples modified was studied in relation to activated carbon and nanostructures not modified with APPh.

The data of sorption study revealed that samples modified demonstrate the best sorption ability (Figure 4.8). In case of phosphorus containing Cu/C nanocomposite the sorption ability after 120hrs was ~40% maximum, but the sorption ability of the sample not containing phosphorus was ~7% after the same time interval (after 20hrs—20% maximum).

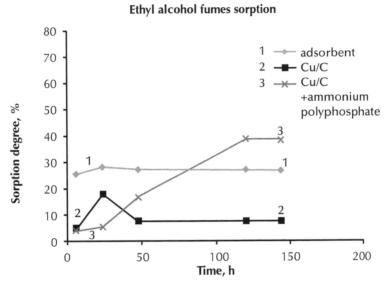

FIGURE 4.8   Diagram of ethyl alcohol fumes sorption by copper/carbon nanocomposites modified with APPh in comparison with the sorption of non-modified samples and activated carbon.

Sorption ability can indicate the nanoproduct activity degree during the modification of polymeric materials.

4.2.8   Preparation of Fine Suspension

Prior to the modification of different materials the preparation of nanocomposites is required. According to reference sources [18], the most advantageous and urgent decision is the application of fine suspensions of nanostructures. They provide the fine modifier uniform distribution through the volume of the material modified.

The liquid medium should be selected in accordance with the composition of the material modified not to distort the structuring processes and not to act as additional inclusions in it. Thus for each material the liquid media should be either the basic binder.

The technique for producing the suspension on ethyl alcohol basis consists in mechanical treatment of nanocomposite and ethyl alcohol following the mode selected with further treatment in ultrasound bath (the treatment time found experimentally—10min). The necessary concentration of nanostructures in suspension was chosen based on alcohol volume and varied from 0.001 up to 0.1%.

4.2.9 The IR Spectrum of Nanocomposite Alcohol Suspension

Nanocomposites, contained Phosphorus atoms, were introduced into the alcohol medium. At the same time, the final content of nanostructures by mass was 0.001%. The suspensions were studied on IR Fourier-spectrometer (Figure 4.9).

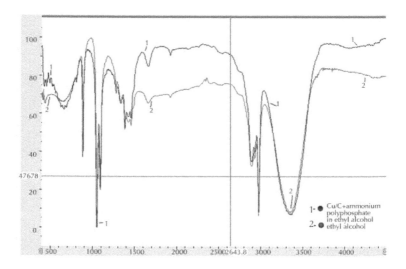

FIGURE 4.9 The IR spectrum of alcohol suspension containing copper/carbon phosphorus containing nanocomposites.

The investigation revealed that IR spectrum of alcohol suspension based on Cu/C NC(P) is more similar to IR spectrum of ethyl alcohol, but the bands, in particular in the range 1,200–1,700, differ by intensity and correlation between each other due to the presence of phosphoryl groups in nanocomposite.

The change in IR spectrum by absorption intensity indicates the availability of nanocomposite in the suspension. All the previous investigations [27–29] describe the composition activity increase at its inclusion into the composition of nanostructures, thus resulting in the improvement of required properties and characteristics of the composition. The interaction of nanocomposites with ammonium polyphosphate results in grating phosphoryl groups to them that allowed using these nanocomposites to modify intumescent fireproof coatings.

The sorption ability of functionalized nanocomposites was studied. It was found out that phospholyrated nanostructures have higher sorption ability than nanocomposites not containing phosphorus. Since the sorption ability indicates the nanoproduct activity degree, it can be concluded that nanocomposite activity increases in the presence of active medium and modifier (ammonium polyphosphate).

The IR spectra of suspensions of nanocomposites on alcohol basis were investigated. It was found that changes in IR spectrum by absorption intensity indicate the presence of excitation source in the suspension. It was assumed that nanostructures are such source of band intensity increase in suspension IR spectra.

The chapter presents the data for improving the quality of nanocomposites and suspensions on their basis after their modification with ammonium phosphates. In accordance with investigation results phosphorus introduction leads to more complete metal reduction, nanoproduct structure improvement, appearing of additional functional groups which increase the quality of suspensions on composite basis. When nanostructures, including those modified with ammonium phosphates, are introduced into the glue composition, the decrease in material flammability is observed thus giving the possibility to use the given composition as fireproof coatings.

4.3   THE FUNCTIONALIZATION OF METAL/CARBON NANOCOMPOSITES BY NITROGEN CONTAINING COMPOUNDS

This decade has been heralded by a large-scale replacement of conventional metal structures with structures from polymeric composite materials (PCM). Currently we are facing the tendency of production growth of PCM with improved operational characteristics. In practice, PCM characteristics can be improved when applying modern manufacturing technologies, for example, the application of "binary" technologies of prepreg production [30], as well as the synthesis of new polymeric PCM matrixes or modification of the existing polymeric matrixes with different fillers.

The most cost-efficient way to improve operational characteristics is to modify the existing polymeric matrixes; therefore, currently the group of polymeric materials modified with nanostructures (NS) is of special interest. The NS are able to influence the super molecular structure, stimulate self-organization processes in polymeric matrixes in super small quantities, thus contributing to the efficient formation of a new phase "medium modified—nanocomposite" and qualitative improvement of the characteristics of final product—PCM. This effect is especially visible when NS activity increases which directly depends on the size of specific surface, shape of the particle and its ultimate composition [31]. Metal ions in the NS used in this work also contribute to the activity increase as they stimulate the formation of new bonds.

The increase in the attraction force of two particles is directly proportional to the growth of their elongated surfaces, therefore the possibility of NS coagulation and decrease in their efficiency as modifiers increases together with their activity growth. This fact and the fact that the effective concentrations to modify polymeric matrixes are usually in the range below 0.01 mass % impose specific methods for introducing NS into the material modified. The most justified and widely applicable are the methods for introducing NS with the help of fine suspensions (FS). This introduction method allows most uniformly distributing particles in the volume of the medium modified, decreasing the possibility of their coagulation and preserving their activity during storage.

4.3.1   Quantum-Chemical Investigation of Fragments Interaction in Epoxy Polymers Modified with Metal/Carbonic Nanocomposites

In the process of quantum-chemical modeling the fragments imitating the initial reagents are optimized: epoxy diane resin (EDR), polyethylene polyamine (PEPA), cobalt/carbon nanocomposite (Co/C NC), nickel/carbon nanocomposite (Ni/C NC), copper/carbon nanocomposite (Cu/C NC) with the inclusion of Co^{2+}, Ni^{2+}, Cu^{2+} ions (Figure 4.10). For each of these fragments the absolute value of binding energy is defined E_{EDR}, E_{PEPA}, $E_{NC(ME)}$, where Me—cobalt, nickel or copper ion.

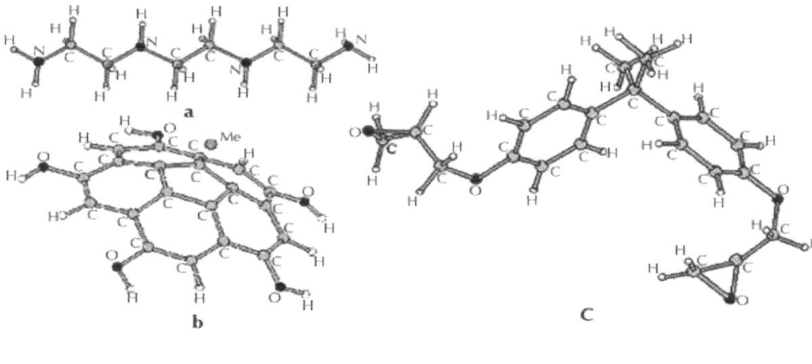

FIGURE 4.10   Fragments of initial substances: (a) PEPA fragment; (b) Me/C NC fragment; (c) EDR fragment.

As the modification process initially assumed the production of fine suspensions (FS) of NC on PEPA basis, the fragments imitating the behavior of the corresponding suspensions of Co/C, Ni/C, Cu/C nanocomposites were optimized (Figure 4.11) and their absolute values of binding energy $E_{PEPA\ NC(Co)}$, $E_{PEPA\ NC(Ni)}$, $E_{PEPA\ NC(Cu)}$ were defined.

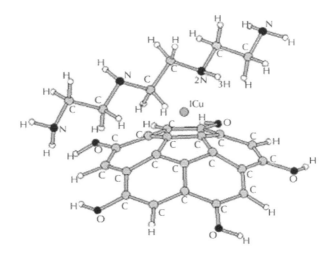

FIGURE 4.11 Fragment of Cu/C NC FS on PEPA Basis.

The next step was to model the influence of fine suspensions of nanocomposites on epoxy resin. The complexes formed similarly with the previous ones were optimized, and the absolute values of binding energy were found for each of them $E_{EDR-PEPA-NC(CO)}$, $E_{EDR-PEPA-NC(Ni)}$, $E_{EDR-PEPA-NC(Cu)}$. The absolute values of binding energy are given in Table 4.1.

TABLE 4.1 Absolute binding energies of the fragments.

	$E_{NC(Me)}$, KJ/mol	$E_{PEPA-NC(Me)}$, KJ/mol	$E_{EDR-PEPA-NC(Me)}$, KJ/mol
Co^{2+}	−19116.50	−29992.05	−51486.96
Ni^{2+}	−18562.38	−29098.04	−50621.15
Cu^{2+}	−18340.32	−28764.72	−50315.94
E_{EDR}, KJ/mol	−21424.25		
E_{PEPA}, KJ/mol	−10131.36		

Using the data from Table 4.1 by the following formula:

$$E_1 = E_{EDR-PEPA-NC(Me)} - E_2 - E_{EDR}$$
$$E_2 = E_{PEPA-NC(PEPA)} - E_{PEPA}$$

(4.2)

the relative interaction energies of molecular complexes E_1 are calculated and the diagram is arranged (Figure 4.12).

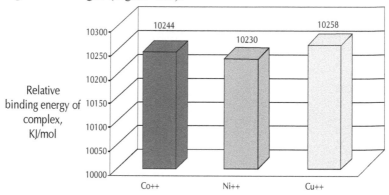

FIGURE 4.12   Diagram of relative interaction energies of molecular complexes.

From the diagram it is seen that the relative interaction energy of molecular complexes with Cu/C NC is higher in comparison with the complexes with Co/C NC and Ni/C NC content. As the polymer forms the strongest complexes with Cu/C NC, therefore after the modification it will be the most effective. The detailed analysis of the lengths of the bonds formed and effective charges before and after the optimization of fragments imitating PEPA interaction with Cu/C NC (Table 4.2 and 4.3) indicates that stable coordination bonds were formed between NC fragment and PEPA (between copper ion and nitrogen atom of amine group NH of PEPA). It was found that after the interaction of two fragments studied a part of electron density of N atom participating in the bond shifted to Cu atom, thus resulting in NH bond weakening.'

TABLE 4.2   Bond lengths.

Bond Designation	Bond length before optimization, Å	Bond length after optimization, Å
Cu-N	2.82	1.95
Cu-C	2.65	2.25

TABLE 4.3 Effective charges.

Atom number (see Figure 4.2)	Atom designation	Effective charge before optimization	Effective charge after optimization
1	Cu (copper)	0.138	−0.250
2	N (nitrogen)	−0.066	0.420
3	H (hydrogen)	0.045	0.027

The bond weakening is indirectly confirmed by the increase in the effective charge of H atom and slight change in the wave number in oscillatory spectra calculated. The wave number of NH bond before the optimization was $3,360 cm^{-1}$, and after—$3,354 cm^{-1}$, which correlates with the data of IR spectra obtained with the help of IR Fourier spectrometer. For instance, in the spectrum of PEPA and NC suspension on PEPA basis with nanocomposite concentration 0.03% the shift of wave numbers of peaks of amine groups is observed from $3,280 cm^{-1}$ to $3,276 cm^{-1}$.

The modeling of hardening process with the participation of epoxy resin, polyethylene polyamine and Cu/C nanocomposite is given in Figure 4.13a, b respectively. From the geometry of optimized molecular systems it is seen that the introduction of Cu/C NC into the system leads to its self-organization (Figure 4.13b) and formation of presumably coordination polymer. The formation of coordination polymer due to the introduction of nanocomposite active particles can result in increasing the adhesive strength and thermal stability as the total number of bonds in polymer grid grows and more energy will be required for its thermal destruction. At the same time, the formation of nanocomposite metal coordination bond with PEPA nitrogen can increase the stability of fine suspension formed. Such coordination results in polyethylene polyamine activity growth during the hardening of epoxy diane resin.

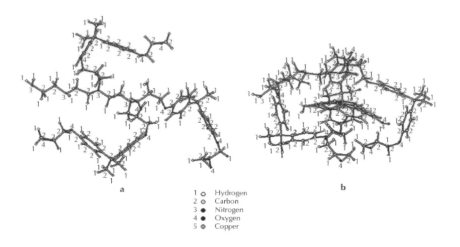

a

1 ○ Hydrogen
2 ◉ Carbon
3 ● Nitrogen
4 ● Oxygen
5 ◉ Copper

b

FIGURE 4.13   Influence of Cu/C NC on epoxy resin (ER) hardening: (a) ER hardening without Cu/C NC; (b) ER hardening in the presence of Cu/C NC.

Thus, quantum-chemical modeling allows predicting the interaction processes of components when hardening the epoxy resin with polyethylene polyamine and active participation of copper/carbon nanocomposite.

4.3.2 Materials

Currently there is a huge need in modern epoxy systems with improved operational characteristics [32] that can be reached, as previously mentioned, when modifying them with nanostructures. Therefore in this work the modification processes of cold-hardened model epoxy composition containing epoxy diane resin ED-20 State Standard (GOST) 10587-84 in the amount of 100 weight fractions and polyethylene polyamine (PEPA) grade A Technical Condition (TU) 2413-357-00203447-99 in the amount of 10 weight fractions were considered as the research object. The modification was carried out when introducing Metal/C nanocomposite into the epoxy resin.

The PEPA represents the mixture of linear branched ethylene polyamines with average molecular mass 200–250 and very wide molecular-mass distribution. The general structural formula of PEPA is as follows:

$$NH_2[-CH_2CH_2NH-]_n-H, \text{ where } n = 2 - 8.$$

Synthesized Co/C, Ni/C, Cu/C nanocomposites were studied with the help of transmission electron microscopy (TEM) and electron microdiffraction (EMD) in the shared centers of research institutions in Ekaterinburg and Moscow.

The investigations of Cu/C and Ni/C nanocomposites revealed that their average sizes (r) differ approximately in two times: r of Cu/C NC is 25nm, r of Ni/C NC is about 11nm. The average size of Co/C nanocomposite practically equals the average size of Ni/C NC. Such correlation of average sizes of nanocomposites is apparently connected with the ability to form metal containing clusters. Atom magnetic moments of nanocomposites are greater than the corresponding moments of microparticles of the same metals. The availability of magnetic properties of nanoproducts widens their application possibilities [33].

4.3.3 Processing of Fine Suspensions

The fine suspension was prepared in a number of stages:
- Preliminary grinding of Cu/C NC in mechanical mortar.
- Mechanical and chemical activation of suspension components when combining PEPA and Cu/C nanocomposite.
- Ultrasound processing for complete and uniform dispergation of Cu/C NC particles in PEPA volume.

For thermo gravimetric investigation also two sets of samples were produced, including the reference and modified FS of Cu/C NC with concentrations 0.001, 0.01, 0.03, 0.05%.

4.3.4 Investigation Technique

- *Quantum-Chemical Modeling:* In the frameworks of this method the fragments imitating the behavior of Co/C, Ni/C, Cu/C nanocomposites, fine suspensions of nanocomposites and systems being epoxy resins modified with nanocomposites were constructed and optimized with the help of software HyperChem v. 6.03 and semi-empirical methods. At the same time, the absolute binding energies of certain components of the molecular systems formed were found and their relative interaction energies were calculated. The oscillatory spectra of fine suspensions of Cu/C nanocomposite were calculated with the help of semi-empirical method PM3.
- *The IR spectroscopy:* For IR spectra IR Fourier-spectrometer FSM 1201 was used. The IR spectra of liquid films of PEPA, NC FS with concentrations 0.001, 0.01 and 0.03% were taken. The spectra were taken on KBr glasses in wave number range 399–4,500cm^{-1}.

- *Optical spectroscopy*: Spectral photometer KFK-03-01 was used to define the optimal time period of NC FS ultrasound processing. The optimal time period value was the interval when the optical density (D) was at the maximum that is, corresponded to the maximum saturation of FS with NC. The optical density of NC FS samples with the concentration 0.01% processed in ultrasound bath Sapfir UZV 28 within 0, 3, 7, 10, 15, 20, 30min, respectively, at US power 0.5kW and frequency 35kHz was measured in 5-ml quartz cuvettes. The work wavelength was found when defining the optical density in the range $\lambda = 320-920$nm.

- *Determination of relative viscosity*: The relative viscosity of PEPA and FS on its basis was found with the help of viscometer VZ-246 in accordance with State Standard (GOST) 8420-74. The relative viscosity values were translated into the kinematic following GOST 8420-74. The viscosity was measured for PEPA processed and not processed with ultrasound, and also for fine suspensions of NC with the concentrations 0.001, 0.01 and 0.03% processed and not processed with ultrasound.

4.3.5   Optimal Time Period of Ultrasound Processing of Fine Suspensions of COPPER/Carbon Nanocomposite

In the process of selecting the work wavelength (λ) on spectral photometer KFK-03-01 to define the optimal time period of ultrasound processing of FS on PEPA basis the optimal wavelength 413nm was found. This wavelength was selected as the optical density of FS with the concentration of Cu/C NC 0.01% not processed with ultrasound was 0.800 that corresponded to the middle of the operational range of this instrument (operational range D = 0.001–1.5). At the wavelength $\lambda = 413$nm the optical densities of all FS investigated were defined. This is reflected in the graphic dependence of optical density D upon the time period of ultrasound processing τ_{us} (Figure 4.14).

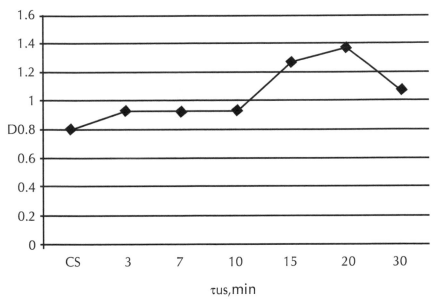

FIGURE 4.14 Dependence of the change in optical density (D) of Cu/C NC FS upon the time period of ultrasound processing τus.

The analysis of optical density dependence upon the time period of ultrasound processing (Figure 4.14) demonstrates that the optimal processing time of Cu/C NC fine suspension on PEPA basis is 20min. Further ultrasound processing is useless as in 20min the maximum optical density 1.37 is reached.

Taking into account the above data for defining the optimal time period of ultrasound processing, the preparation of fine suspension for IR investigations, for finding the viscosity and for producing the modified epoxy resin was carried out with preliminary ultrasound processing within 20min.

4.3.6 Viscous Properties of Fine Suspensions of Copper/Carbon Nanocomposite

The following diagram was prepared based on the results of viscosity measurements of PEPA and fine suspension (FS) with different Cu/C NC content before and after US processing (Figure 4.15).

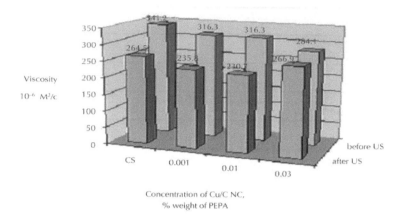

FIGURE 4.15 Diagram of the dependence of FS kinematic viscosity on Cu/C NC concentration.

The decrease in FS viscosity with the increase in Cu/C NC concentration on the diagram (Figure 4.15) is explained by the fact that the system consisting of two phases, in our case (PEPA—dispersion medium and Cu/C NC—disperse phase) tends to surface energy decrease. This tendency is expressed by self-decrease in interface surface due to sorption. The PEPA molecules start sorbing on Cu/C NC surface, probably producing the interface between Cu/C NC particles being the obstacle for coagulation. Due to such localization the interaction between medium particles diminishes, resulting in intermolecular friction decrease, and, consequently, decrease in system viscosity (Figure 4.16a). When fine suspensions are processed with ultrasound, the sizes of disperse phases decrease.

Thus the growth of surface energy increases the sorption ability and leads to the action region on medium increases (Figure 4.16b).

The action regions of Cu/C nanocomposite in fine disperse suspension (FS) with concentrations 0.001 and 0.01% do not overlap that is confirmed by viscosity decrease, and for FS with concentration 0.03% the action regions overlap and, consequently, the increase in intermolecular friction and viscosity are observed.

Taking into account that the action degree on the medium mainly depends on the action time of Cu/C NC on the medium modified, FS were studied for 48hrs. After this time interval FS layered into floccular structures with no sediment on the bottom observed (Figure 4.16c).

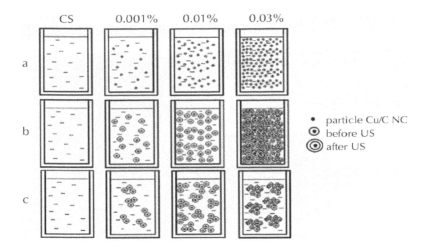

FIGURE 4.16 Distribution of Cu/C NC in PEPA: a before ultrasound processing; b after ultrasound processing; c after 48hrs.

Mechanical action resulted in FS recovery, which is indirectly confirmed by the availability of PEPA stable layer between Cu/C nanocomposite particles preventing the coagulation and possibility of suspension production with the ability to recover the distribution of nanocomposite particle distribution.

4.3.7 The IR Spectroscopic Investigation

The comparative IR spectroscopic investigation of bis(2-aminoethyl) amine—one of PEPA olygomers demonstrates the consistency of wave numbers appropriate for these compounds.

In both spectra the wave numbers appropriate for the oscillations of amine groups $v_s(NH_2)$ 3,352cm^{-1} and asymmetric $v_{as}(NH_2)$ 3,280cm^{-1} are available, there are wave numbers that refer to symmetric $v_s(CH_2)$ 2,933cm^{-1} and asymmetric valence $v_{as}(CH_2)$ 2,803cm^{-1}, deformation wagging oscillations $v_d(CH_2)$ 1,349cm^{-1} of methylene groups, deformation oscillations v_d (NH) 1,596cm^{-1} and $v_d(NH_2)$ 1,456cm^{-1} of amine groups, and also the oscillations of skeleton bonds are vivid $v(CN)$ 1,059–1,273cm^{-1} and $v(CC)$ 837cm^{-1}.

The IR spectra of PEPA and Cu/C NC FS were taken (Figure 4.17).

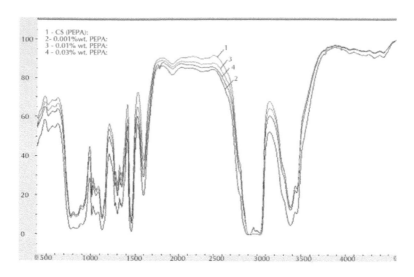

FIGURE 4.17   The IR spectra of PEPA and FS of Cu/C nanocomposite.

The comparison of IR spectra of polyethylene polyamine and fine suspension of Cu/C NC on PEPA basis (Figure 4.17) indicate that practically all changes of wave numbers in the spectra are within the error ± 2cm^{-1}. However in FS spectra the vivid increase in peak intensity corresponding to deformation oscillations of NH bonds is observed. These changes can spread onto the vast areas arranging a certain super molecular structure, apparently involving the adjoining amine groups into the process, which is demonstrated by the intensity change of these peaks.

Quantum-chemical modeling methods allow quite precisely defining the typical reaction of interaction between the system components, predict the properties of molecular systems, decrease the number of experiments due to the imitation of technological processes. The computational results are completely comparable with the experimental modeling results.

The optimal composition of nanocomposite was found with the help of software HyperChem v. 6.03. Cu/C nanocomposite is the most effective for modifying the epoxy resin. Nanosystems formed with this NC have higher interaction energy in comparison with nanosystems produced with Ni/C and Co/C nanocomposites. The effective charges and geometries of nanosystems were found with semi-empirical methods. The fact of producing stable fine suspensions of nanocomposite on PEPA basis and increasing the operational characteristics of epoxy polymer was ascertained. It was demonstrated that the introduction of Cu/C nanocomposite into PEPA facilitates the formation of coordination bonds with nitrogen of amine groups, thus resulting in PEPA activity increase in epoxy resin hardening reactions.

It was found out that the optimal time period of ultrasound processing of copper/carbon nanocomposite fine suspension is 20min.

The dependence of Cu/C nanocomposite influence on PEPA viscosity in the concentration range 0.001–0.03% was found. The growth of specific surface of NC particles contributes to partial decrease in PEPA kinematic viscosity at concentrations 0.001%, 0.01% with its further elevation at the concentration 0.03%.

The IR investigation of Cu/C nanocomposite fine suspension confirms the quantum-chemical computational experiment regarding the availability of NC interactions with PEPA amine groups. The intensity of these groups increased in several times when Cu/C nanocomposite was introduced.

KEYWORDS

- **Functionalization**
- **IR Spectroscopy**
- **Nanostructure**
- **Phosphorylation**
- **Redox**

REFERENCES

1. Blokhin, A. N. (2012). *Development of process of composites nanocarbon modification on base epoxy resins and mechanization of its process.* Thesis of Canada Technical Science. Tambov: Tambov State Technical University.
2. Postnikov, P. S., Trusova, M. E., Feduschak, T. A., et al. (2010). Aryl diazoniy tonzilates as new effective agents of covalent grafting aromatic groups to carbon shells of metallic nanoparticles. *Russian Nanotechnologies, 5*(7–8), 49–50.
3. Kulakova, I. I., Korol'kov, V. V., Yakovlev, R. Yu., & Lisichkin, G. V. (2010). Structure of particles of chemical modified diamond obtained by knocking. *Russian Nanotechnologies, 5*(7–8), 66–73.
4. Aparcin, E. K., Novohashina, D. S., Nastaushev, Yu. V., & Ven'yaminova, A. G. (2012). Fluorescent tagging UNT and their hybrids with oligonucleotides. *Russian Nanotechnologies, 7*(3–4), 38–45.

5. Demin, A. M., Uiimin, M. A., Schegoleva, N. N., et al. (2012). Surface modification of magnetic nanoparticles on the basis Fe_3O_4 by (S)-naproxen. *Russian Nanotechnologies, 7*(3–4), 66–70.

6. Ebbesen, T. W., Ajayan, P. M., Hiura, H., & Tanigaki, K. (1994). Purification of nanotubes. *Nature, 367,* 519.

7. Kuznetsova, A., Popova, I., & Yates, J. T. (2001). Oxygen-containing functional groups on single-wall carbon nanotubes: NEXAFS and vibrational spectroscopic studies. *Journal of the American Chemical Society, 123,* 10699.

8. Kneller, J. M., Soto, R. J., & Surber, S. E. (2000). TEM and laser-polarizes 129Xe NMR characterization of oxydatively purified carbon nanotubes. *Journal of the American Chemical Society, 122,* 10591.

9. Mawhinney, D. B., Naumenko, V., Kuznetsova, A., Yates, J. T., Liu, J., & Smalley, R. E. (1999). Infrared Spectral Evidence for the etching of carbon nanotubes: Ozone oxidation at 298K. *IBID,* 2383–2384.

10. Geng, H. Z., Rosen, R., & Zheng, B., et al. (2002). Fabrication and properties of composites of poly (ethylene oxide) and Functionalized carbon nanotubes. *Advanced Materials, 14,* 1387.

11. Zhu, J., Peng, H., & Rodriguez-Macias, F., et al. (2004). Reinforcing epoxy polymer composites through covalent integration of functionalized nanotubes. *Advanced Functional Materials, 14,* 643.

12. Wong, S. S., Joselevich, E., & Wooley, A. T., et al. (1998). Covalently functionalized nanotubes as nanometre-sized probes in chemistry and biology. *Nature, 394,* 52.

13. Michael, G. C., & Wong, S. (2002). Solubilization of oxidized single-walled carbon nanotubes in organic and aqueous solvents through organic derivatization. *Nano Letters, 2,* 1215–1218.

14. Micklson, E. T., Huffman, C. B., & Rinzler, A. G., et al. (1998). Fluorination of single wall carbon nanotubes. *Chemical Physics Letters, 296,* 188.

15. Mickelson, E. T., Chiang, I. W., & Zimmerman, J. L., et al. (1999). Solvation of fluorinated single wall carbon nanotubes in alcohol solvents. *The Journal of Physical Chemistry B, 103,* 4318.

16. Allongue, P., Delamar, M., & Desbat, B., et al. (1997). Covalent modification of carbon surfaces by aryl radicals generated from the electrochemical reduction of diazonium salts. *Journal of the American Chemical Society, 119,* 201.

17. Delamar, M., Hitmi, R., Pinson, J., & Saveant, J. M. (1992). Covalent modification of carbon surfaces by grafting of functionalized aryl radicals produced from electrochemical reduction of diazonium salts. *Journal of the American Chemical Society, 114,* 5883.

18. Kodolov, V. I., & Khokhriakov, N. V. (2009). *Chemical physics of formation and transformation processes of nanostructures and nanosystems* (Vol. 1. p. 365 and Vol. 2, p. 4153). Izhevsk: Izhevsk State Agricultural Academy.

19. Bulgakov, V. K., Kodolov, V. I., & Lipanov, A. M. (1990). *Modeling of polymeric materials combustion* (p. 238). Moscow: Chemistry.

20. Shuklin, S. G., Kodolov, V. I., Larionov, K. I., & Tyurin, S. A. (1995). Physical and chemical processes in modified two-layer fire- and heat-resistant epoxy-polymers under the action of fire sources. *Physics of Combustion and Explosion, 31*(2), 73–79.

21. Glebova, N. V., & Nechitalov, A. A. (2010). Surface functionalization of multi-wall carbon nanotubes. *Journal of Technical Physics, 36*(19), 12–15.
22. Rakov, E. G. (2006). *Nanotubes and fullerenes: Students' book for higher educational institutions* (p. 376). Moscow: University Book.
23. Kodolov, V. I., Vasilchenko, Yu. M., Akhmetshina, L. F., Shklyaeva, D. A., Trineeva, V. V., Sharipova, A. G., Volkova, E. G., Ulyanov, A. L., & Kovyazina, O. A. (2010). *Patent 2393110 Russia Technique of obtaining carbon metal containing nanostructures,* declared on 17.10.2008.
24. Kodolov, V. I., Kodolova, V. V. (Trineeva), Semakina, N. V., Yakovlev, G. I., & Volkova, E. G., et al. (2008). *Patent 2337062 Russia Technique of obtaining carbon nanostructures from organic compounds and metal containing substances,* declared on 28.08.2006.
25. Kodolov, V. I., & Trineeva, V. V. (2012). Perspectives of idea development about nanosystems self-organization in polymeric matrixes. In *The problems of nanochemistry for the creation of new materials* (pp. 75–100). Poland, Torun: IEPMD.
26. Kodolov, V. I., Trineeva, V. V., Kovyazina, O. A., & Vasilchenko, Yu. M. (2012). Production and application of metal/carbon nanocomposites. In *The problems of nanochemistry for the creation of new materials* (pp. 17–22). Poland, Torun: IEPMD.
27. Akhmetshina, L. F., & Kodolov, V. I. (2012). Influence of metal/carbon nanocomposites on silicate composition properties. In *The problems of nanochemistry for the creation of new materials* (pp. 23–36). Poland, Torun: IEPMD.
28. Akhmetshina, L. F., Koreneva, E. Yu., Kodolov, V. I., et al. (2010). Nanostructures interaction with silicate compositions. *Nanotechnics, 3,* 13–16.
29. Akhmetshina, L. F., Kodolov, V. I., Tereshkin, I. P., & Korotin, A. I. (2010). The influence of carbon metal containing nanostructures on strength properties of concrete composites. *Nanotechnologies in Construction, 6,* 35–46.
30. Panfilov, B. F. (2010). Composite materials: production, application, market tendencies. *Polymeric materials,* (2–3), 40–43.
31. Kodolov, V. I., Khokhriakov, N. V., Trineeva, V. V., & Blagodatskikh, I. I. (2008). Activity of nanostructures and its expression in nanoreactors of polymeric matrixes and active media. *Chemical Physics and Mesoscopy, 10*(4), 448–460.
32. Bobylev, V. A., & Ivanov, A. V. (2008). New epoxy systems for glues and sealers produced by CJSC "Chimex Ltd". *Glues Sealers Paints,* (2), 162–166.
33. Kodolov, V. I., Kovyazina, O. A., Trineeva, V. V., Vasilchenko, Yu. M., Vakhrushina, M. A., & Chmutin, I. A. (2010). On the production of metal/carbon nanocomposites, water and organic suspensions on their basis. *VII International Scientific-Technical Conference "Nanotechnologies to the production". Proceedings. Fryazino,* 52.

CHAPTER 5

THE INVESTIGATION OF METAL OR CARBON NANOCOMPOSITES ELECTRON STRUCTURE BY X-RAY PHOTOELECTRON SPECTROSCOPY

I. N. SHABANOVA, N. S. TEREBOVA, V. I. KODOLOV,
I. I. BLAGODATSKIKH, V. V. TRINEEVA, L. F. AKHMETSHINA,
and M. A. CHASHKIN

CONTENTS

5.1 THE INVESTIGATION OF NANOSTRUCTURE SYNTHESIS PROCESSES BY X-RAY PHOTOELECTRON SPECTROSCOPY

5.1.1 The Method of Investigation of Nanosystem Electron Structure

The nanoproducts obtained are investigated by X-ray photoelectron spectroscopy (XPES), transmission electron microscopy and electron microdiffraction methods. The samples for XPE spectra are applied to indium plate. The spectra are obtained using X-ray photoelectron magnetic spectrometer (AlK_α radiation) and X-ray electron spectrometer ES-2401 (MgK_α radiation). The microphotographs of the samples are obtained on electron microscope JEM-200CX with the accelerating voltage 160 kV.

The spectra, given on Figure 5.1 and 5.2, are obtained on X-ray photoelectron magnetic spectrometer (Siegbahn type) with the apparatus resolution of 0,1eV. The metal carbon tubulenes are nonconducting samples. It is known that the positive electric charge, accumulated on the sample during the process of radiation owing to electrons can shift the atomic levels by some electron-volts. The aluminum grid of several mkm thickness was applicated just to avoid the influence of these effects on XPS data obtaining. The grid was settled on the way of X-ray radiation between cathode of the X-ray tube and the sample. The absence of the charging was controlled by C1s, O1s lines. The spectra shift was not observed.

The comparative spectra of Cls lines (XPE spectra) of the samples, containing Cu and Mn, are given as an example (Figure 5.1 and 5.2).

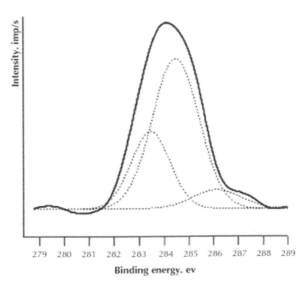

FIGURE 5.1 C1s XPE spectrum of manganese containing carbon tubulenes.

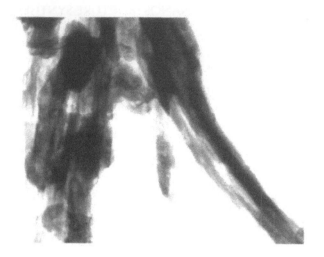

FIGURE 5.2   Microphotography of dendrite-like scrolls, including cobalt (×110,000).

Apparently, in our case, the variant of "scroll" formation from carbon melts is more preferable in micro cavities between salt and metal particles. Therefore, extensive cylindrical structures have rather large diameters, and can branch similarly to dendrites. This is confirmed while using transmission electron microscopy (Figure 5.3). The products extracted from the mixtures of anthracene, cobalt (or nickel), aluminum and sodium chlorides taken in molar ratio as 1:1:10:10, are investigated using X-ray photoelectron spectroscopy, transmission electron microscopy with electron microdiffraction determination.

In accordance with electron microscopy data, lamellar structures are formed around the particles which can refer to metal(cobalt)carbon clusters obtained. The structures appear as a result of joints by side surfaces of tubes and scrolls formed (Figure 5.4).

The same structures, but more clear, are observed in the samples obtained in the presence of nickel chloride. Large quantities of point reflexes, which form concentric circles, are observed in the electronograms of these samples (Figure 5.4).

As it is seen, the "scroll" formed consists of rolled graphite—like shells on the surface of a comparatively extensive metal particle (dark area). The presence of a metal phase and the rolling up in a spiral, that is tubulenes formation, is confirmed by the electron microdiffraction splitting.

FIGURE 5.3   Microphotography of tubulene sample, including nickel
(×150,000).

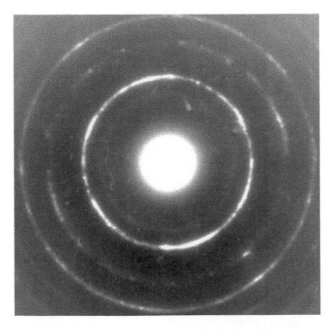

FIGURE 5.4   Microdiffraction picture of tubulene sample, including nickel.

The estimation of lattice parameters shows that the reflections observed belong to graphite and diamond (it is characteristic for tubulenes being formed [1–17]).

According to XPE spectra (Figure 5.5), graphite-like carbon structures are formed in reactive media containing cobalt. The maximum corresponding to carbon with binding energies 284 and 285.3eV (which is attributed to carbon in graphite-like structures and polyaromatic compounds), as well as a small wide peak of about 7% from the main maximum intensity, which can be attributed to carbon carbonyl and carbonate in accordance with binding energies (286.7 and 288.8eV), are observed in Cls spectrum. Co2p³ spectrum corresponds to the presence of oxidized cobalt, apparently connected with carbon particles, and cobalt which has interacted with the aromatic rings.

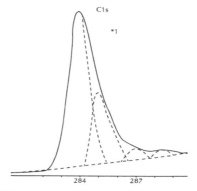

FIGURE 5.5   The XPE spectrum (C1s) of tubulene containing product including cobalt.

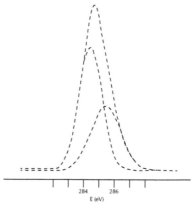

FIGURE 5.6   The XPE spectrum (C1s) of tubulenes containing product including nickel.

In accordance with XPS spectra (C1s) of samples intercalated by nickel (Figure 5.6), there are more carbon structures corresponding to carbon in graphite-like substances and there are no substances containing C–O bonds in comparison with tubulenes including cobalt.

The difference in the morphology of nanostructures being formed which contain cobalt and nickel, is apparently explained by the dispersion of metal particles due to $H-C^{\delta-}$—$Met^{\delta+}$ redox reaction. Such supposition agrees with the data [2] about the formation of nanotubes on metal catalysts.

The mechanism of tubulenes formation in salt and polyphosphoric acid melts needs further investigation, and the character of competing interactions of aluminum and transition metal ions determines the results obtained. Aluminum absence in the reaction products formed is noted, this is connected with the stronger interaction of transition metals with π-electron systems of aromatic hydrocarbons.

The above scheme of tubulenes obtaining is complicated because usually polyvinyl alcohol macromolecules form lamellar aggregates of a certain expansion, and the geometry distortion of the aggregates mentioned is possible, if a lamellar active mineral medium is a melt or water solution. The medium, which has a considerable defectiveness, accelerates tubulenes formation and increases the number of their growth centers. Therefore, the formation of dendrite-like structures and small-crystal structure is possible. In accordance with electron microscopy data the samples, obtained in polyphosphoric acid melt containing manganese ions, have a very small structure which produces incredible pictures of microdiffractions comprising bright and weak spots, the combination of which form pictures like "six-pointed stars" (Figure 5.7).

Here, it is mentioned that such structures are formed like dendrite on needle-like formations (Figure 5.8).

FIGURE 5.7 Microdiffraction picture of small crystal structure formed from PVA (in polyphosphoric acid).

FIGURE 5.8   Microphotography of dendrite-like tubular structure obtained in polyvanadic acid intercalated by Mo.

Three methods of carbon- metal containing tubulenes formation are considered. It is found that the "scroll" with metal inclusions and defective regions can be formed in active media having lamellar structure (in this investigation, in metal chloride melts or at the boundaries of aromatic hydrocarbon melts with metal chloride, as well as polyphosphoric acid melts or in polyvanadic acid solution which contain metal ions). The character of tubulene defects depends on the method of their obtaining, the active media structure, and the method of product isolation from the medium. Carbon and carbon-metal-containing groups in tubulenes, including the row of Mn-Co–Ni–Cu metals, are determined by X-ray photoelectron spectroscopy. Quantum-chemical calculating experiment shows the increase of metal activity from copper to nickel in dehydropolycondensation reaction and stimulated carbonization, which are stimulated by tubulenes formation. At the same time, the absence of metal lines in the spectra is explained by the masking of carbon atoms and by the small sizes of metal inclusions in comparison with the scroll size. This is observed in the microphotographs obtained by the transmission electron microscopy and is confirmed by the microdiffraction pictures. The metal ions play a greater role in the tubulenes formation in comparison with metal powder particles as it has been established.

5.1.2 X-ray Photoelectron Spectroscopy as a Method to Control the Received Metal-Carbon Tubules

In 1991 the idea of nano-tubular forms of substances started to develop in connection with the progress made in the sphere of studying structures and properties of carbon nano-objects such as a new allotropic form of carbon—fullerenes.

As one of the possible metastable nano-forms of carbon a quasi-one-dimensional tubular structure—an extended cylinder—was offered [18], formed by rolling up an atomic "ribbon" cut out of graphen grid. This structure was called nanotubulene or tubulene.

It is known that tubules can be received by the destruction of carbon electrodes in the electric arc in the presence of super dispersed metal particles or by the destruction of carbon electrodes in the process of electrolysis of salt melts [1–3]. In [19], some feasible mechanisms of the formation of tubules from aromatic hydrocarbons or polymers, containing functional groups, were offered.

The studies of metal-carbon systems conducted previously showed that the tubulene formation depends on the content of the initial components: anthracene—$NaCl$—$AlCl_3$—$MeCl_2$. In this connection, the dependence of the formation of C–C bonds on the content of the initial components in the mixture prepared from fine powders of anthracene and metal chloride was analyzed.

The studies were performed by means of the X-ray photoelectron spectrometer with an apparatus resolution 0.1eV at excitation of AlKα- lines [20]. The investigated systems were non-conducting samples. In the process of photoemission, a positive charge accumulated on their surfaces; the charge created a retarding field and shifted the atom levels by several electron-volts [21, 22]. To exclude the influence of the effect of charging the aluminum foil was set across the X-ray radiation path. When the foil was irradiated by X-ray radiation, additional electrons were knocked out, which neutralized the surface positive charge. The absence of the charging effects was determined by C1s—and O1s—lines. The displacement of the spectrum of these lines was not observed.

The XPS data processing was done in accordance with the procedure described in [23]. The spectra were smoothed, and the background was subtracted from them. The analysis of C1s spectra was carried out by means of the program based on the method of least squares together with the use of the known chemical shifts of the interior electron levels taking into account the width of the peaks at their half-heights characteristic to the reference patterns. The analysis was carried out with the help of Gauss function with the maximal approximation of the envelope curve to the experimental curve. The accuracy in determination of the positions of the peaks made up 0.1eV.

In the given work, the investigation of metal-carbon tubules was carried out. Metal-carbon tubules were obtained by the method of low-energy synthesis from multiring aromatic hydrocarbons in active media of a lamellar structure, which turn into salt melts on the exposure to the temperature. Duration of the synthesis was 3hrs [24]. Among polyaromatic compounds anthracene was selected. As active media, the eutectic melts containing aluminum and sodium chlorides together with chlorides of manganese, cobalt and nickel were used. The selection of these metal chlorides is due to their stimulating action in the reactions of dehydropolycondensation resulting in the formation of metal-carbon tubules as it was shown in [19]. The theoretical basis of the process of tubulene formation is presented in [25].

In Figure 5.9 C1s spectra of metal-carbon systems are displayed. These systems were obtained from the mixture of fine powders of anthracene and manganese chloride; molar ratio of the components was different (anthracene: salt— a) 1:1; b) 1:2; c) 1:3). The spectra have complex shapes. The spectra a) and b) consist of three components, which correspond to graphite-like bonds C–C (284.3eV), C–H (285.0eV), and C–O–C and (C = O)–O–C (286.1–287.0eV) by their positions [26]. The analysis of the spectra shows that two-fold increase of the salt content causes very insignificant changes in the structures of the spectra. The portion of C–C bonds grows from 53.7 to 67.7%, C–H bonds decreases from 27.3 to 26.4%. The shapes of the spectra are practically the same. The spectrum displayed in Figure 5.1c is a little different. At further increase of manganese chloride content insignificant carbide bonds appear (282.7–283.0eV). This fact seemingly indicates the excess of the content of salt. Manganese ions start to form bonds with aromatic rings. From the spectrum it follows, that the relation between the intensities of C–C and C–H bonds changed insignificantly.

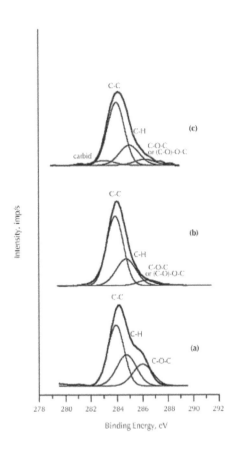

FIGURE 5.9 C1s-spectra of nanostructures obtained from the mixtures of fine powders of anthracene and manganese chloride with different mole ratios of the initial components: a) 1:1; b) 1:2; c) 1:3.

In Figure 5.10 C1s spectra of the systems obtained from the mixture containing fine powders of anthracene and cobalt chloride are shown. At the ratio anthracene: salt equal to 1:1, four components corresponding to carbide (283.0eV), C–C (284.3eV), C–H (285.0eV), and C–O–C and (C = O)–O–C (286.1–287.0eV) bonds emerge. The emergence of the carbide component is seemingly connected with the presence of super dispersed particles of nickel in the mixture. Two-fold increase of the content of cobalt chloride led to a great change of the intensity of the components in C1s spectrum (see Figure 5.10b). The portion of C–C bonds grew approximately in two times, from 28 to 60%, and the portion of C–H bonds decreased from 51.2 to 32.6%. As in the given sample nickel particles were absent, the carbide component is not observed. The increase of the content of salt in the mixture up to 3 moles (see Figure 5.2c) led to further growth of the portion of C–C bonds (72.9%) and further decrease of C–H bonds (20.3%).

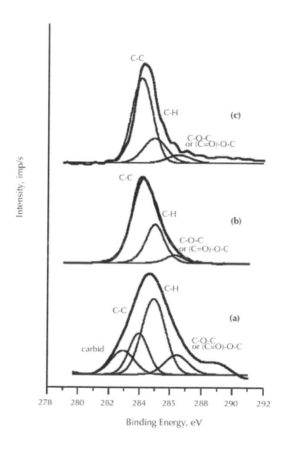

FIGURE 5.10   C1s-spectra of nanostructures obtained from the mixtures of fine powders of anthracene and cobalt chloride with different mole ratios of the initial components: a) 1:1; b) 1:2; c) 1:3.

In Figure 5.11 the XPS C1s spectra of metal-carbon systems are presented; these systems were obtained from anthracene and nickel chloride mixtures with different contents of the initial substances. Depending on the nickel chloride content the percentage of C–C and C–H bonds changes. The comparison of the spectra shows that at the increase of nickel chloride content in the mixture from 1 mole to 2 moles the growth of the portion of C–C bonds from 49.1 to 72.3% and the decrease of the portion of C–H bonds to 15.4% are observed. The formation of analog of carbide bonds (4.6%) can also be seen. The further increase of nickel chloride content does not result in further growth of the portion of C–C bonds, as it is indicated in Figure 5.9c. When the content of $NiCl_2$ is equal to three moles, the portion of C–C bonds makes up 59.8%, C–H bonds—12.2%, and due to the excess of nickel ions the carbide bonds' portion grows (~21.9%).

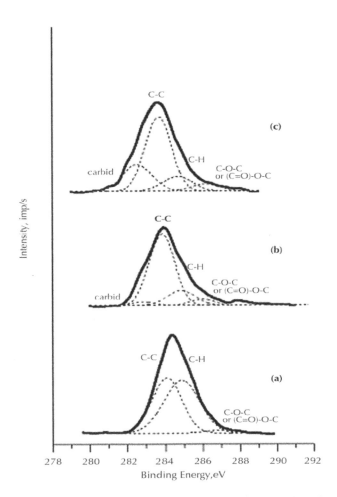

FIGURE 5.11 C1s-spectra of nanostructures obtained from the mixtures of fine powders of anthracene and nickel chloride with different mole ratios of the initial components: a) 1:1; b) 1:2; c) 1:3.

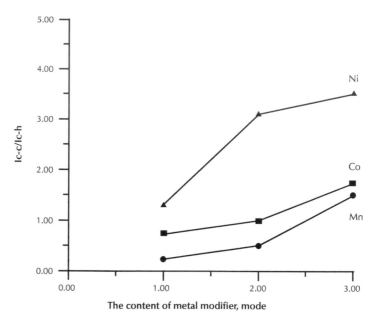

FIGURE 5.12   Dependence of IC–C/IC–H on the contents of metal chlorides.

In Figure 5.12 dependences of I_{C-C}/I_{C-H} on the content of 3d-metal chlorides are presented. Curve 1 corresponds to nickel chloride, curve 2—to cobalt chloride, curve 3—to manganese chloride. It seems that such behavior of dependences could be connected with the ability of the atoms of 3d-metals to form chemical bonds. As far as, the electron density in the vicinity of the atoms of 3d-metals decreases in case of the elements belonging to the row Ni–Co-Mn, and the energy of interaction of the metals with chlorine ions increases, then the dependences displayed in Figure 5.13 behave differently. When fine powders of nickel chloride and anthracene are mixed, the saturation by C–C bonds is observed even at the salt amount equal to 2 moles. In the case of cobalt chloride, an abrupt increase of the portion of C–C bonds is observed, however, judging by the shape of the curve it is impossible to say that the saturation by C–C bonds is taking place; and for manganese chloride, a slow growth of the curve of the relation I_{C-C}/I_{C-H} is observed. Different positions of the initial points in Figure 5.12 are connected with different oxidizing abilities of the ions of 3d-metals under discussion.

In order to use XPS as a method to control the formation of tubules, it was necessary to investigate the microstructure of the obtained samples. The investigation of the microstructures was carried out by means of the transmission electron microscope JEM-200CX at the accelerating voltage of 160 kV (the microstructures are shown in Figure 5.13).

FIGURE 5.13 Microstructures of metal-carbon systems obtained from the mixture of fine powders of anthracene and metal chloride: $MnCl_2$ (a); $CoCl_2$ (b); $NiCl_2$ (c).

The following was stated based on the TEM data. Mainly the microstructures of the samples are the rolled up perforated films, which are butted together by their lateral surfaces. As a result of thermal destruction of an edge of the sample on the exposure to the electron beam the spatial cross-section of the sample, lamellar arrangement of films in it and cellular texture of the films themselves were clearly manifested. The sizes of perforations in the films were from 0.5 to 2.5μm.

Also, in the samples the presence of nano-particles is observed; their electron diffraction patterns are similar to the ones characteristic to the amorphous state. The sizes of such particles are from 50 to 200nm; these particles can be polyhedral, or they can have spherical form. The microcrystalline structure state, which forms also on larger objects, was observed for all the samples. The TEM data are given in work [25].

The results of the XPS and TEM studies allow stating that it is possible to control the formation and growth of tubules with the help of the data on the portions of C–C and carbide bonds in the XPS spectrum. The increase of the portions of these bonds in the spectrum correlates with the growth of metal-carbon tubules according to the TEM data.

It is believed that different behavior of the curves in Figure 5.12 allows drawing a conclusion on the mechanisms of formation and growth of metal-carbon tubules. Amorphous carbon serves as a base for the formation of tubules. Amorphous carbon contains embedded nano-particles of metal, which can be considered as "embryos growing up" into tubules. The growth of dendrite-like structures, the so-called "dendrites", occurs on the particles of metal. In this work different stages of the formation of tubules are presented. For example, in case of manganese chloride only two structures are formed representing the base for the further growth of tubules. The second stage of tubulene formation is shown in the example of cobalt chloride. The dendrite-like structures are formed on the particle of the metal. The formation and growth of nano-tubes take place in the case of nickel chloride. These assumptions based on the data of XPS are fully confirmed by the TEM results. The obtained regularities may facilitate the development of new trends in the synthesis of tubules with unique properties.

Thus, based on the given study one may draw the following conclusions:

- There is a possibility of receiving multilayer carbon tubes from the mixtures of fine-dispersed powders of aromatic hydrocarbons and metal salts. The direct observation reveals that the formed tubes (tubules) have the sizes from 50 to 200nm, and they are the result of the rolling up of graphite planes in a spiral on the surfaces of metallic particles. Mainly the ends of these tubes are closed and there is amorphous carbon on their surfaces.
- The study of the structure of metal-carbon systems by XPS together with TEM allows using XPS to control the formation of metal-carbon tubules.
- Based on the XPS spectra it is possible to conclude that the increase of the content of 3d-metal salt in the mixture of fine powders leads to the growth of the portion of C–C bonds.
- Depending on the oxidizing abilities of the ions of 3d-metals the curves $I_{C?C}/I_{C/H}$ for various metals behave differently. In Figure 5.4, curves 1, 2, 3 correspond to different stages of formation and growth of nanotubes.

5.1.3   X-ray Photoelectron Study of Carbon-Nickel-Containing Nanostructures Obtained in Nanoreactors in Polyvinyl Alcohol Gels and in Gels Prepared from a Mixture of Polyvinyl Alcohol and Polyethylene Polyamine

The need for ecologically pure productions opens wide prospects of application of synthesis of chemical materials in nanoreactors of different matrices. However, it is difficult to implement directed syntheses in nanoreactors without preliminary computational experiment, which predicts the behavior of chemical particles, and development of diagnostic methods, which would allow controlling the intermediate and final products of occurring processes. Application of nanostructures in nanoreactors of polymer matrices for synthesis seems to be expedient due to the decrease in the energy consumed in the process and, in the ideal case, reduced output of minor products into the environment. In this context, development of X-ray photoelectron spectroscopy (XPS) for monitoring synthesis and testing obtained nanostructures is of particular importance. An essential feature of this method is its nondestructive character, since the X-rays used to excite photoelectrons do not damage most materials. This does not hold true for the methods of surface analysis based on ion or electron bombardment. In this chapter, along with the XPS data, the results obtained by transmission electron microscopy (TEM), electron diffraction (ED), and IR spectroscopy are used.

The purpose of this chapter is to develop a model of formation of carbon-metal-containing nanostructures in active media and a technology of their fabrication.

The XPS investigation was performed on an X-ray electron magnetic spectrometer with the following characteristics: automated control system, double focusing, resolution 0.1eV, excitation by the Al $K\alpha$ line (1,486.6eV), and vacuum 10^{-6}–10^{-7} Pa. This spectrometer has a number of advantages over electrostatic spectrometers, specifically: constant aperture ratio, independence of resolution on the electron energy, and high spectral contrast.

The XPS method is used to monitor the fabrication of nanostructures from the relative content of C–C and C–H bonds, type of hybridization of valence s and p electrons of carbon atoms and the satellite structure of the C1s spectrum, and the type of interaction between metal atoms with carbon. This method makes it possible to identify C1s spectra and determine the chemical bonds of elements, nearest environment of atoms, and the type of sp hybridization of valence electrons in nanostructures with the use of the satellite structure of the C1s spectra [26]. The XPS data make it possible to monitor the formation and growth of nanostructures. It has been established that, in the C1s spectrum, the satellite spaced at a distance of 22eV from the main peak, is characteristic of the C–C bond with sp^2 hybridization of valence electrons; its relative intensity is 10% of the main spectrum intensity. The satellite can be caused by plasmon losses. If the C1s spectrum exhibits a satellite at a distance of 27eV, the C–C bonds are characteristic of the sp^3 hybridization of valence electrons of carbon atoms (diamond, fullerene). One might suggest that the satellite ratio should change, depending on the diameter and shape of the nanostructures obtained. For rolls and multilayer tubulenes of large diameter, one might expect a much higher intensity of the satellite corresponding to the sp^2 hybridization in comparison with the satellite characteristic of the sp^3 hybridization.

The C1s spectrum of the single-wall nanotubes contains two satellites near 306 and 313eV. Therefore, the spectrum of carbon single-wall nanotubes contains a component characteristic of the C–C bonds with sp^2 and sp^3 hybridization of valence electrons at carbon atoms. The intensity of the component characteristic of the sp^3-hybridized C–C bond of valence electrons is lower by a factor of 2 in comparison with that of the component characteristic of the s^2-hybridized C–C bond. Hence, single-all nanotubes are formed as a result of roll-up of a graphite plane. A similar situation is observed in the C1s spectrum of multilayer nanotubes of small diameter; however, in this case, the intensity of the component characteristic of the sp^3-hybridized C–C bonds of valence electrons is higher; apparently, due to the screening effect.

In this chapter, the results of the XPS study of the nanoproducts obtained in nanoreactors of polymer matrices are reported in the form of gels of polyvinyl alcohol (PVA); PVA and polyethylenepolyamine (PEPA); and PVA, PEPA, and acetylacetone (AA) in the presence of nickel chloride $NiCl_2$.

The PVA, PVA–PEPA, and PVA–PEPA–AA polymer matrices were obtained by mixing solutions of the corresponding components, according to the technique described in [26]. Then, the gels formed during predrying were treated with a nickel chloride solution. According to the IR spectra, interaction of nickel ions with oxygen of hydroxyl and ketone groups and nitrogen of amine groups occurs in the color films formed. In this case, the coordination number of Ni^{2+} is 4 in PVA gels and 6 in PVA–PEPA and PVA–PEPA–AA gels. Upon subsequent thermal treatment (step heating to 400°C), the film color changed to black and a porous semiproduct was formed, which was rinsed and dispersed with subsequent isolation and drying of the nanoproduct. The formation of nanostructures is a redox process, which implies metal reduction and carbonization of the organic components that form nanoreactor walls.

In the XPS study, the C1s and Ni3p spectra of the nanoproducts obtained from nickel chloride and PVA, PVA–PEPA, and PVA–PEPA–AA gels were analyzed. Figure 5.13a shows the results of the calculations of the multiplet structure related to the 3p–3d interaction of unoccupied shells in nickel complexes. The Ni3p spectrum consists of three strong peaks: the main one and two peaks located at distances of 2.5 and 5.0eV from the main peak; in addition, a series of weak peaks is observed at a distance of 10eV [27].

The XPS data on the samples are presented in the Table 5.1, Figure 5.14 and 5.15.

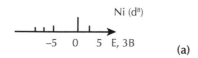

(a)

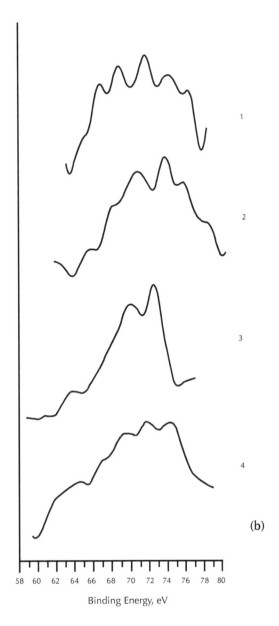

58 60 62 64 66 68 70 72 74 76 78 80

Binding Energy, eV

FIGURE 5.14   (a) Results of the calculation of the multiplet structure of electron spectra of the 3 p of nickel in complexes [3] and (b) the Ni$_3$ p X-ray photoelectron spectra of the samples obtained from the (1) 2PVA +1NiCl2, (2) 1PVA + 1PEPA + 1NiCl$_2$, (3) 2PVA + 2PEPA + 1NiCl$_2$, and (4) 1AA + 2PVA +2PEPA +1NiCl$_2$ mixtures.

TABLE 5.1   Relative Ni and C contents in the samples represented for investigation.

No.	Composition	Ratio of carbon bonds in the C 1s spectrum	Ratio of nickel bonds in the Ni 3p spectrum
1	2ПВС+1NiCl$_2$	C–C(sp^2):C–H:C–O = (50:40:10)%	Ni–O(H):Ni–CNi–Cl = (55:45)%
2	1ПВС+1ПЭПА+1NiCl$_2$	Ni–C:C–C(sp^2):C–C (sp^3):C–H = (21:42:11:26)%	Ni–C:Ni(N$^+$):Ni–CNi–Cl = (12:35:53)%
3	2ПВС+2ПЭПА+1NiCl$_2$	Ni–C:C–C (sp^2):C–C (sp^3):C–H = (15:28:14:43)%	Ni–C:Ni(N$^+$):Ni–CNi–Cl = (23:50:27)%
4	1АА+2ПВС+2ПЭПА+1NiCl$_2$	Ni–C:C–C (sp^2):C–C (sp^3):C–H = (10:39:11:40)%	Ni–C:Ni(N$^+$) = (33:67)%

The C1s XPS spectrum of the nanoproduct obtained from the 2PVA–NiCl$_2$ mixture showed the presence of the sp^2-hybridized C–C bond; that is, similar to graphite, as is evidenced by the satellite structure at 306 eV, as well as the presence of hydrocarbons and carbon–oxygen (C–O) bond. The sp^3-hybridized C–C bond is weakly pronounced for this mixture. The O1s spectrum indicates the presence of adsorbed and bound oxygen. The percent ratio of all these components is 50:40:10 (Table 5.1, row 1; Figure 5.2a).

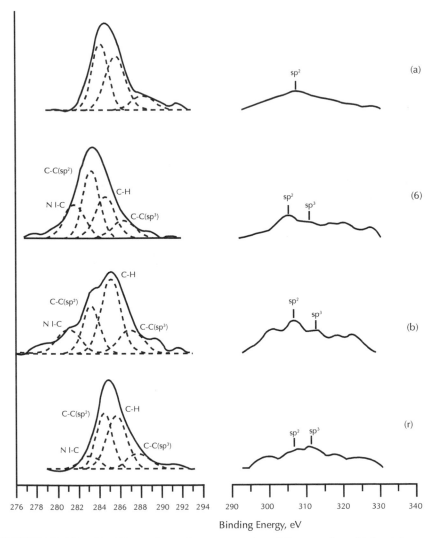

FIGURE 5.15 C1s X-ray photoelectron spectra of the samples obtained from the (1) 2PVA + 1NiCl$_2$, (2) 1PVA + 1PEPA + 1NiCl$_2$, (3) 2PVA + 2PEPA + 1NiCl$_2$, and (4) 1AA + 2PVA + 2PEPA + 1NiCl$_2$ mixtures.

Analysis of the chemical shifts in the Ni3p spectrum (Figure 5.14b, curve 1) indicates the presence of nickel bonds: Ni–O or Ni–O(H), as well as Ni–Cl or (H)O–Ni–Cl with approximately the same percent ratio. Reduced Ni atoms are absent in the analyzed layer. However, according to the X-ray diffraction (XRD), ED, and TEM data, metal (nickel) is present in multilayer tubular nanostructures, which form dense bundles ("aggregates") [29]. This fact can explain the sp^2 hybridization of carbon. The TEM images show that the formed carbon films roll up into nickel-containing "rolls". In this case, the lateral surfaces of such nanostructures can be fairly active to form aggregates.

The instant of roll-up of a nanofilm, as well as the product of the process (aggregates of nickel-containing tubulenes), are shown in Figure 5.16.

FIGURE 5.16   The TEM image demonstrating the instant of roll-up of a nanofilm and aggregates formed.

The $C1s$ spectrum of the nanoproduct (Table 5.1, row 2; Figure 5.15b) obtained from the mixture of PVA, PEPA, and $NiCl_2$ in the ratio 1:1:1 contains components that are due to the Ni–C interaction, C–C bond with s^2 hybridization of the valence electrons of carbon, C–C bond with sp^3 hybridization of the valence electrons of carbon, and C–H bond. These components are in the following percent ratio: Ni–C: C–C (sp^2): C–C (sp^3): C–H = 21:42:11:26.

Analysis of the Ni$3p$ spectrum (Figure 5.14b, curve 2) showed the presence of nickel (12%), N–O or Ni–O(H) (35%), and Ni(N$^+$) (53%) in the nanoproduct. Comparison of the IR and XPS N$1s$ spectra indicates the presence of =N$^+$= groups in the nanoproduct, which have an electronegativity of 3.3 [30]. This fact explains the corresponding chemical shift in the Ni$3p$ spectrum. The formation of =N$^+$ = groups was noted in the analysis of the IR spectrum of the obtained color films (xerogels). These groups are evidenced by the existence of bands in the ranges 2,280–2,130 and 1,710–1,570 cm^{-1} [31]. The spectra revealed bands associated with the C=N and N=N bonds. Analyzing the spectra, one can conclude that conjugation chains are formed; this process indicates possibility of forming of a thermostable coordination polymer with participation of Ni atoms. In this case, the coordination number of nickel, as in inorganic analogs, can be 6. The presence of C–C bonds with sp^2 and sp^3 hybridization of valence electrons of carbon in the ratio 2:1 indicates the presence of nanotubes; however, since the number of sp^2-hybridized C–C bond is larger, it can be suggested that graphite inclusions are present, which arise due to the crystallization of the polymer precursor. This fact is confirmed by the TEM data. One might suggest that the presence of PEPA in the mixture leads to reduction of nickel, partially oriented to carbon, and increases the probability of breaking nanofilms at phase boundaries with formation of single-layer nanostructures of smaller diameter. This mechanism can be illustrated by the TEM data (Figure 5.17). The micrograph demonstrates nanotubes with a diameter of about 10nm and a length of about 200nm against the background of amorphous thin "rumpled" nanofilms and small graphite particles with metal-containing nanocrystals.

With an increase in the content of the polymer phase and a decrease in the content of $NiCl_2$ (Table 5.1, row 3; Figure 5.15c) in the C1s spectrum of the nano-product obtained from the 2PVA–2PEPA–$NiCl_2$ mixture, the component ratio changed as follows: Ni–C:C–C (sp^2):C–C (sp^3):C–H = 15:28:14:43.

Analysis of the Ni3p spectrum (Figure 5.14b, curve 3) showed the presence of nickel in the sample and the Ni(N^+) interactions in the ratio Ni(C):Ni(N^+) = 23:77, which is indicative of coordination of nickel with positively charged nitrogen. In this case, the amount of interacting nitrogen greatly exceeds the amount of oxygen that can be present in the nanoproduct. The decrease in the content of Ni ions in the initial composition changed the character of the coordination interaction, which enhanced nickel reduction and increased the content of the heat-resistant polymer phase. The intensity ratio of the satellites reflecting the sp^2 and sp^3 hybridizations decreased. Therefore, one can state that the content of small single-wall nanotubes increased, all the more that, according to the TEM and ED data, graphite inclusions and a large number of small tubular nanostructures were observed. This fact can be explained by the formation of more stable crystalline polymer phases with an increase in the PEPA content.

During formation of nanostructures in nanoreactors of the PVA–PEPA–AA gel (Table 5.1, row 4; Figure 5.15d), the satellite ratio increased, apparently, due to the change in the mechanism of nickel coordination to the nitrogen and oxygen present in the walls of nanoreactors. Therefore, the fraction of the hydrocarbon part of the heat resistant polymer phase increased and the content of reduced carbon-saturated nickel somewhat decreased, because, during thermal action and catalytic process, stresses in the nanofilms increase, leading to their break and roll-up of pieces of amorphous film fragments under the action of Ni ions or atoms. The roll-up of a nanofilm is shown in the TEM image in Figure 5.18.

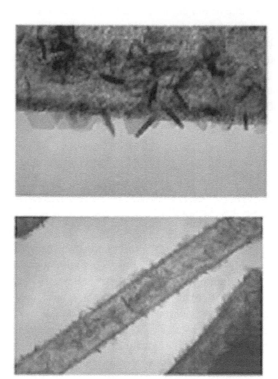

FIGURE 5.17 The TEM images demonstrating the formation of small nanotubes and nickel nanocrystals on nanofilms and nanoribbons.

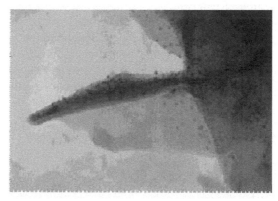

FIGURE 5.18 The TEM image demonstrating the instant of roll-up of a nanofilm with small graphite pieces and nickel nanocrystals.

The ratio of components in the C1s spectrum is as follows: Ni(C):C–C(sp^2):C–C(sp^3):C–H = 10:39:11:40. From the ratio of the C–C(sp^2) and C–C(sp^3) components, one can judge about the shape and sizes of the nanostructures obtained. The TEM and ED data confirm the presence of graphite films in the sample and increase in the nanostructure diameter; that is, they correlate with the XPS data (the contribution of the sp^3- hybridized component decreased). In addition, individual aggregates of nanostructures are observed; their presence explains the increase in the intensity of the satellite assigned to the sp^2 hybridization. Analysis of the Ni3p spectrum (Figure 5.14b, *curve* 4) showed the presence of reduced nickel, Ni–O(H) bonds, and Ni(N$^+$) interactions in the ratio 22:34:44.

Based on the results of the XPS study and the TEM, ED, and IR spectroscopy data, one can propose the following model of formation of carbon nickel-containing nanostructures.

Carbon nanotubes or tubulenes containing nickel nanoclusters (in some cases, nickel nanocrystals) are formed during the redox process, in which nickel compounds play the role of an oxidant and hydrocarbon or amine groups serve as reducers.

During this process, chlorine and oxygen are removed from the sphere of interactions, and carbonization occurs with the formation of corresponding nanostructures. In this case, amorphous nanofilms are formed first; under the action of Ni ions or atoms, they roll up into cylindrical nanostructures of certain diameter. In the case of nanoreactors in the PVA gel, multilayer tubulenes are formed, which are prone to formation of aggregates due to the presence of sp^2-hybridized C–C bonds. Formation of gels of complex composition, including polyethylene polyamine or polyethylene polyamine and acetylacetone, facilitates metal reduction and graphitization. The formation of crystalline phases in carbon nanofilms is accompanied by an increase in the internal stress and break of the nanofilm with isolation of its amorphous part. Pieces of amorphous parts of the film roll up into single-wall nanotubes or form shells for nickel nanocrystals.

On the basis of the results obtained, one can draw the following conclusions.

- In contrast to the PVA gel, complex PEPA-containing gels exhibit an increase in the amount of reduced nickel, coordinated to the e=e bond, during the process under study.
- With a decrease in the nickel content in a nanoreactor, the amount of thermostable coordination polymer increases, and small nanotubes are formed.
- Addition of AA to PVA and PEPA increases the content of graphite inclusions without decreasing the content of the coordination polymer.

Thus, XPS can be successfully used to establish the character of nanostructures, their sizes, and characteristics of the processes of formation of metal nanoclusters.

5.1.4 The XPS Studies of the Carbon Nanostructures Obtained in Nano-reactors and Polymer Matrices Using Polyvinyl Alcohol as an Example

During last decade, a scientific field under the name nanotechnology has emerged and is intensively evolving. Within the framework of nanotechnology, disperse systems consisting of objects of nanometer size are considered. Carbon nanostructures such as fullerenes, carbon nano-tubes, multi-layer carbon nanoparticles and nanoporous carbon occupy a special place among others. The use of carbon nanotubes is mainly restricted due to the energy consumption and low efficiency of the existing methods for their synthesis. In this connection, works on low-energy synthesis of carbon nanotubes are undoubtedly actual [32, 33].

This work is concerned with the control over the synthesis of metal nanoparticles in a carbon shell using the method of X-ray electron spectroscopy (XES) and the investigation of the possibility of introducing the developed methods with the purpose of developing nano-metallurgy. The objective of the X-ray electron investigation is the determination of conditions for obtaining nanostructures from metal salts in nano-reactors and from the polymer matrices of polyvinyl alcohol.

In accordance with the above objective, the interaction of 3d metal chlorides (Cu, Ni, Co, and Fe) and their mixtures with an active polymer medium (polyvinyl alcohol) was studied. In all the cases, aqueous solutions of salts and polyvinyl alcohol (PVA) were formed. After drying, films were formed. These films were being heated to 300°C for 2–3 hrs. The product was rinsed with hot water and dried. For the control over the intermediate and final results: 1) the investigations of the solutions of PVA and metal salts and their mixtures were carried out by means of photocolorimetry and optical microscopy. These methods allow assessing the activity of the interaction of PVA and the ions of the metal salts, the shapes of the nanoreactors and of anticipated nanostructures; 2) the investigations of the product obtained with the use of the X-ray electron spectroscopy (XES), atomic force microscopy (AFM) and transmission electron microscopy (TEM) methods were conducted.

The analysis was carried out on an electron magnetic spectrometer with AlKα-line excitation. The vacuum in the spectrometer chamber was 10^{-3} Pa. The spectrometer resolution was 1.2 eV and the accuracy of measuring the positions of the peaks was 0.1 eV [3].

In the PVA C1s-spectrum (Figure 5.1), a satellite is observed at the distance of 22 eV from the main peak in the high-energy region, which is characteristic of the

C–C bond with the *sp²*- hybridization of the carbon atoms [20]. The C1s spectrum consists of three components. Because of its energy position (284.3 eV) and the presence of the satellite, the maximum of the low-energy component corresponds to the atoms of carbon in the C–C groupings having the *sp²*- hybridization of the carbon atoms. The component with the binding energy $E_{bind} = 285$ eV corresponds to the atoms of carbon in the C–H groupings, and the third component with the binding energy $E_{bind} = 287$eV corresponds in its position to the C–OO–H bonds.

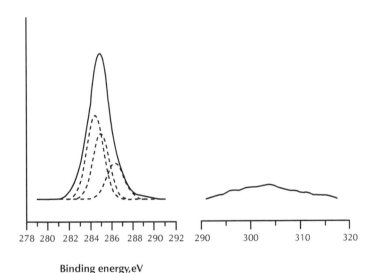

Binding energy, eV

FIGURE 5.19   The X-ray electron C1s spectrum of polyvinyl alcohol (PVA).

The O1s spectrum for this sample consists of one component, the maximum of which corresponds in its position to the O–H bond ($E_{bind} = 532$eV).

From the analysis of the X-ray electron spectra of PVA, one can draw a conclusion that in vacuum during the X-ray radiation, some changes are taking place from the very beginning in the PVA film, and this explains the appearance of the satellite in the high-energy region and of the low-energy component in the main spectrum.

When the spectrum is being taken from the PVA + NiCl₂ (1:2) sample (T = 20°C, a film), at the beginning the C1s spectrum (Figure 5.20a) consists of four components, that is, C–C ($E_{bind} = 284.3$ eV) with a satellite at the distance of 22 eV in the high-energy region that is indicative of *sp²*- hybridization of the carbon atoms; C–H ($E_{bind} = 285$ eV); C–C ($E_{bind} = 286$ eV) having *sp³* - hybridization of the carbon atoms and C–O ($E_{bind} = 287$eV). In 40min, two components remain in the spectrum of this sample (Figure 5.2b): C–C ($E_{bind} = 284.3$ eV) with a high-energy satellite that corresponds to the *sp²*- hybridization of the electrons of the carbon atoms, and C–O. The O1s line of this sample is of low contrast and corresponds to oxygen adsorbed on the surface.

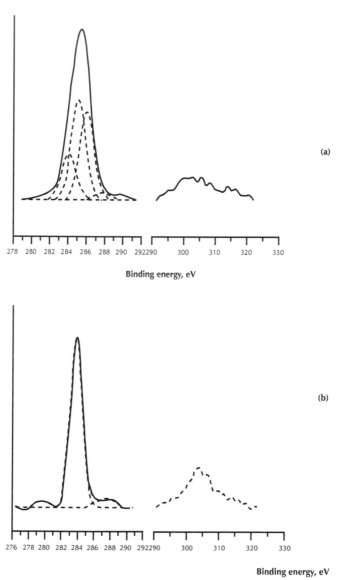

FIGURE 5.20   The C1s spectrum of the sample obtained from PVA and NiCl₂ at room temperature: a) at the initial moment; b) in 40 min.

The C1s spectrum of the PVA + NiCl$_2$ (1:2) sample (T = 300°C, powder) is stable and does not change with time (Figure 5.21). In the C1s spectrum, there are four components with binding energies characteristic of the C–C bond (E$_{bind}$ = 284.3 eV) with the sp^2- hybridization of the valence electrons on the carbon atoms, C–H (E$_{bind}$ = 285 eV), C–C (E$_{bind}$ = 286.1 eV) with the sp^3- hybridization of the valence electrons on the carbon atoms, and C–O (E$_{bind}$ = 287 eV). In the high-energy region there are satellites characteristic of the C1s spectrum of graphite, that is, the sp^2- hybridizations of the valence electrons of the carbon atoms and the sp^3- hybridizations. In the O1s spectrum, two maxima O–H (E$_{bind}$ = 532 eV) and C–O (E$_{bind}$ = 534 eV) are observed; chlorine is not found.

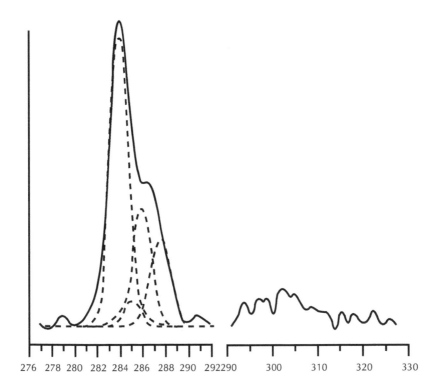

Binding energy, eV

FIGURE 5.21   The C1s spectrum of the sample obtained from PVA and NiCl$_2$ (1:2) at the temperature of 300°C.

Based on the XES results, the following regularities have been obtained: 1) in the very beginning, there are only two components characteristic of C–H and C–O bonds in the C1s spectrum of the samples under study. During long X-ray irradiation of the sample surface, a component characteristic of C–C bonds with the sp^2- hybridization of valence electrons appears in the carbon spectrum. The appearance of carbon in the sp^2- state in the C1s spectrum of PVA and gels PVA + MeCl$_2$ is apparently related to the breaking of the polymer shell of a nano-reactor on exposure to X-ray radiation; 2) the gels from the PVA and CoCl$_2$ mixture dried at room temperature are more stable on exposure to vacuum and X-ray radiation than the gels from the PVA and NiCl$_2$ mixture; 3) the heat-treated gels at 300°C are stable to the action of X-ray radiation in vacuum. In the C1s spectra of these samples, there are three components with the binding energies characteristic of the C–C bonds with sp^2- and sp^3- hybridizations in the ratio of 2:1, which is characteristic of nanotubes [35]. In addition, a significant component of the bond of carbon with hydrogen remains (35–40%).

The investigation with the use of TEM and AFM have resulted in the finding of globe-shaped and cylindrical nanostructures in the products of heat treatment, which are metal nano-particles in carbon shells. From the micro-diffraction patterns, it has been established that metals are divided into nanostructures, which are different in size and shape.

The investigations show the regularities in the formation of carbon nanostructures from the metal salts and the polymer matrices of polyvinyl alcohol. It is found that the films obtained at room temperature are not stable on exposure to vacuum and X-ray radiation in contrast to the films obtained at the temperature of 300°C.

Moreover, the experimental results obtained with the use of the XES and AFM methods allow concluding that in the gels obtained at 300°C from the mixture of polyvinyl alcohol and the salts of transition metals, the origination of differently shaped nanostructures takes place.

5.2 THE INVESTIGATION OF PROPERTIES OF METAL/CARBON NANOCOMPOSITES OBTAINED

5.2.1 Application of the X-ray Photoelectron Spectroscopy Method for Studying the Magnetic Moment of 3d Metals in Carbon-Metal Nanostructures

One of the main characteristics of magnetic materials is the magnitude of the atomic magnetic moment. Known methods for the determination of this magnitude give an averaged value. Such investigations are informative for the materials with homogeneous crystalline structure. However, there are heterogeneous objects, the metal atoms of which have different nearest surrounding and chemical bond and they can differ in the value of the atomic magnetic moment. In this case the averaged magnetic moment value gives only indirect information.

Thus, it is obvious that it is necessary to conduct a detail investigation of the interconnection of the electronic structure and varying magnetic characteristics of materials based on 3d-transition metals at changing temperature and composition using modern spectroscopy methods for structure investigations.

One of the most powerful direct methods for investigating electronic structure, chemical bond, nearest surrounding of the atoms of substances is X-ray photoelectron spectroscopy (XPS). The choice of an electron magnetic spectrometer is due to a number of advantages in comparison with electrostatic spectrometers, namely, its constant luminosity and resolution, which do not depend on the energy of electrons and high contrast of spectra. In addition, the constructional separation of the energy analyzer of magnetic type from the spectrometer vacuum chamber allows using various ways of action on a sample in vacuum right at the moment of spectra taking. Neither heating nor cooling of samples nor mechanical cleaning of the sample surface from contaminations deteriorates the spectrometer resolution.

At present, the unique physical properties of nanoparticles are the subject of intensive studies. Special attention is paid to magnetic properties, in which the difference between massive material and nanoscale material is exhibited more clearly. By changing the size, form, composition and structure of nanoparticles it is possible to regulate magnetic characteristics of the nanoparticle-based materials to a certain extent.

The method for the preparation of the nanoscale structures under study was grinding of superdispersed metal-containing particles (d-metal oxides) and mixing them with an active polymer medium (polyvinyl alcohol). In all the cases aqueous solutions with polyvinyl alcohol (PVA) were formed. After drying, films were formed. The films were heated at 400°C for 2–3 hrs.

In the synthesis of the carbon and carbon-polymer metal-containing nanostructures, the approach was based on the reactions of coordination and redox processes in the nano-reactors of polymer matrices, which allowed significantly to simplify the synthesis and made it unnecessary to create a special atmosphere and to use appropriate catalysts and reagents.

The main investigation method was the method of X-ray photoelectron spectroscopy (XPS). The analysis was conducted on an electron magnetic spectrometer with the excitation of the AlK-line. The vacuum in the spectrometer chamber was 10^{-9} torr. The instrument resolution was 1.2eV and the accuracy of the measurement of the peak positions was 0.1 eV.

In contrast to conventional methods which give the information integrated over a studied sample, the X-ray photoelectron spectra present the information on the local characteristics of the substance structure.

The XPS 3s-spectra of transition metals are known to correlate with the magnetic moment of metal atoms [36]. The interpretation of 3s-spectra of transition metal within the single electron theory describes the $3s^{-1}$-multiplet in 3d metals as two groups of components well separated by energy, the relative intensities of which are associated with the total spin of the 3d-subshell and, consequently, with the magnetic moment on an atom of a transition metal.

The presence of two maxima is characteristic of the 3s-spectra of reference metals, in which multiplet splitting is reflected [36]. Based on the model offered by Van Vleck, the values of atomic magnetic moments for the reference systems of 3d transition metals are obtained from the spectra parameters with the use of the approximation:

$$I_2/I_1 = S/(S+1) \qquad (5.1),$$

where S is spin quantum number.

The distance between the peaks determines the exchange interaction energy:

$$\Delta = (2S+1)I_{SD} \qquad (5.2),$$

where I_{SD} is intra-atomic exchange integral between 3s and 3d electrons depending on the overlapping of 3d- and 3s-shells.

The magnetic moment of atoms is determined from the expression:

$$\mu_S = 2\mu_B\sqrt{(S+1)S} \qquad (5.3),$$

where μ_B is Bohr magneton, S is the spin moment of unpaired 3d electrons.

Based on our experimental results and the literature data [37, 38] on multiplet splitting in the 3s-spectra, a systematic investigation has been conducted in order to find a correlation of the intensity ratio of the maxima of the multiplet splitting lines in the 3s-spectrum with the spin magnetic moment of metal atoms. For the binary systems M-M (M is metal) and M-X (X is metalloid) it is shown that there is a dependence, which is nearly linear: $I_2/I_1 = 0.9 \times S/(S+1)$ (Figure 5.22). The maximal error in the determination of I_2/I_1 is $\approx 15\%$. For the systems, in which the charge transfer between a transition metal atom and a ligand atom is absent, the model predicts correctly the tendency of the spin state variation (the error is 5%).

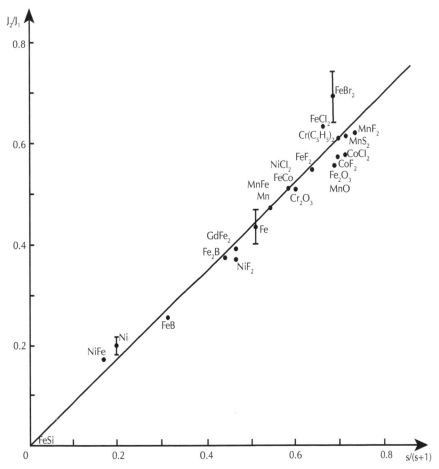

FIGURE 5.22 The dependence of the intensity ratio of the multiplet splitting maxima on the ratio characterizing spin moment.

The XPS spectra parameters are connected with the spin state or the atomic magnetic moment in the following manner:

- The relative intensity of the 3s-spectra multiplet maxima correlates with the magnitude of the magnetic moment of the atoms in the 3d-metal systems;

- The distance between the multiplet maxima yields the information about the exchange interaction of 3s–3d shells; based on the changes taking place in the 3d-shell, the information about the changes in the distance between neighboring atoms is obtained;

- The shape of the valence bands, which is the energy distribution of the density of d-states, determines their localization and presents the information about the nearest surrounding of the atoms;
- The presence of variations in the 3s spectra shapes and the valence bands gives the information about the changes in the alloy structure.

For the low-temperature synthesis of carbon metal-containing nanostructures, superdispersed powders of transition metal oxides were used. The method of photoelectron spectroscopy shows that the metal reduction is possible at the interaction of polyvinyl alcohol with the metal oxide powders. The presence of reduced metal leads to the formation of carbon metal-containing nanostructures [39].

The objective of the present work was the X-ray photoelectron study of the spin state of the nanostructures based on Ni, Co, and Cu.

The X-ray photoelectron spectra are obtained for the reference samples from Fe, Mn, Co, and Ni pure metals and the nanostructures based on Ni, Co, Cu (Figure 5.23, 5.24 and 5.25).

For all the samples studied, the O1s spectrum is low-intensity and characteristic of adsorbed oxygen. The Ni2p, Co2p and Cu2p spectra do not contain an oxide phase. In the C1s spectra of the studied nanostructure samples there is a Me–C component, which indicates the formation of carbon-metal nanostructures [39]. The investigation of the specific features of the Me3s spectra structure of the studied samples shows the change of the relative intensity of the multiplet splitting maxima and the distance between them compared to the 3s spectra of pure metals.

Using expressions 1–3 and the curve of the dependence of the intensity ratio of the multiplet splitting maxima on the ratio value characterizing the atomic magnetic moment of various 3d metal-based systems (Figure 5.22), the atomic magnetic moment have been calculated for the nanostructures containing Co, Ni, and Cu.

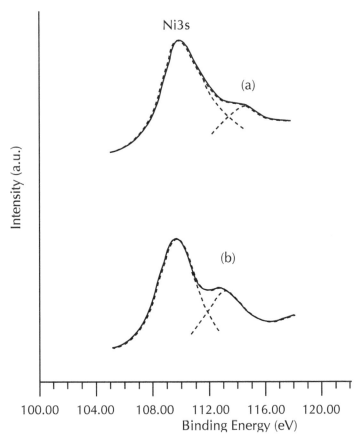

FIGURE 5.23   The X-ray photoelectron 3s-spectra of Ni in the crystalline (a) and nanoscale (b) states.

The comparison with a massive sample shows the increase in the atomic magnetic moment from 0.5 to $1.8\mu_B$ in nanostructures (in the Ni-based sample); the distance between the multiplet maxima decreases from 4.3 to 3.0eV compared to pure nickel.

The results obtained show the increase of the atomic magnetic moment by $1.3\mu_B$ in the Ni-based nanostructures compared to the Ni massive sample.

The decrease of the distance between the multiplets by $1.3\mu_B$, in comparison with the Ni massive sample, indicates a sharp weakening of the 3s–3d interaction on the Ni atoms because the number of unpaired electrons increases due to the formation of the bond between some Ni d electrons and the p electrons of the carbon atoms and the formation of a strong hybridized chemical bond in the nanostructures obtained.

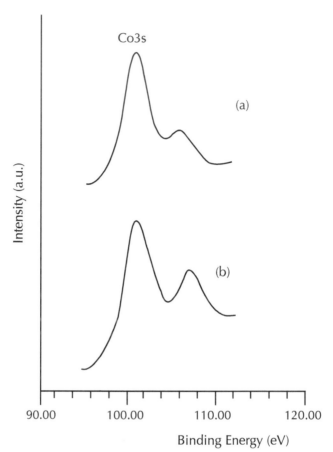

FIGURE 5.24   The X-ray photoelectron 3s-spectra of Co in the crystalline (a) and nanoscale (b) states.

In the case with the cobalt-based nanostructures, the situation is different: though the atomic magnetic moment increases by 0.7_B as compared to that of the massive sample, in contrast to nickel-based nanostructures, the distance between the multiplets increases by 0.8 eV. In [5], a similar effect was observed for the cobalt atom at the hybridization of d electrons of the Co atoms with d electrons of the Mn atoms. The result obtained in this case can be explained by the overlapping of d wave functions of Co atom electrons with p wave functions of C atom electrons, and consequently, by the formation of a strong chemical bond between these atoms. The difference in the behavior of the 3s–3d shell interaction is associated with the distances between the atoms and the average atomic volume in the nanoscale forms of Co and Ni, that is the formation of different nanoscale forms.

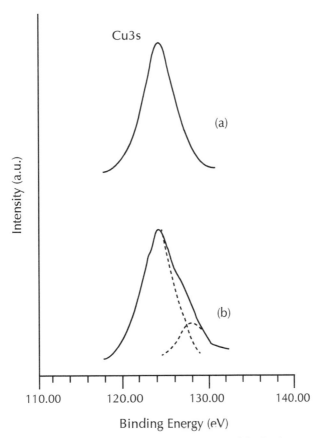

FIGURE 5.25   The X-ray photoelectron 3s-spectra of Cu in the crystalline (a) and nanoscale (b) states.

The appearance of the atomic magnetic moment in the Cu-based nanostructures indicates a strong chemical bond of d electrons of the copper atoms with p electrons of the carbon atoms. The appearance of the atomic magnetic moment on the Cu atoms is due to the fact that some electrons of the copper atoms participate in the hybridized bond with p electrons of the carbon atoms, which leads to the appearance of uncompensated d electrons on the Cu atoms. A similar situation is observed in liquid copper [40–41].

In Table 5.2, the parameters of the 3s spectra multiplet splitting are given for the Ni, Co massive samples and the Ni-, Co-, and Cu-based nanostructures obtained with the use of the low-temperature synthesis method.

TABLE 5.2 The parameters of the 3s spectra multiplet splitting in reference samples and the nanostructures.

Sample	I_2/I_1	E_{max}, eV	$E_{sat,}$ eV	Δ, eV	μ_{Co}, μ_B	μ_{Ni}, μ_B	μ_{Cu}, μ_B
Ni3s$_{massive}$	0.15	110	114.3	4.3		0.5	
Ni3s$_{nanostructure}$	0.32	110	113	3		1.8	
Co3s$_{massive}$	0.29	101	105.6	4.6	1.6		
Co3s$_{nanostructur}$	0.48	101	106.4	5.4	2.3		
Cu3s$_{nanostruc-ture}$	0.24	124	127.5	3.5			1.5

I_2/I_1 –the ratio of the intensities of the maxima of the multiplet splitting lines;

E_{max}, eV—the energy position of the main peak

$E_{sat,}$ eV—the energy position of the satellite peak

Δ—the energy distance between the multiplet splitting maxima in the 3s-spectra of pure metals Co, Ni and the Ni-, Co- and Cu-based nanostructures;

μ—the atomic magnetic moment of the alloy components calculated by the authors

Thus, the investigations conducted show that:

1.   The atomic magnetic moment in 3d-metals in nanostructures increases in comparison with the atomic magnetic moment in massive samples from the same metals.

2.   The nanostructures formed as the result of the interaction of the transition metal oxides with polyvinyl alcohol are not of the same type. The differences are associated with the participation of transition metals in their formation, which have different d-shell filling.

In the present work the variation of the spin magnetic moment is shown for the atoms of transition metals in metal-containing nanostructures in comparison with the atomic magnetic moment of the same metals in massive samples. The atomic magnetic moment in nanostructures increases on the Ni and Co atoms. The appearance of the multiplet splitting in the Cu3s spectrum indicates the appearance of uncompensated d electrons on the Cu atoms, and consequently, the appearance of the magnetic moment on the Cu atom.

5.2.2 X-ray Photoelectron Study of the Functionalization of Carbon Metal-Containing Nanotubes with Phosphorus Atoms

Unique mechanical, thermal, electrical, magnetic and other properties of carbon metal-containing nanostructures find different applications. For improving the material properties it is necessary to modify them with nanostructures. Since the nanostructure surface has low reactivity as a binding agent between the nanostructure surface and a material, the functionalization of the nanostructure surface is used, that is the addition of certain chemical groups of sp-elements in it, which form a covalent bond with the atoms on the nanostructure surface. In this case, the dispersive ability and solubility of nanostructures improve and their aggregation into bundles is prevented due to the repulsion of sp-elements attached on lateral areas. The variation of the synthesis parameters allows obtaining nanostructures of a specified dimension and imperfection. Functionalization creates additional conditions for improving nanostructure properties.

The objective of the conducted investigation is the study of the mechanism of the functionalization of the surface of carbon metal-containing nanostructures and the influence of functional groups on the change of the nanostructure properties.

In the present chapter, the X-ray photoelectron spectroscopy method is used for studying the functionalization of the carbon metal-containing (Me: Cu, Ni, Fe) nanotubes with phosphorus atoms (ammonium polyphosphate) and the influence of the functionalization on the change of the metal atomic magnetic moment.

The investigation was conducted by X-ray photoelectron spectroscopy on an X-ray electron magnetic spectrometer with resolution 10^{-4}, luminosity—0.085% at the excitation by the AlKα line in vacuum 10^{-8}–10^{-10}Pa. The spectrometer used has a number of advantages over electrostatic X-ray electron spectrometers, which are the constancy of the luminosity and resolution independent of the energy of electrons, high contrast of spectra and the possibility of external actions on a sample during measurements without deteriorating the spectrometer resolution and luminosity [42]. The important feature of the method is its non-destructive character. It cannot be said about other methods for surface analysis which are based on ion or electron bombardment of a surface.

For studying the formation of the covalent (hybridized) bond between the atoms of carbon metal-containing nanostructures and sp-elements' functional groups, the investigation of X-ray photoelectron Me3s-spectra was conducted, in which the parameters of the multiplet splitting correlated with the number of noncompensated d electrons of the metal atoms and its spin magnetic moment.

In [43] it is proved that the relative intensity of the peaks of the multiplets in 3s-spectra correlates with the number of noncompensated d electrons of atoms in the 3d-metal systems, and the distance between the multiplet peaks provides the information about the exchange interaction of 3s- and 3d-shells. The changes in the 3d-shell (localization or hybridization) provide the information about the change of the distance between the neighboring atoms and the change of the structure of the nearest surrounding of the atoms of the 3d-metals.

A model developed in [43] was used for the determination of the atomic magnetic moment in carbon metal-containing nanotubes in comparison with massive samples of metals [44]. The change of the relative intensity of the peaks of the multiplet splitting and the distance between them in the nanostructures is shown in comparison with massive samples. The results obtained indicate the increase of the number of noncompensated d electrons in the Fe, Co and Ni metal atoms and their appearance on the Cu atoms in the nanostructures.

The increase of the number of noncompensated d electrons is explained by the participation of d electrons of the metal atoms in the chemical hybridized bond with p electrons of the carbon atoms. Thus, in comparison with pure metal, the atomic magnetic moment in carbon metal-containing (Ni, Co) nanostructures increases and appears on the Cu atoms.

The model [43] is used for the study the change of the atomic magnetic moment on the Fe, Ni and Cu atoms of the functionalized carbon metal-containing nanotubes. Chemical groups containing different P concentrations are used for the functionalization. The Table 5.2 shows the Me3s spectra parameters and atomic magnetic moments on the metal atoms in the nanotubes functionalized with the chemical groups containing P (ammonium polyphosphate) and in the nanotubes in which the concentration of ammonium polyphosphate is increased by half (Figure 5.26).

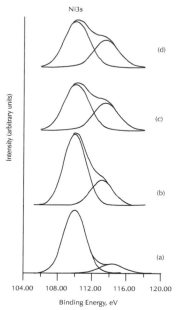

FIGURE 5.26   The X-ray photoelectron Ni3s-spectra: The Ni3s-spectrum of crystalline nickel; The Ni3s-spectrum of carbon nickel-containing nanotubes prepared by the low-energy synthesis method; The Ni3s-spectrum of carbon

nickel-containing nanotubes prepared by the low-energy synthesis method and functionalized with chemical groups containing P (ammonium polyphosphate).

The Ni3s-spectrum of carbon nickel-containing nanotubes prepared by the low-energy synthesis method and functionalized with chemical groups containing P (ammonium polyphosphate) with the concentration larger by half.

In the functionalized carbon nickel-containing nanotubes, the atomic magnetic moment of Ni increases in comparison with the Ni atomic magnetic moment in non-functionalized carbon nickel-containing nanotubes, that is the number of noncompensated d electrons increases. Consequently, the nearest surrounding of the Ni atoms and their chemical bond are changed. With growing number of phosphorus atoms in the functional group, the Ni atomic magnetic moment is increasing insignificantly that is additional phosphorus atoms do not form bonds with the nanostructure atoms, and the excess of phosphorus is in the oxidized state.

Similar situation is observed in the functionalized carbon copper-containing nanotubes. The Cu atomic magnetic moment increases in comparison with that in the non-functionalized nanostructures. The two-fold increase of the phosphorus amount does not change the atomic magnetic moment value and additional phosphorus forms the bond with oxygen.

The process of the reduction of carbon and iron becomes difficult and the nanoproduct yield is small. The synthesis activity grows at the functionalization with phosphorus-containing chemical groups. In this case, iron and carbon are reduced and the formation of carbon iron-containing nanostructures takes place; therefore the Fe atomic magnetic moment increases with an increase in the phosphorus content.

The C1s-spectra are similar for all studied nanotubes and consist of two components C–C with sp^2- and C–C with sp^3-hybridization of carbon atoms in the ratio of 1:0.5 (Figure 5.26). Thus, the phosphorus atoms form bonds with the metal atoms rather than with the carbon atoms in the carbon metal-containing nanostructures. This is also indicated by the change of the number of noncompensated d electrons because of the change in the nearest surrounding of the metal atoms and the covalence degree of the Me–P bond which is higher than that of Me–C [44] because phosphorus has a larger covalence radius that is close to the radius of the atoms of d-metals. Thus, the Me–P chemical bond is stronger than the Me–C chemical bond. The presence of the phosphorus atoms liberates a certain amount of the carbon atoms from the bond with the metal. The carbon atoms form the bond with the material atoms at the material modification with nanostructures. In contrast to the result obtained in [45], it is shown that in the case when the functional group consists of fluorine, the covalent bond of fluorine and carbon is formed on the surface of a carbon nanostructure, and at the material modification the bond with the material atoms is realized via the fluorine atoms.

C1s

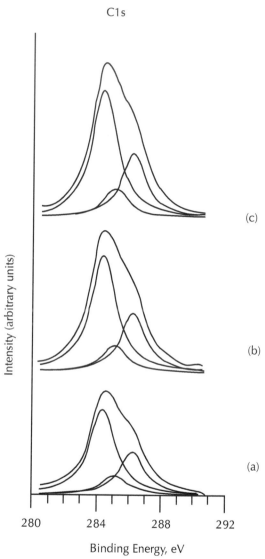

FIGURE 5.27   The X-ray photoelectron C1s-spectra.

- The C1s-spectrum of carbon nickel-containing nanotubes prepared by the low-energy synthesis method;
- The C1s-spectrum of carbon nickel-containing nanotubes prepared by the low-energy synthesis method and functionalized with chemical groups containing P (ammonium polyphosphate);

- The C1s-spectrum of carbon nickel-containing nanotubes prepared by the low-energy synthesis method and functionalized with chemical groups containing P (ammonium polyphosphate) with the concentration larger by half.

Thus it can be concluded that the formation of the bond between the atoms of functional groups and nanostructures is influenced by the following factors: electronegativity of atoms and the possibility of the formation of a strong covalent bond between atoms. Therefore in the present chapter phosphorus atoms form a strong chemical bond with metal atoms rather than carbon atoms.

In the present investigation the mechanism of functionalization of the surface of carbon metal-containing nanostructures with phosphorus atoms and the influence of such functionalization on the change of the metal atomic magnetic moment of carbon metal-containing nanotubes are studied.

It is shown that

- The value of the atomic magnetic moment in the functionalized carbon metal-containing (Cu, Ni, Fe) nanotubes increases in comparison with non-functionalized nanotubes at the growth of the phosphorus content up to a certain amount.
- When the carbon iron-containing are functionalized with phosphorus, the yield of the synthesized nanostructures and the value of the iron atomic magnetic moment are increased in comparison with the case of non-functionalized carbon iron-containing nanotubes.
- It is shown that the covalent bond of Me atoms with P atoms is stronger than that between metal and carbon atoms, which increases the activity of the nanotube surfaces that is necessary for the material modification with nanotubes.
- The formation of the chemical bond between the atoms of the functional sp-groups and nanostructures is influenced by the following factors: electronegativity of the elements and the possibility of the formation of a strong covalent bond between the atoms.

5.2.3   Functionalization of Copper/Carbon Nanocomposite with Nitrogen Atoms

An effective method for improving material properties is the material modification with nanostructures. Studying mechanisms of the material modification and methods for obtaining required material service properties is very important in the field of the creation of new materials.

The surface of nanostructures has low reactivity. To increase the nanostructure reactivity for providing better cohesion between a nanostructure surface and a material, the method of the functionalization of the surface of nanostructures is used, that is the attachment of the atoms of sp-elements to the surface of a nano-structure, which form a covalent bond with the atoms of the nanostructure surface.

In addition, functionalization provides additional conditions for improving the properties of nanostructures [46].

The objective of the present chapter is the investigation of the mechanism of the functionalization of the surface of carbon metal-containing nanostructures with nitrogen atoms, the influence of the functionalization on the formation of stable nanocomposites from carbon copper-containing nanostructures and atoms of nitrogen obtained from polyethylene-polyamine, and the chemical bond be-tween the atoms.

The number of the experimental methods for the investigation on the atomic level of the chemical structure of nanostructures is limited. In the present chapter, the investigation by the X-ray photoelectron spectroscopy (XPS) method is pre-sented. The XPS allows study the electronic structure, interatomic interaction and nearest environment of an atom.

The study was conducted on a unique automated X-ray electron magnetic spectrometer EMS-3 with double focusing, the main performance specifications of which are: the resolution—10^{-4} and luminosity—0.185% [47].

The X-ray electron magnetic spectrometer has a number of advantages over electrostatic spectrometers, namely, the constancy of luminosity and resolution independent of electron energy, high contrast of spectra, the possibility of external action on a sample during an experiment without deteriorating the resolution and luminosity. The XPS is a nondestructive method which is especially important for investigating metastable systems.

The samples of carbon metal-containing nanostructures were prepared by low-temperature synthesis from CuO and polymers. Carbon copper-containing nanotubes were treated with nitrogen (polyethylene-polyamine $H_2N[CH_2-CH_2-NH]_mH$, where m = 1–8) with the use of mechanoactivation.

The investigations were conducted in the X-ray Photoelectron Spectroscopy Laboratory of the Physicotechnical Institute of the Ural Branch of the Russian Academy of Sciences.

For studying the mechanism of the formation of the chemical bond between the atoms of carbon, nitrogen and oxygen in the considered system, the spectra of the core levels C1s, N1s, and O1s were investigated. The decomposition of the spectra into components was performed with the use of a program based on the least-squares method. For the spectra decomposition, the energy position, the width of spectra components and the components' intensities based on the data obtained for the spectra of reference samples were entered in the program. The accuracy in the determination of the peak positions is 0.1eV. The error in the determination of the contrast of the electron spectra was less than 5%.

In the C1s-spectrum (Figure 5.28) of the nanostructures functionalized with nitrogen atoms, the following components are found: C–H (285eV), C–N(-H) (286.4 eV) and insignificant peak of the adsorbed component C–OH (289 eV) (Figure 5.28). The C–H bond is relatively weak and the C–N (H) with high degree of covalence is stronger since the radii of carbon and nitrogen are close. The appearance of the C–N(H) bond between the carbon atoms of carbon copper-containing nanotubes and nitrogen atoms of amino groups of polyethylene-polyamine indicates the formation of a stable nanocomposite.

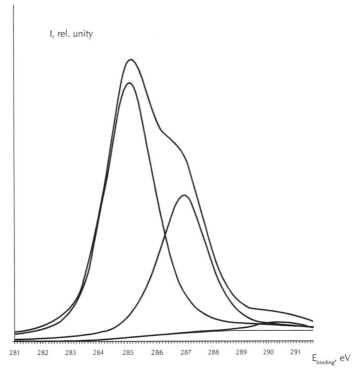

FIGURE 5.28 The C1s-spectrum of functionalized nanostructures.

The presence of the C–N(H) bond in the N1s-spectrum (Figure 5.29) consisting of two components NH (397 eV) and C–N(H) (398.8 eV) also confirms the formation of a stable composite with the covalent C–N(H) bond.

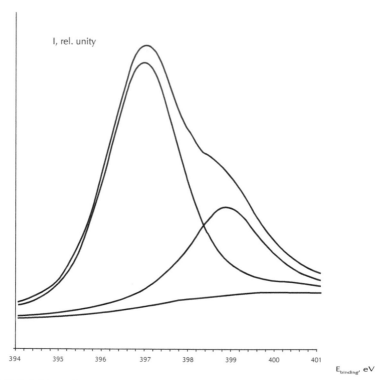

FIGURE 5.29   The N1s-spectrum of functionalized nanostructures.

In the O1s-spectrum, the low-intensity peak corresponds to adsorbed oxygen.

As it is shown in [46] that when nanostructures are functionalized with fluorine atoms, a strong covalent bond between the atoms of fluorine and the atoms of nanostructure carbon is formed in the system and the bond of nanostructures with the atoms of the material modified is realized through fluorine atoms. In our case, the bond between carbon metal-containing nanostructures and material is realized through the nitrogen atoms. Such bond is not form between the metal (Cu) atoms and nitrogen atoms.

It can be concluded that the formation of the strong covalent chemical bond between the atoms of functional sp-groups and atoms of nanostructures is influenced by the following factors: electronegativity of the component atoms and closeness of their covalent radii.

The results of this investigation are following:

- It is shown that the formation of a strong covalent bond between the carbon atoms of carbon copper-containing nanotubes and nitrogen atoms (functional sp-group) leads to an increase in the activity of the nanostructure surface, which is necessary for effective material modification.

- The formation of the strong covalent chemical bond between the atoms of functional sp-groups and nanostructure atoms is influence by the following factors: electronegativity of the component atoms and the closeness of their covalent radii.
- The X-ray photoelectron spectroscopy method can be used for monitoring the quality of the functionalization of nanostructures.

5.2.4   Study of the Nanostructures Obtained from a Mixture of Metallurgical Dust (Ni, Fe) and Raw Polymer Materials

Reprocessing of not only slags but also all other wastes (including powders), which makes the use of raw materials much more comprehensive, plays an important role in metallurgy. One of the methods for obtaining nanostructures is the synthesis of metal containing carbon nanostructures during carbonization of raw polymer material.

The study by X-ray photoelectron spectroscopy (XPS) was performed on an X-ray electron magnetic spectrometer with double focusing (resolution 0.1 eV, excitation in the Al$K\alpha$ line (1486.6eV), residual pressure 10–6Pa) [47–48]

The XPS method was used to monitor the preparation of nanostructures from the relative content of C–C and C–H bonds, hybridization type of valence s and p electrons of carbon atoms and the satellite structure of the C1s spectrum, and the type of interaction of metal atoms with carbon. This method makes it possible to identify the C1s spectra and determine the chemical bonding of the elements, the nearest atomic environment, and the type of sp hybridization of valence electrons in nanostructures using the satellite structure of the C1s spectra. Application of $3d$ transition metal salts as catalysts in synthesis of nanostructures leads to the formation of C–C bonds with sp^2 and sp^3 hybridizations, whose ratio depends on the degree of occupying the d shell of the metal [49].

A possible way of utilizing the metallurgical powder is its treatment by a polymer medium (polyvinyl alcohol (PVA)) and secondary polyvinylchloride (PVC) [50].

The investigation was performed on the following three samples. Sample 1 was the initial fine-grained metallurgical Ni-containing powder; Sample 2 was repaired by mixing two solid components, fine-grained metallurgical Ni-containing powder and PVA, and heating the mixture to 400°C; and Sample 3 was prepared by mixing a PVC solution in HCl and metallurgical Fe-containing powder, with subsequent heating to 400°C.

X-ray photoelectron spectra were measured for all samples. The full-range spectra of Sample 1 of the initial fine-grained metallurgical Ni-containing powder exhibited C1s, O1s, and Ni3p lines.

The C1s spectrum (Figure 5.30) has two components: C–H (285.0eV) and C–O (287.0eV), which indicate the presence of hydrocarbon contaminants in the metallurgical powder. The O1s spectrum also has two components, characteristic of metal oxides Me–O (529.5eV) and adsorbed oxygen (532.0eV). The Ni3p spectrum exhibits oxidized states of metals, that is, the metal enters the composition of gas-dust jections mainly in the form of oxides.

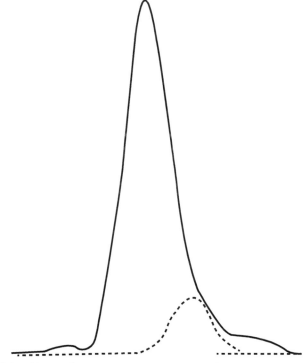

FIGURE 5.30   X-ray photoelectron C1s spectra of Sample 1.

The spectrum C1s of Sample 2 (Figure 5.31) demonstrates the presence of carbon not only in the C–H bond (285eV) but also in the C–C bond with sp hybridization (284.3 eV) and C–C bond with sp^3 hybridization (286.1 eV).

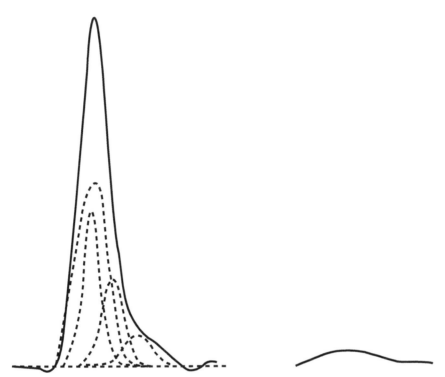

FIGURE 5.31   X-ray photoelectron C1*s* spectra of Sample 2.

The changes in the Ni2*p* spectrum are indicative of partial reduction of nickel. According to the data in the literature [51] on ultra dispersed Ni powders with different particle sizes (from several tens to several hundred A), the thermal stability of NiO on particles of different size is different. On particles with sizes of several tens A, NiO decomposes even at 200–250°C. On large (300 A) particles, NiO is retained at temperatures up to $T = 400$°C. Such a difference is related to the smaller thickness of the oxide film (15 A) on small particles; the film thickness on particles of several thousand A in size increases by a factor of 1.5–2. Therefore, reduction of Ni from NiO requires lower temperatures in the case of nanoparticles.

Since iron is not reduced in PVA, secondary polyvinylchloride (PVC) dissolved in HCl as a raw polymer material was used. This allows obtaining graphite films, as evidenced by the presence of the component with sp^2 hybridization of electrons in the C1*s* spectrum (Figure 5.3). In addition, the C1*s* spectrum contains C–H and C–O components. The Fe2*p* spectrum has a low intensity, which indicates that only iron traces are present in the samples under study. Iron is in the oxidized state. In contrast to the Ni-containing powder, study of the Fe-containing powder did not reveal metal reduction.

Thus, the X-ray photoelectron study of the samples obtained by heating mixtures of metal powder and raw polymer materials showed that metal reduction is observed only in the nickel-containing powder.

Reduction of Ni from NiO requires lower temperatures in the case of nanoparticles. In contrast to the Ni-containing powder, metal is not reduced from the iron-containing powder.

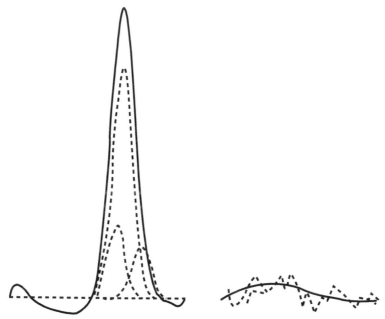

FIGURE 5.32   X-ray photoelectron C1s spectra of Sample 3.

The X-ray photoelectron study of the possibility of using metallurgical powder to form carbon nanostructures (with a metal compound as a necessary component involved in their formation) showed that an important factor is the metal ability to be reduced at certain temperatures. The interaction of PVA with metallurgical Ni-containing powder leads to metal reduction with the formation of both sp^2 and sp^3 hybridized structures, a fact indicating that carbon nanostructures are produced.

As a result of interaction of PVC with metallurgical Fe-containing powder, iron fails to be reduced to pure metal; however, the C1s spectrum contains a component with sp^2 hybridization of electrons, which is indicative of the formation of graphite-like structures.

KEYWORDS

- Ketone
- Metallurgical Dust
- MgK$_\alpha$ Radiation
- Photoelectron
- Tubulenes

REFERENCES

1. Ijima, S. (1991). *Nature, 354,* 56.
2. Ebbsen, T. V., & Ajayan, P. M. (1992). *Nature, 358,* 220.
3. Hsu, W. K., Terrones, M., & Hare, J. P., et al. (1996). *Chemical Physics Letters, 262,* 161.
4. Ajayan, P. M., & Ijima, S. (1993). *Nature, 361,* 333.
5. Barker, R. T. (1989). *Carbon, 27,* 315.
6. McAlister, P., & Wolf, E. E. (1992). *Carbon, 30,* 189.
7. Mordkovich, V. Z., Baxendale, M., & Yudasaka, M., et al. (1998). *Molecular Crystal and Liquid Crystal, 310,* 159.
8. Ebbesen, T. W. (1996). *Physics Today, 273,* 26.
9. Baxendale, M., Mordkovich, V. Z., & Yoshimura, S. (1997). *Physical Review B, 56,* 2161.
10. Kodolov, V. I., & Turin, S. A. (1992). *Chemical Physics (Russian Academy of Sciences), 11,* 1275.
11. Bulgakov, V. K., Kodolov, V. I., & Lipanov, A. M. (1990). *Modeling of polymer materials combustion.* Moscow: Chemistry.
12. Fischer, E. O., & Hafner, W. Z. (1955*). Naturforschung, 10b,* 665.
13. Wachters, A. J. H. (1970). *Journal of Chemical Physics, 52,* 1033.
14. Dupuiis, M., et al. (1989). *Computer Physics Communication, 52,* 415.
15. Schmidt, M. W., Baldridge, K. K., & Boatz, J. A., et al. (1993). *Journal of Computational Chemistry, 14,* 1347.
16. Molina, J. M., & Melchor, S. F. *Private Communication.*
17. Bethune, D. S. Klang, C. H., & de Vries, M. S., et al. (1993). *Nature, 363,* 605.
18. Mintmire, J. W. Dunlap, B. I., & White, C. T. (1992). *Physical Review Letters, 68,* 631.
19. Kodolov, V. I. Boldenkov, O. Yu., Khokhriakov, N. V., & Babushkina, S. N., et al. (1999). *Analytics and Control Technology, 4,* 18–25.
20. Shabanova, I. N., Sapozhnikov, V. P., Bayankin, V. Ya., & Bragin, V. G. (1981). *Pribori i tekhnika eksperimenta, 1,* 138.

21. Nefyodov, V. I. (N. A.). *X-ray photoelectron spectroscopy of chemical compounds.* Moscow (in Russian).
22. Briggs, D., & Seach, M. P. *(1983). Practical surface analysis by Auger and X-ray photoelectron spectroscopy.*
23. Kodolov, V. I., Shabanova, I. N., & Makarova, L. G., et al. (2001). *Zhurnal strukturnoi khimii, 42*(2), 260.
24. Kodolov, V. I., Kuznetsov, A. P., & Nicolayeva, O. A., et al. (2001). *Surface and Interface Analysis, 32,* 10–14.
25. Beamson, G., & Briggs, D. (1992). *HRXPS of Organic Polymers. The Scienta ESCA A300 Database.* Chichester: Wiley
26. Makarova, L. G. Shabanova, I. N., & Terebova, N. S. (2005). *Zavodskaya Laboratoriya, Diagnostic Materials, 71*(5), 26.
27. Blagodatskikh, I. I., Volkova, E. G., & Kodolov, V. I. (2006). *Trudy mezhdunarodnoi nauchno-prakticheskoi konferentsii "Nanotekhnologii proizvodstvu".* Proceedings of International. Scientific and Practical Conference "Nanotechnologies for Production" Fryazino, Russia (p. 358). Moscow: Yanus-K.
28. Demekhin, V. F. Nemoshkalenko, V. V., & Aleshin, V. G. (1975). *Metallofizika, 60,* 27.
29. Didik, A. A. (2004). *Study of the formation of carbon metal-containing nanostructures upon carbonization of polyvinyl alcohol.* Canadian Science (Chemical), Dissertation. Izhevsk: IPM UrO RAN.
30. Siegbahn, K. Nordling, C., & Fahlman, A., et al. (1971). *ESCA: Atomic, Molecular, and Solid State Structure Studied by Means of Electron Spectroscopy,* Amsterdam: North Holland, 1967. Translated under the title Electronnaya Spectroskopiya. Moscow: Mir.
31. Bellamy, L. T. (1963). *The infra-red spectra of complex molecules,* London: Methuen 1954 (Translated under the title Infrakrasnye spektry slozhnykh molekul). Moscow: Inostr. Lit.
32. Kodolov, V. I., Khokhryakov, N. V., Kusnetsov, A. P., & Lipanov, A. M. (2001). *Khimicheskaya fizika and mezoscopiya, 3*(2), 162.
33. Didik, A. A., Kodolov, V. I., Volkov, A. Yu., Volkova, Ye. G., & Halmayer, K. H. (2003). *Neorganicheskiye Materially, 39*(6), 693.
34. Kolobova, K. M., Shabanova, I. N, & Kulyabina, O. A., et al. (1981). *FMM, 5*(4), 890.
35. Makarova, L. G., Shabanova, I. N., Kusnetsov, A. P., & Terebova, N. S. (2004). Izvestia Akedemii Nauk. *Seriya fiziko Tekhnicheskikh, 68*(5), 653.
36. Briggs, D., & Seah, M. P. (Eds). (1987). *Analysis of surface by the methods of Auger- and X-ray photoelectron spectroscopy* (p. 60). Moscow: Mir.
37. van Acker, J. F., Stadnik, Z. M., Fuggle, J. C., Hoekstra, H. J. W. M., Buschow, K. H. J., & Stroink, G. (1988). *Physical Review Letters, 37,* 6827–6834.
38. Kozo, O., & Akio, K. (1992). *Journal of the Physical Society of Japan, 61*(12), 4619–4637.
39. Makarova, L. G., Shabanova, I. N., & Kodolov, V. I., et al. (2008). *Izvestia RAN, seria Physicheskaya, 72*(4), 461–495.
40. Lomova, N. V., Shabanova, I. N., & Terebova, N. S., et al. (2005). *Izvestia RAN, seria Physicheskaya, 69*(7), 1015–1019.
41. Shabanova, I. N., & Mitrokhin, Yu. S. (2004). *Electron Spectroscopy and Related Phenomena, 137–140,* 569–571.

42. Trapeznikov, V. A., Shabanova, I. N., & Varganov, D. V., et al. (1986). New automated X-ray electron spectrometers: spectrometers with technological adapters and manipulators, a spectrometer for studying melts. *Izvestia Akedemii Nauk SSSR, seria fiziko, 50*(9), 1677–1682.
43. Lomova, N. V., & Shabanova, I. N. (2004). The study of the electronic structure and magnetic properties of invar alloys based on transition metals. *Journal of Electron Spectroscopy and Related Phenomena, 137–140,* 511–517.
44. Shabanova, I. N., & Terebova, N. S. (2010). *Surface and interface analysis, 42*(6–7), 846–849.
45. Shabanova, I. N., Terebova, N. S., Mitrokhin, Yu. S., & Nebogatikov, N. M. (2002). *SIA, 34,* 606–609.
46. Khabashesku, V. N. (2011). Covalent functionalization of carbon nanotubes: synthesis, properties and application of fluorinated derivatives. *Uspehi khimii, 80*(8), 739–760.
47. Shabanova, I. N., & Terebova, N. S. (2011). Dependence of the value of the atomic magnetic moment of d-metals on the chemical structure of nanoforms. *Polymers Research Journal, 5,* 7–13.
48. Shabanova, I. N., Varganov, D. V., & Dobysheva, L. V., et al. (1986). *Izv. Akad. Nauk SSSR, Ser Fiz, 50*(9), 1677.
49. Makarova, L. G, Shabanova, I. N., & Terebova, N. S. (2005). *Zavod. Laboratory Diagnosis Materials, 71*(5), 26.
50. Kodolova, V. V., Makarova, L. G., & Volkova, E. G. (2007). Proceedings of XIX all-Russia Science School–Seminar "X-ray and electronic spectra and chemical bonding" (pp. 99). Izhevsk.
51. Shabanova, I. N. (1990). X-ray photoelectron spectroscopy of disordered systems based on transition metals, *Doctoral (Phyics–Mathematics) Dissertation,* Izhevsk: FTI UrO.

CHAPTER 6

COMPUTATION MODELING OF NANOCOMPOSITES ACTION ON THE DIFFERENT MEDIA AND ON THE COMPOSITION MODIFICATION PROCESSES BY METAL OR CARBON NANOCOMPOSITES

V. I. KODOLOV, N. V. KHOKHRIAKOV, V. V. TRINEEVA,
M. A. CHASHKIN, L. F. AKHMETSHINA, YU. V. PERSHIN, and
YA. A. POLYOTOV

CONTENTS

6.1   THE QUANTUM CHEMICAL INVESTIGATION OF METAL OR CARBON NANOCOMPOSITES INTERACTION WITH DIFFERENT MEDIA AND POLYMERIC COMPOSITIONS

6.1.1   The Quantum Chemical Investigation of Hydroxyl Fullerene Interaction with Water

Currently various sources contain experimental data that prove radical changes in the structure of water and other polar liquids when super small quantities of surface active nanoparticles (according to a number of papers—about thousandth of percent by weight) are introduced. When the water modified with nanoparticles is further applied in technological processes, this frequently results in qualitative changes in the properties of products, including the improvement of various mechanical characteristics.

The investigation demonstrates the results of quantum-chemical modeling of similar systems that allow making the conclusion on the degree of nanoparticle influence on the structure of water solutions. The calculations were carried out in the frameworks of *ab initio* Hartree–Fock method in different basis sets and semi-empirical method PM3. The program complex GAMESS was applied in the calculations [1]. The equilibrium atomic geometries of fragments were defined and the interaction energy of molecular structures E_{int} was evaluated in the frameworks of above energy models. The interaction energy was calculated by the following formula:

$$E_{\text{int}} = E - E_a - E_b,$$

(6.1)

where, E—energy of the molecular complex with an optimized structure, E_a and E_b—bond energies of isolated molecules forming the complex; when calculating E_a and E_b the molecule energy was not additionally optimized.

At the first stage quantum-chemical investigation of ethyl alcohol molecule interaction with one and few water molecules was carried out. The calculations were carried out in the frameworks of different semi-empirical and *ab initio* models. Thus, based on the data obtained we can make a conclusion on the error value that can occur due to the use of simplified models with small basis sets. The calculation results are given in Table 6.1.

TABLE 6.1   Equilibrium parameters of the complex formed by the molecules of ethyl alcohol and water

Calculation method	E_{int}, kcal/mol	R_{9-10}	R_{3-9}	Q_3	Q_9	Q_{10}
PM3	1.88	2.38	0.95	−0.33	0.20	−0.37
3-21G	11.29	1.80	0.97	−0.74	0.41	−0.72
6-31G	7.52	1.89	0.96	−0.81	0.48	−0.83
TZV	6.90	1.91	0.96	−0.64	0.42	−0.83
3-21G MP2	13.17	1.78	1.00	−0.64	0.35	−0.63
TZV**++ MP2	6.27	1.90	0.97	−0.49	0.33	−0.65
TZV MP2	7.52	1.84	0.99	−0.58	0.39	−0.79
TZV+ MP2 for the geometry obtained by the method 6-31G	6.27	1.89	0.96			

The geometrical structure of the complex formed by the ethyl alcohol and water molecules are given in Figure 6.1.

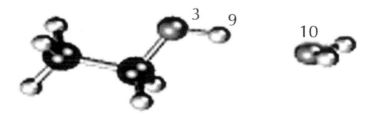

FIGURE 6.1   Equilibrium geometric structure of the complex formed by the molecules of ethyl alcohol and water. Optimization by energy is carried out in the basis 6-31G.

The energy minimization of the structure demonstrated was carried out on the basis 6-31G. The atom numbers in the figure correspond to the designations given in the tables.

The calculations demonstrate that equilibrium geometrical parameters of the complex change insignificantly in the calculations by different methods. It should be noted that in the structures obtained without electron correlation the atoms 3, 9, 10, and hydrogen of water molecule are in one plane. The structure planarity is distorted when considering electron correlation by the perturbation theory MP2. Even more considerable deviations from the planarity are observed when calculating by the semi-empirical method PM3. In this case, the straight line connecting the atoms 3 and 9 become practically perpendicular to the water molecule plane. Simultaneously, the semi-empirical method gives the hydrogen bond length exceeding *ab initio* results by over 20%.

The analysis of energy parameters indicates that the energy of interaction of molecules inside the complex calculated by the formula (1) is adequately predicted in the frameworks of Hartree–Fock method in the basis 6-31G. If for the geometrical structure obtained by this method the calculations of energy are made in the frameworks of more accurate models, the calculation results practically coincide with the successive accurate calculations.

The atom charges obtained by different methods vary in a broad range, in this case no regularity is observed. This can be connected with the imperfection of the methodology of charge evaluation by Mulliken.

The investigation of water interaction with the cluster of hydroxyfullerene $C_{60}[OH]_{10}$ was carried out in the frameworks of *ab initio* Hartree–Fock method in the basis 6-31G. To simplify the calculation model the cluster with the rotation axis of 5th order of fullerene molecule C_{60} was applied. Figure 6.2a demonstrates the fullerene molecule structure optimized by energy, Figure 6.2b is optimized structure of the cluster $C_{60}[OH]_{10}$.

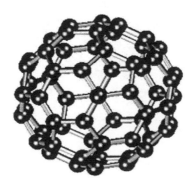

FIGURE 6.2   *(Continued)*

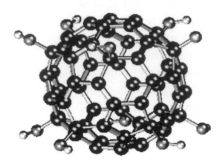

FIGURE 6.2   Equilibrium geometric structure of the molecule C_{60} (a) and cluster $C_{60}[OH]_{10}$ (b) Optimization by energy is carried out in the basis 6-31G.

All the calculations were carried out preserving the symmetry C_s of molecular systems. The pentagon side length calculated in the fullerene molecule is 1.452Å, and bond length connecting neighboring pentagons—1.375 Å. These results fit the experimental data available and results of other calculations [2].

To evaluate the energy of interaction of the cluster $C_{60}[OH]_{10}$ with water, the complex $C_{60}[OH]_{10} \cdot 10\ H_2O$ with the symmetry C_s was studied. The complex geometry obtained in the basis 6-31G is given in Figure 6.3.

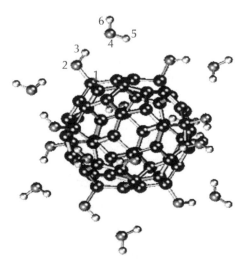

FIGURE 6.3   Molecular system formed by the cluster $C_{60}[OH]_{10}$ and 10 water molecules. Optimization by energy is carried out in the basis 6-31G.

The complex geometrical parameters and atom charges are shown in Table 6.2.

TABLE 6.2   Equilibrium parameters of the complex $C_{60}[OH]_{10} \cdot 10\,H_2O$ (calculations by ab initio method in the basis 6-31G).

Charges					
Q_1	Q_2	Q_3	Q_4	Q_5	Q_6
0.17	−0.78	0.50	−0.86	0.46	0.42
Bond lengths					
$R_{1\text{-}2}$	$R_{2\text{-}3}$	$R_{3\text{-}4}$	$R_{4\text{-}5}$	$R_{4\text{-}6}$	
1.42	0.97	1.86	0.95	0.95	

The atom numeration in the table corresponds to the numeration given in Figure. 6.3.

TABLE 6.3   Characteristics of the complexes formed by the molecules $C_{60}[OH]_{10}$, C_2H_5OH with water. *Ab initio* calculations in the basis 6-31G.

	E_{int}, kcal/mol	L,A	Q_H	Q_O (OH)	Q_O (H$_2$O)
$C_{60}(OH)_{10}+10H_2O$	12.69	1.86	0.5	−0.78	−0.86
$C_2H_5OH+H_2O$	7.52	1.89	0.47	−0.81	−0.83
H_2O+H_2O	8.37	1.89	0.42	−0.82	−0.76

Table 6.3 contains the comparison of the breaking off energy of water molecule from the complexes $C_{60}[OH]_{10} \cdot 10\,H_2O$, $C_2H_5OH \cdot H_2O$, $H_2O \cdot H_2O$. Besides, the table gives the lengths of hydrogen bonds in these complexes L and charges of Q atoms of oxygen and hydrogen of OH-group and oxygen atom of water molecule. All the calculations are carried out by *ab initio* Hartree–Fock method in the basis 6-31G. The calculations demonstrate that the energy of interaction of ethyl alcohol molecule with water is somewhat less than the energy of interaction between water molecules. At the same time, nanoparticle $C_{60}[OH]_{10}$ interacts with water molecule about two times more intensively. When comparing the atom charges of the complexes it can be seen that the molecule π—fullerene electrons in the nanoparticle act as a reservoir for electrons. Thus the molecule OH-group is practically electrically neutral. This intensifies the attraction of water molecule oxygen. Therefore, the strength of hydrogen bond in the complex $C_{60}[OH]_{10} \cdot 10\,H_2O$ increases.

To evaluate the cluster sizes with the center in nanoparticle $C_{60}[OH]_{10}$ that can be formed in water, semi-empirical calculations of energies of attachment of water molecules to the nanoparticle $C_{60}[OH]_2$ with the chain formation are carried out (Figure 6.4). Semi-empirical method PM3 was applied in the calculations. Water molecules are attached to the chain successively; after each molecule is attached, the energy is minimized.

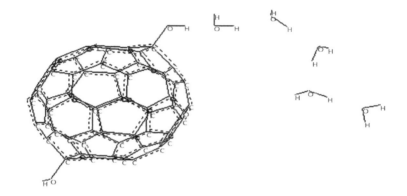

FIGURE 6.4 Model system formed by the cluster $C_{60} [OH]_{10}$ and chain from five water molecules. Optimization by energy is carried out by semi-empirical method PM3.

Table 6.4 contains the results obtained for the chain growth from the nanoparticle, alcohol molecule, and water molecule. Bond energies of water molecules with the chain growing from $C_{60} [OH]_{12}$ over two times exceed similar energies for the chain growing from OH-group of alcohol molecule and water molecule.

TABLE 6.4 Energies of attachment of water molecule to the chain connected with the molecules $C_{60} [OH]_{10}$, $C_{60} H_5OH$. Semi-empirical calculations by the method PM3.

n	$(H_2O)_n$	$C_2H_5OH+n\ H_2O$	$C_{60}(OH)_2+nH_2O$
1		−1.40	−4.06
2	−1.99	−2.19	−4.83
3	−2.38	−2.51	−5.02
4	−2.54	−2.58	−4.88
5	−2.56	−2.68	−4.71
6	−2.66		−3.04

At the same time, the energy of hydrogen bond first increases till the attachment of the third molecule, and then gradually decreases. When the sixth molecule is attached, the energy decreases in discrete steps.

Thus, the investigations demonstrate that hydroxyfullerene molecule forms a stable complex in water, being surrounded by six layers of water molecules. Obviously, the water structure is reconstructed at considerably larger distances. The introduction of insignificant number of nanoparticles results in complete reconstruction of water medium, and, therefore, in considerable change in the properties of substances obtained on its basis. Thus, the results are reported on the considerable change in the properties of constructional materials obtained on the water basis after the insignificant additions of active graphite-like nanoparticles with chemical properties similar to the ones of hydroxyfullerene.

6.1.2 Influence Comparison of Hydroxyl Fullerenes and Metal or carbon Nanoparticles with Hydroxyl Groups on Water Structure

Quantum-chemical investigations of the interactions between hydroxylated graphite-like carbon nanoparticles, including metal containing ones, and water molecules are given. Hydroxyfullerene molecules as well as carbon envelopes consisting of hexangular aromatic rings and envelope containing a defective carbon ring (pentagon or heptagon) are considered as the models of carbon hydroxylated nanoparticles. The calculations demonstrate that hydroxyfullerene molecule $C_{60}(OH)_{10} \cdot (H_2O)_{10}$ forms a stronger hydrogen bond with water molecule than the bond between water molecules. Among the unclosed hydroxylated carbon envelopes, the envelope with a pentagon forms the strongest hydrogen bond with water molecule. The presence of metal nanocluster increases the hydrogen bond energy in 1.5–2.5 times for envelopes consisting of pentagons and hexagons.

At present, we can find in the literature experimental data confirming radical changes in the structure of water and other polar liquids when small quantities (about thousandth percent by mass) of surfacants are introduced [3–5]. When further used in technological processes, the water modified by nanoparticles frequently results in qualitative changes of product properties including the improvement of different mechanical characteristics.

The paper is dedicated to the investigation of water molecule interactions with different systems modeling carbon enveloped nanoparticles, including those containing clusters of transition metals. All the calculations were carried out following the same scheme. At the first stage, the geometry of complete molecular complex was optimized. At the second stage the interaction energy of the constituent molecular structures E_{int} was assessed for all optimized molecular complexes. The interaction energy was calculated by the formula

$$E_{int} = E - E_a - E_b$$

where, E—bond energy of the optimized structure molecular complex, E_b and E_b—bond energies of isolated molecules forming the complex. When calculating E_a and E_b, the molecule energy was not additionally optimized. The energy of chemical bonds was assessed by the same scheme. All the calculations were carried out with the help of software PC-GAMESS [6]. Both semi-empirical and *ab initio* methods of quantum chemistry with different basis sets were applied in the calculations. Besides, the method of density functional was used. The more detailed descriptions of calculation techniques are given when describing the corresponding calculation experiments.

At the first stage, the quantum-chemical investigation of the interaction of α-naphthol and β-naphthol molecules with water was carried out within various quantum-chemical models. Thus, based on the data obtained the conclusion on the error value can be made, which can be resulted from the application of simplified methods with insufficient basis sets. There are three isomers of naphthol molecule named α- naphthol (complex with water is shown on Figure 1a), trans rotamer and cis rotamer of β-naphthol (Figure. 6.5b, c).

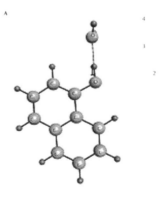

FIGURE 6.5   *(Continued)*

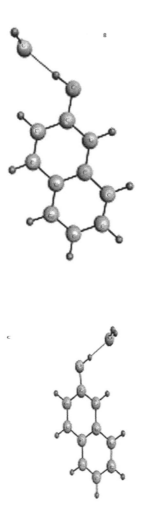

FIGURE 6.5 Equilibrium geometric structure of the complex formed by the molecules of α-naphthol (a), trans-β-naphthol (b) and cis- β-naphthol (c) with water. The optimization by energy was carried out at HF/6-31G level.

In what follows, we denote them as α-naphthol, trans-β-naphthol, and cis-β-naphthol correspondingly. The calculations within all methods applied, except for the semi-empirical one, showed that cis-β-naphthol is the most stable isomer, α-naphthol has almost equal stability, and trans-β-naphthol is the least stable structure. This conclusion is in accordance with available experimental data [7, 8].

The results for the complexes with α-naphthol are given in the Table 6.5.

TABLE 6.5 Equilibrium parameters of the complex formed by α-naphthol molecules and water. In the column E_{int} the interaction energy between water molecules in dimer is given in brackets for comparison.

Calculation method	E_{int}, kJ/mol	r_{1-2}, nm	r_{2-3}, nm	r_{3-4}, nm	q_1	q_2	q_3
PM3	16.27(8.32)	0.1364	0.0963	0.1819			
HF 3-21G	68.66(46.11)	0.1361	0.098	0.1701	0.45	-0.82	0.44
HF 6-31G	43.91(33.13)	0.1365	0.0961	0.1805	0.39	-0.85	0.50
HF 6-31G*	31.91(23.61)	0.1345	0.0954	0.194	0.42	-0.81	0.52
HF 6-31G* MP2	42.65(30.75)	0.1367	0.0984	0.1855	0.31	-0.72	0.48
DFT 6-31G* B3LYP	42.58(32.33)	0.136	0.0983	0.1821	0.28	-0.69	0.44

The geometry structure of the complex formed by naphthols and water molecules is shown in Figure 6.5. The structures given in Figure 6.5 were obtained by minimizing the energy using Hartree–Fock method in the basis 6-31G. The numbers of atoms in Figure 6.5 correspond to those given in the table. The calculations demonstrate that equilibrium geometric parameters of the complex vary insignificantly when applying different calculation methods. It should be pointed out that in the structures obtained at HF/6-31G level of theory without taking into account of electron correlation, atoms 2, 3, 4, and hydrogens of water molecule are almost in one plane. The structure planarity is broken with the inclusion of polarization functions into the basis set and electron correlation consideration. The semi-empirical calculation forecasts the planarity disturbance of the atom group considered. The hydrogen bond for β-naphthol complexes is longer than one for α-naphthol. These values are 0.1812nm for cis-β-naphthol and 0.1819nm for trans-β-naphthol calculated at HF/6-31G level. The values obtained at MP2/6-31G* level of theory are 0.1861 and 0.1862nm, respectively. They are in accordance with the results of more accurate calculations 0.1849 and 0.1850nm [9].

The second column in Table 6.1 demonstrates the energy of interaction between α-naphthol and water molecule calculated using the equation (1). The interaction energies between water molecules in water dimer are given in brackets for comparison. Although all energy characteristics for different methods vary in a broad range, the ratio of energies E_{int} for the complex of water with α-naphthol and water dimer is between 1.3 and 1.4. Thus, we can assume that Hartree–Fock method in the basis 6-31G provides qualitatively correct conclusions for the energies of intermolecular interaction. It should be pointed out that energy E_{int} calculated in the basis 6-31G agrees numerically with the results of more accurate calculations. Nevertheless, the interaction energies between water and α-naphthol exceed experimental value of 24.2 ± 0.8kJ/mol [5]. The energies of interaction between water molecules in dimer are overestimated as well. The errors are known to be caused by the superposition of basis sets of complex molecules (BSSE) and zero-point vibration energy (ZPVE). Calculations have shown that taking into account BSSE decreases the interaction energy by 10–20% depending on the basis, but does not lead to qualitative changes. Taking into account, ZPVE results in additional decrease in the energies of intermolecular interaction. Thus, taking into account ZPVE and BSSE considerably improves the accordance with the available data. Interaction energy between water and α-naphthol is 30kJ/mol in the basis 6-31G after correction. In general, analysis of energy parameters shows that all the models considered give the correct correlation for energies and demonstrate that the energy of interaction between water and α-naphthol molecules exceeds the energy of interaction between water molecules. Numerical accordance with experimental value is achieved only after taking into account electronic correlation at the MP2 level. In this model, the interaction energy after all the corrections mentioned becomes 23.7 kJ/mol; Cis-β-naphthol is known to be more stable than trans-isomer. The energy difference calculated in 6-31G basis set is 3.74 kJ/mol that is equal to experimental value [8]. HF/6-31G binding energy for water dimer is 18.9kJ/mol after corrections (experimental value is 15±2 kJ/mol [10]). As in the case of α-naphthol complex MP2 calculation is necessary to obtain value that is equal to experimental value.

The vibrational frequencies of naphthols and their complexes with water molecule were calculated and multiplied by a scaling factor of 0.9034 [8]. IR-active frequencies corresponding to naphthol OH-group stretching are 3,660cm^{-1} for trans-β-naphthol and 3654cm^{-1} for cis-β-naphthol. They are in accordance with experimental frequencies 3,661cm^{-1} and 3,654cm^{-1} [8]. The influence of hydrogen bond on naphthol OH-group stretching calculated in the present article is overestimated. Calculated frequencies at HF/6-31G level of theory in the presence of water molecule are 3,479cm^{-1} for trans-β-naphthol and 3,466cm^{-1} for cis-β-naphthol (for comparing experimental values are 3,523cm^{-1} and 3,512cm^{-1}, respectively). The same result was observed in research [8]. The atom charges obtained within different methods vary in a broad range without any regularity. It can be connected with the imperfectness of Mulliken's charge evaluation technique.

Thus, comparison of calculation results with experimental data demonstrates *ab initio* Hartree–Fock method in the basis 6-31G basis set provides qualitatively correct conclusion on hydrogen bond energies for systems considered. This model was applied to investigate the interaction between water and hydroxyfullerene cluster $C_{60}[OH]_{10}$. Hydroxyfullerene is a proper object for quantum chemical modeling of interaction between activated carbon nanoparticle and water. Simultaneously, hydroxyfullerene is of great interest because of its possible applications in medicine, for water disinfection and for polishing nanosurfaces.

The cluster with the rotation axis C5 was used in our research of hydroxyfullerene to simplify the calculation model. Figure 6.6a demonstrates the structure of C_{60} molecule optimized by energy, Figure 6.6b—the optimized structure of cluster $C_{60}[OH]_{10}$. All the calculations were carried out conserving the symmetry C_5 of molecular systems.

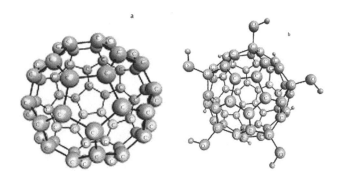

FIGURE 6.6 Equilibrium geometric structure of C_{60} molecule (a) and cluster $C_{60}[OH]_{10}$ (b).

The optimization by energy is carried out at HF/6-31G level

TABLE 6.6 Equilibrium parameters of the complex $C_{60}[OH]_{10}.10H_2O$ (HF/6-31G calculation).

Charges					
q_1	q_2	q_3	q_4	q_5	q_6
0.17	−0.78	0.50	−0.86	0.46	0.42
Bond lengths, nm					
r_{1-2}	r_{2-3}	r_{3-4}	r_{4-6}	r_{4-6}	
0.142	0.097	0.186	0.095	0.095	

TABLE 6.7 Characteristics of the complexes formed by C60[OH]10 and α-naphthol molecules and water HF/ 6-31G calculation.

Complex	E_{int}, kJ/mol	L, nm	q_H	$q_O(OH)$	$q_O(H_2O)$
$C_{60}[OH]_{10} \cdot 10H_2O$	53.09	0.186	0.5	−0.78	−0.86
$C_{10}H_7OH_2O \cdot H_2O$	43.91	0.181	0.5	−0.85	−0.84
$H_2O . H_2O$	33.13	0.189	0.47	−0.88	−0.83

The length of pentagon side in the fullerene molecule was 0.1452 nm, and the length of bond connecting the neighboring pentagons –0.1375 nm. These results agree with the experimental data available and results of other calculations [11, 12].

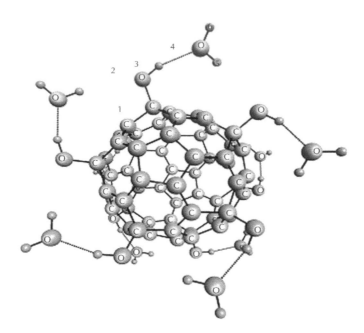

FIGURE 6.7   Molecular system formed by the cluster C60[OH]10 and 10 water molecules. The optimization by energy is carried out at HF/6-31G level.

Complex $C_{60}[OH]_{10} \cdot 10\ H_2O$ with symmetry C_5 was considered to obtain the interaction energy of cluster $C_{60}[OH]_{10}$ and water. The geometric structure of the complex optimized by energy is given in Figure 6.7. The complex geometric parameters and atom charges are given in Table 6.6. The atoms are enumerated in the same way as in Figure 6.7. Table 6.7 contains the water molecule isolation energies from complexes $C_{60}[OH]_{10} \cdot 10\ H_2O$, $C_{10}H_7OH \cdot H_2O$, and $H_2O \cdot H_2O$. Besides, the table contains the lengths of hydrogen bonds in these complexes L and charges Q of oxygen and hydrogen atoms of fullerene OH-group and oxygen atom of water molecule. The calculations demonstrate that the energy of interaction between nanoparticle $C_{60}[OH]_{10}$ and water molecule nearly twice as big as the interaction energy of water molecules in dimer. Basis set superposition error was also analyzed. The calculation indicates that the account of this effect decreases the energies of intermolecular interaction by 15%. The qualitative conclusions from the calculations do not change the gain in the interaction energy for the complex $C_{60}[OH]_{10} \cdot 10\ H_2O$ increases insignificantly in comparison with other complexes. When comparing the atom charges of the complexes, it can be concluded that fullerene π-orbitals in nanoparticle act as a reservoir for electrons. Thus, the charge of hydroxyfullerene OH group decreases. This increases the attraction of water molecule oxygen and the strength of hydrogen bond in complex $C_{60}[OH]_{10}$. $10\,H_2O$.

To estimate the size of the cluster with the center in nanoparticle $C_{60}[OH]_{10}$ that can be formed in water, the semi-empirical calculations of energies of water molecules adding to nanoparticle $C_{60}[OH]_2$ with chain formation were made (Figure 6.8). The semi-empirical method PM3 was used in calculations. Water molecules were bonded to the chain consequently; the energy was minimized after each molecule was added. Table 6.8 demonstrates the comparative results obtained for the chain growth both from the nanoparticle and water molecule. The binding energy of water molecule with the chain growing from $C_{60}[OH]_2$ over twice as big as similar energies for the chain growing from the water molecule. At the same time, the energy of hydrogen bond first increases before the addition of the third molecule, and then slowly decreases. When the sixth molecule is added, the energy decreases step-wise. Thus, the research demonstrates that hydroxyfullerene molecule forms a stable complex in water, being surrounded with six layers of water molecules. It is clear that the water structure changes at much longer distances. The introduction of insignificant number of nanoparticles results in complete reconstruction of aqueous medium, and, consequently, to sufficient changes in the properties of substances obtained on its base.

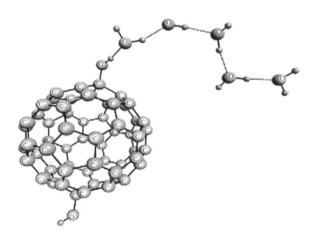

FIGURE 6.8 Model system formed by cluster $C_{60}[OH]_2$ and 5 water molecule chain. The optimization by energy was carried out with semi-empirical method PM3.

TABLE 6.8   Energies of binding water molecules to the chain connected with molecules $C_{60}[OH]_2$ and water chain (kJ/mol). Semi-empirical calculations with PM3 method.

n	$[H_2O]_n$	$C_{60}[OH]_2 + n \cdot H_2O$
1		16.99
2	8.33	20.21
3	9.96	21.00
4	10.63	20.42
5	10.71	19.71
6	11.13	12.72

In some experimental works carbon nanoparticles are obtained in gels of carbon polymers in the presence of salts of transition metals [13]. Simultaneously, nanocomposites containing metal nanoparticles in carbon envelopes are formed. The envelope structure contains many defects, but it mainly has an aromatic character. The nanoparticles considered are widely used for the modification of different substances; therefore, it is interesting to investigate the influence of metal nuclei on the interaction of carbon enveloped nanoparticles with polar fluids.

The systems demonstrated in Figures 6.9, 6.10, and 6.11 were considered as the nanoparticle model containing a metal nucleus and hydroxylated carbon envelope. In all the figures of the paper carbon and hydrogen atoms are not signed. At the same time, carbon atoms are marked with dark spheres, and hydrogen atoms with light ones. Clusters comprising two metal atoms are introduced into the system to model a metal nucleus. Carbon envelopes are imitated with 8-nucleus polyaromatic carbon clusters. The carbon envelope indicated in Figure 6.9 contains one pentagonal aromatic ring, and in the envelope in Figure 6.11 one heptagon. The envelope in Figure 6.10 contains only hexagonal aromatic rings. The corresponding numbers of hydrogen atoms are located on the boundaries of carbon clusters. Besides, one OH-group is present in each of the carbon envelopes. Thus, the chemical formula of hydroxylated carbon envelope with pentagonal defect $C_{23}H_{11}OH$ hydroxylated with defectless carbon envelope, $C_{27}H_{13}OH$, and for the hydroxylated carbon envelope with heptagonal defect, $C_{31}H_{15}OH$. Copper and nickel are taken as metals. The interaction of the considered model systems with water molecules is investigated in the paper.

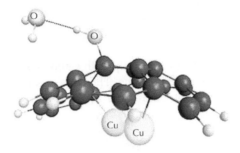

FIGURE 6.9 Complex $Cu_2C_{23}H_{11}OH \cdot H_2O$ with hexagonal defect in carbon envelope.

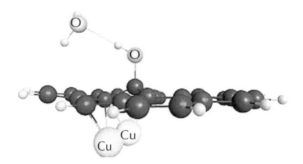

FIGURE 6.10 Complex $Cu_2C_{27}H_{13}OH \cdot H_2O$.

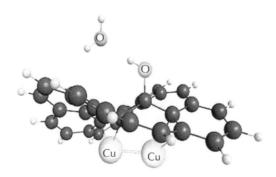

FIGURE 6.11 Complex $Cu_2C_{31}H_{15}OH \cdot H_2O$ with heptagonal defect in carbon envelope.

The calculations are made in the frameworks of density functional with exchange-correlative functional B3LYP. For all the atoms of the model system, apart from oxygen, the basis set 6-31G was applied. Polarization and diffusion functions were added to the basis set for more accurate calculations of hydrogen bond energy and interaction of hydroxyl group with carbon envelope. Thus, the basis set 6-31+G* was used for oxygen atoms. The basis sets from [14] were used for atoms of transition metals; the bases from software PC-GAMESS were applied for rest of the elements.

Hydrogen bond energy in water dimers calculated in this model exceeded 22 kJ/mol being a little more than the value obtained in the frameworks of the more accurate method—19.65 kJ/mol [15]. The experimental value is 14.9 ± 2 kJ/mol [15]. If we consider the energy of zero oscillations of dimer and water molecules in the frameworks of the calculation model, hydrogen bond energy decreases to 12.57 kJ/mol and becomes similar to the experimental value. When the basis extends due to polarization functions on hydrogen atoms, hydrogen bond energy goes down by 2%.

TABLE 6.9 Bond lengths (in angstrems).

Compound	C – O	O – H	O ··· H
$C_{23}H_{11}OH$	1.492	0.973	–
$Cu_2C_{23}H_{11}OH$	1.450	0.973	–
$Ni_2C_{23}H_{11}OH$	1.454	0.973	–
$C_{27}H_{13}OH$	1.545	0.974	–
$Cu_2C_{27}H_{13}OH$	1.469	0.973	–
$Ni_2C_{27}H_{13}OH$	1.464	0.972	–
$C_{31}H_{15}OH$	1.506	0.973	–
$Cu_2C_{31}H_{15}OH$	1.470	0.973	–
$Ni_2C_{31}H_{15}OH$	1.455	0.972	–
H_2O	–	0.969	–

TABLE 6.9 (*Continued*)

$C_{23}H_{11}OH \cdot H_2O$	1.470	0.981	1.967
$Cu_2C_{23}H_{11}OH \cdot H_2O$	1.439	0.984	1.869
$Ni_2C_{23}H_{11}OH \cdot H_2O$	1.440	0.985	1.893
$C_{27}H_{13}OH \cdot H_2O$	1.524	0.979	2.119
$Cu_2C_{27}H_{13}OH \cdot H_2O$	1.451	0.984	1.911
$Ni_2C_{27}H_{13}OH \cdot H_2O$	1.449	0.981	1.964
$C_{31}H_{15}OH \cdot H_2O$	1.490	0.979	2.024
$Cu_2C_{31}H_{15}OH \cdot H_2O$	1.453	0.981	1.964
$Ni_2C_{31}H_{15}OH \cdot H_2O$	1.446	0.979	1.997
$H_2O \cdot H_2O$	–	0.978	1.906

At the first stage of the investigation, we minimized the energy of all the considered structures in the presence of water molecule and without it. In Figures 6.9, 6.10, and 6.11 it can be seen that the optimized geometric structures for complexes containing copper cluster. The second column of Table 6.5 shows the lengths of $C-O$ bond formed by OH-group with carbon envelope atom for the calculated complexes; the third column shows the bond length in hydroxyl group and the forth shows the lengths of hydrogen bond for complexes with a water molecule.

The analysis of geometric parameters demonstrates that in all cases a metal presence results in the decrease of $C-O$ bond length formed by OH-group with the carbon envelope atom. The especially strong effect is observed in the defectless carbon envelope which is explained by a high stability of defectless graphite-like systems. The metal availability results in the increased activity of the carbon envelope due to the additional electron density and destabilization of aromatic rings. The defects of carbon grid have a similar effect. Simultaneously, the pentagonal defect is usually characterized by the electron density excess, and the heptagonal one by its lack. By comparing the two metals clusters considered, it is seen that for the system with pentagonal defect, copper influence is stronger than nickel one, and for rest of the systems it shows the opposite effect. A metal has only an insignificant influence on $O-H$ bond length.

The interaction with water molecule results in the increase of $O-H$ bond length in the hydroxyl group of carbon envelope and decrease of $C-O$ bond length. In the presence of metal the length of hydrogen bond decreases in all the cases, copper clusters cause the formation of shorter hydrogen bonds. In the calculations, it is taken into consideration the basis superpositional error (BSSE). Table 6 shows the interaction energies of the considered model systems with water molecule. In the table the energy of intermolecular interaction in water dimer calculated using the basis 6-31+G* is given in the table for comparison.

From the table, it is seen that the interaction of hydroxylated carbon envelopes with water molecule is significantly weaker than the intermolecular interaction in water dimer. The strongest hydrogen bond is formed in case of the envelope with pentagonal defect $C_{23}H_{11}OH$. This result correlates with the minimal length of $C-O$ bond in the envelope $C_{23}H_{11}OH$ (Table 6.9). Thus, hydrogen bond is getting weaker in the hydroxyl group and it provides the formation of a stronger hydrogen bond. The main conclusion from Table 6 is a considerable strengthening of hydrogen bond in metal systems. For complex $Cu_2C_{27}H_{13}OH \cdot H_2O$ with defectless carbon envelope the energy of hydrogen bond increases in 2.5 times in comparison with a similar complex without metal. The greatest energy of hydrogen bond is obtained for complex $Ni_2C_{27}H_{13}OH \cdot H_2O$. This energy exceeds the energy of hydrogen bond in water dimer by over 40%.

TABLE 6.10   Interaction energy of model clusters with water molecule (kJ/mol).

Compound	E_{int}
$C_{23}H_{11}OH \cdot H_2O$	18.12
$Cu_2C_{23}H_{11}OH \cdot H_2O$	26.06
$Ni_2C_{23}H_{11}OH \cdot H_2O$	31.56
$C_{27}H_{13}OH \cdot H_2O$	11.01
$Cu_2C_{27}H_{13}OH \cdot H_2O$	27.04
$Ni_2C_{27}H_{13}OH \cdot H_2O$	19.65
$C_{31}H_{15}OH \cdot H_2O$	15.08

TABLE 6.10 *(Continued)*

$Cu_2C_{31}H_{15}OH \cdot H_2O$	13.75
$Ni_2C_{31}H_{15}OH \cdot H_2O$	16.88
$C_{60}(OH)_2 \cdot H_2O$	21.98
$H_2O \cdot H_2O$	22.37

It should be noted that in the frameworks of model HF/6-31G the energy of hydrogen bond for hydroxyfullerene $C_{60}(OH)_{10} \cdot (H_2O)_{10}$ is higher than the energy of hydrogen bond in water dimer by 60% (Table 6.7). The semi-empirical calculations for complex $C_{60}(OH)_2 \cdot H_2O$ predict that the energy of hydrogen bond two times exceeds the one in water dimer. When calculating by the method described in the final part of this paper, the energy of hydrogen bond for hydroxyfullerene $C_{60}(OH)_{10} \cdot (H_2O)_{10}$ is higher by 37% than the bond energy in water dimer. Thus, taking into account diffusion and polarization functions we considerably change the results. The additional reason for the difference in the calculation of complex $C_{60}(OH)_{10} \cdot (H_2O)_{10}$ is the use of the method of hydrogen bond energy evaluation with breaking off all water molecules while preserving the system symmetry.

The main conclusion from the calculation results presented is a considerable increase in hydrogen bond energy with water molecule for carbon graphite-like nanoparticles containing the nucleus from transition metal atoms. Besides, the energy of C – O bond between carbon envelope and hydroxyl group increases in the presence of metal. Thus, the probability of the formation of hydroxylated nanoparticles increases. The nanoparticles considered have a strong orientating effect on water medium and become the embryos while forming materials with improved characteristics. A similar effect is observed when water molecule interacts with hydroxyfullerene. In this case, the interaction energy depends considerably on the number of OH–groups in the molecule. Thus, the interaction energy of water molecules with hydroxyfullerene in cluster $C_{60}(OH)_{10} \cdot (H_2O)_{10}$ exceeds the interaction energy in water dimer by 40–60% depending on the calculation model на 40–60%. At the same time, for complex $C_{60}(OH)_2 \cdot H_2O$ hydrogen bond is a little weaker than in water dimer.

6.1.3 Quantum—Chemical Investigation of Interaction between Fragments of Metal-Carbon Nanocomposites and Polymeric Materials Filled by Metal Containing Phase

There is a strong necessity to define formulations components processes conditions for nanostructured materials filled by metallic additives. Another task is optimization of components, nanocomposites and diluents combination and, in what follows, curing processes with determined temperature mode. The result of these arrangements will be materials with layerwise homogeneous metal particles or nanocomposites distribution formulation in ligand shell.

The result of these arrangements will be homogeneous metal particles or nanocomposites distribution in acetylacetone ligand shell. In the capacity of conductive filler silver nanoclusters or nanocrystals and copper-nickel or carbon nanocomposites can be used. It should be lead to decrease volume resistance from 10^{-4} to 10^{-6} Ωcm and increase of adhesive or paste adhesion.

Silver filler is more preferable due to excellent corrosion resistance and conductivity, but its high cost is serious disadvantage. Hence, alternative conductive filler, notably nickel-carbon nanocomposite was chosen to further computational simulation. In developed adhesive or paste formulations metal containing phase distribution is determined by competitive coordination and cross-linking reactions.

There are two parallel technological paths that consist of preparing blend based on epoxy resin and preparing another one based on polyethylene polyamine (PEPA). Then prepared blends are mixed and cured with diluent gradual removing. Curing process can be tuned by complex diluent contains of acetylacetone (AcAc), diacetonealcohol (DAcA) (in ratio AcAc:DAcA=1:1) and silver particles being treated with this complex diluent. Epoxy resin with silver powder mixing can lead to homogeneous formulation formation. Blend based on PEPA formation comprises AA and copper- or nickel-carbon nanocomposites introducing. Complex diluent removing tunes by temperature increasing to 150–160° C. Initial results formulation with 0.01% nanocomposite (by introduced metal total weight) curing formation showed that silver particles were self-organized and assembled into layer-chained structures. Acetylacetone (AcAc) infrared spectra were calculated with HyperChem software. Calculations were carried out by the use of semi-empirical methods PM3 and ZINDO/1 and *ab initio* method with 6-31G** basis set [16]. Results of calculation have been compared with instrumental measurement performed IR-Fourier spectrometer FSM-1201. Liquid adhesive components were preliminary stirred in ratio 1:1 and 1:2 for the purpose of interaction investigation. Stirring was carried out with magnetic mixer. Silver powder was preliminary crumbled up in agate mortar. Then acetylacetone was added. Obtained mixture was taken for sample after silver powder precipitation.

Sample was placed between two KBr glass plates with identical clamps. In the capacity of comparative sample were used empty glass plates. Every sample spectrum was performed for five times and the final one was calculated by striking an average. Procedure was carried out in transmission mode and then data were recomputed to obtain absorption spectra.

There are four possible states of AA: ketone, ketone-enol, enol A, and enol B (Figure 6.12). Vibrational analysis is carried out with one of semi-empirical of *ab initio* methods and can allow recognizing the different states of reagents by spectra comparison.

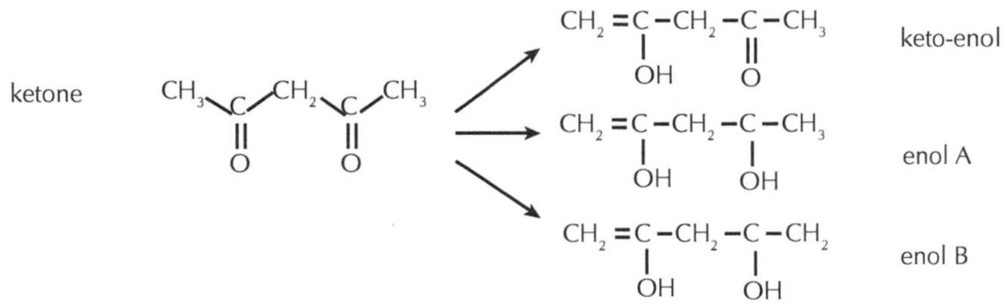

FIGURE 6.12   Different states of acetylacetone.

The vibrational frequencies are derived from the harmonic approximation, which assumes that the potential surface has a quadratic form. The association between transition energy ΔE and frequency ν is performed by Einstein's formula

$$\Delta E = h \cdot \nu \tag{6.2}$$

where, h is the Planck constant. IR frequencies ($\sim 10^{12}$Hz) accord with gaps between vibrational energy levels. Thus, each line in an IR spectrum represents an excitation of nuclei from one vibrational state to another [17].

Comparison between MNDO, AM1, and PM3, methods were performed in [18]. Semi-empirical method PM3 demonstrated the closest correspondence to experimental values.

Figure 6.13 showed experimental data of IR-Fourier spectrometer AA measurement and data achieved from the National Institute of Advanced Industrial Science and Technology open database. Peaks comparison is clearly shown both spectra generally identical. The IR spectra for comparison with calculated ones were received from open AIST database [1]. As it can be concluded from AIST data, experimental IR spectrum was measured for AA ketone state.

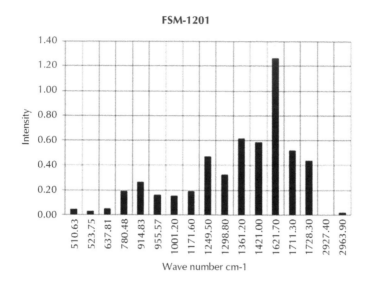

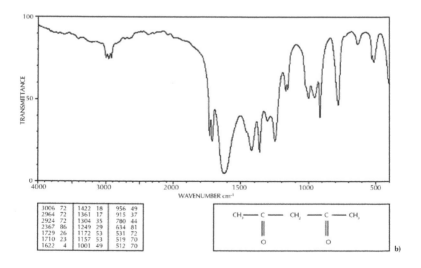

FIGURE 6.13   Experimental IR spectrum of acetylacetone obtained from IR-Fourier FSM-1201 spectrometer (a) and IR spectrum of acetylacetone received from open AIST database (b).

As it is seen, most intensive peak 1,622cm^{-1} belongs to C=O bond vibrations. Oxygen relating to this bond can participate in different coordination reactions. Peaks 915 cm^{-1} and 956cm^{-1} belong to O–H bond vibrations. This bond can also take part in many reactions.

Computational part is comprised of several steps. There were several steps in computational part. Firstly, geometry of AA was optimized to achieve molecule stable state. Then vibrational analysis was carried out. Parameters of the same computational method were used for subsequent IR spectrum calculation. Obtained IR spectra are shown in Figure 6.14.

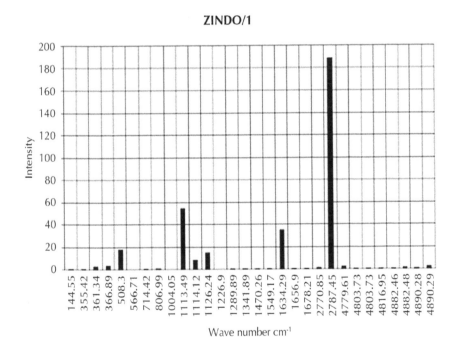

ZINDO/1

FIGURE 6.14   (*Continued*)

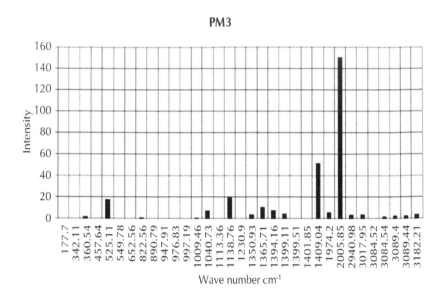

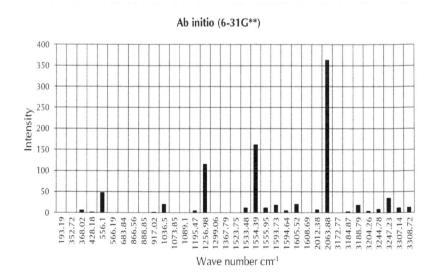

FIGURE 6.14   *(Continued)*

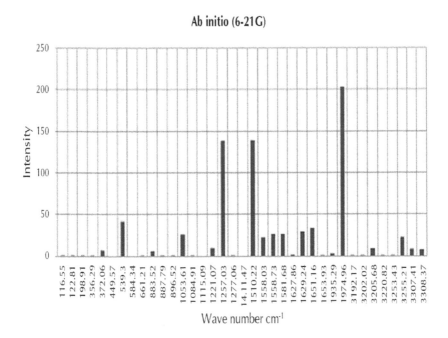

FIGURE 6.14   IR spectra of acetylacetone calculated with HyperChem software: semi-empirical method ZINDO/1 (a); semi-empirical method PM3 (b); ab initio method with 6-31G** basis set (c); ab initio method with 6-21G basis set (d).

Vibrational analysis performed by HyperChem showed that PM3 demonstrates more close correspondence with experimental and *ab initio* calculated spectra than ZINDO/1. As expected, *ab initio* spectra were demonstrated closest result in general. As it is seen in Figure 6.14c, d, peaks of IR spectrum calculated with 6-21G small basis set are staying closer with respect to experimental values than ones calculated with large basis set 6-31G**. Thus, it will be reasonable to use both PM3 and *ab initio* with 6-21G basis set methods in further calculations.

Ab initio (6-21G)

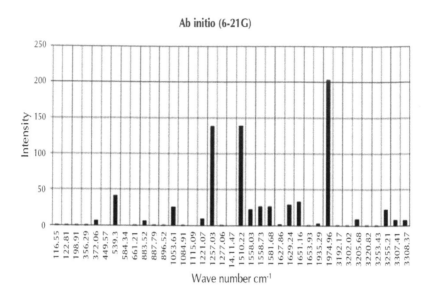

Wave number cm⁻¹

Ab initio (6-21G)

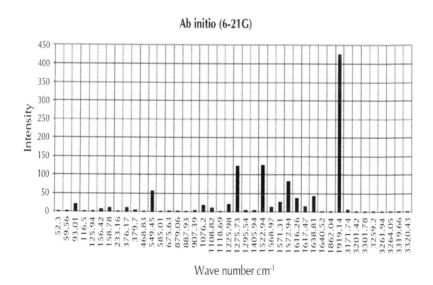

Wave number cm⁻¹

FIGURE 6.15 (*Continued*)

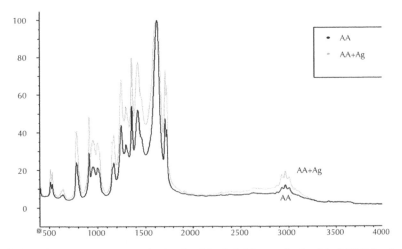

FIGURE 6.15   IR spectra obtained with ab initio method with 6-21G basis set: acetylacetone (a); acetylacetone with silver powder (b); experimental data obtained from IR-Fourier FSM-1201 spectrometer: acetylacetone with silver powder in comparison with pure acetylacetone (c).

As shown in Figure 6.15a, b, peaks are situated almost identical to each other with the exception of peak 1,573 cm^{-1} (Figure 6.15b) which was appeared after silver powder addition. Peak 1,919 cm^{-1} (Figure 6.15b) has been more intensive as compared with similar one on Figure 6.15a.

Peaks are found in range 600–700 cm^{-1} should usually relate to metal complexes with acetylacetone [19, 20]. The IR spectra AA and AA with Ag$^+$ ion calculations were carried out by *ab initio* method with 6-21G basis set. PM3 was not used because this method has not necessary parameterization for silver. In the case of IR spectra, *ab initio* computation unavoidable calculating error is occurred, hence, all peaks have some displacement. Peak 549 cm^{-1} (Figure 6.15b) is equivalent to peak 638 cm^{-1} (Figure 6.15c) obtained experimentally and it is more intensive than similar one on Figure 6.4. Peak 1,974 cm^{-1} was displaced to mark 1,919 cm^{-1} and became far intensive. It can be explained Ag+ influence and coordination bonds between metal and AA formation.

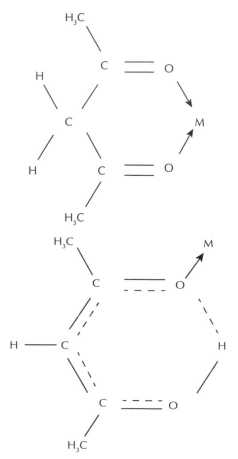

FIGURE 6.16   Different ways of complex between metal and AA formation: kentone with metal (a); enol with metal (b).

Complexes between silver and AA depending on enol or ketone state can form with two different ways (Figure 6.16).

IR spectra of pure acetylacetone and acetylacetone with silver were calculated with different methods of computational chemistry. PM3 method was shown closest results with respect to experimental data among semi-empirical methods. It was established that there is no strong dependence between sizes of *ab initio* method basis set and data accuracy. It was found that IR spectrum carried out with *ab initio* method with small basis set 6-21G has more accurate data than analogous one with large basis set 6-31G**. Experimental and calculated spectra were shown identical picture of spectral changes with except of unavoidable calculating error. Hence, PM3 and *ab initio* methods can be used in metal-carbon complexes formations investigations.

6.2 THE METAL OR CARBON NANOCOMPOSITES INFLUENCE MECHANISMS ON MEDIA AND ON COMPOSITIONS

The modification of materials by nanostructures including metal or carbon nanocomposites consists in the conducting of the following stages:

- The choice of liquid phase for the making of finely dispersed suspensions intended for the definite material or composition.
- The making of finely dispersed suspension with sufficient stability for the definite composition.
- The development of conditions of the finely dispersed suspension introduction into composition.

At the choice of liquid phase for the making of finely dispersed suspension, it should be taken the properties of nanostructures (nanocomposites) as well as liquid phase (polarity, dielectric penetration, and viscosity).

It is perspective if the liquid phase completely enters into the structure of material formed during the composition hardening process.

When the correspondent solvent on the base of which the suspension is obtained is evaporated, the re-coordination of nanocomposite on other components takes place and the effectiveness of action of nanostructures on composition is decreased.

The stability of finely dispersed suspension is determined on the optical density. The time of the suspension optical density conservation defines the stability of suspension. The activity of suspension is found on the bands intensity changes by means of IR and Raman spectra. The intensity increasing testifies to transfer of nanostructure surface energy vibration part on the molecules of medium or composition. The line speeding in spectra testify to the growth of electron action of nanocomposites with medium molecules. Last fact is confirmed by X-ray photoelectron investigations.

The changes of character of distribution on nanoparticles sizes take place depending on the nature of nanocomposites, dielectric penetration, and polarity of liquid phase.

Characteristics of finely dispersed suspensions of metal or carbon nanocomposites are given below.

The distribution of nanoparticles in water, alcohol, and water-alcohol suspensions prepared based on the above technique are determined with the help of laser analyzer. In Figures 6.17 and 6.18 one can see distributions of copper or carbon nanocomposite in the media different polarity and dielectric penetration.

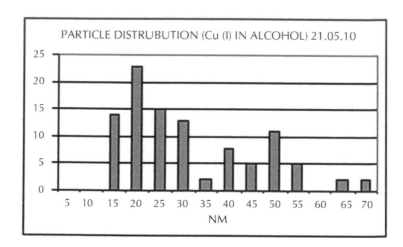

FIGURE 6.17   Distribution of copper or carbon nanocomposites in alcohol.

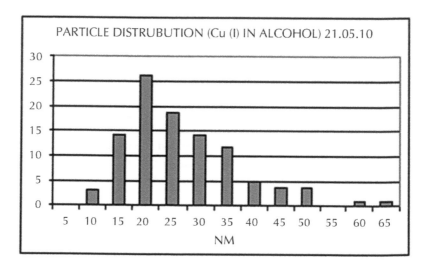

FIGURE 6.18   Distribution of copper or carbon nanocomposites in water.

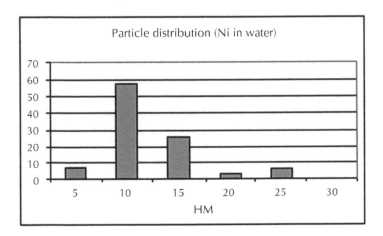

FIGURE 6.19   Distribution of nickel or carbon nanocomposites in water.

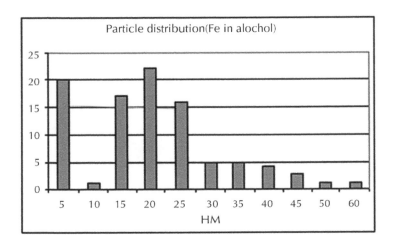

FIGURE 6.20   Distribution of iron or carbon nanocomposites.

When comparing the figures, it is seen that ultrasound dispergation of one and the same nanocomposite in media different by polarity results in the changes of distribution of its particles. In water solution the average size of Cu/C nanocomposite equals 20 nm, and in alcohol medium is greater than 5 nm.

Assuming that the nanocomposites obtained can be considered as oscillators transferring their oscillations onto the medium molecules, it can be determined, to what extent the IR spectrum of liquid medium will change, for example, polyethylene polyamine applied as a hardener in some polymeric compositions, when we introduce small and supersmall quantities of nanocomposite into it.

IR spectra demonstrate the change in the intensity at the introduction of metal or carbon nanocomposite in comparison with the pure medium (IR spectra are given in Figure 6.21). The intensities of IR absorption bands are directly connected with the polarization of chemical bonds at the change in their length, valence angles at the deformation oscillations, that is, at the change in molecule normal coordinates.

When nanostructures are introduced into media, it shows the changes in the area and intensity of bands that indicates the coordination interactions and influence of nanostructures onto the medium (Figures 6.21, 6.22, and 6.23).

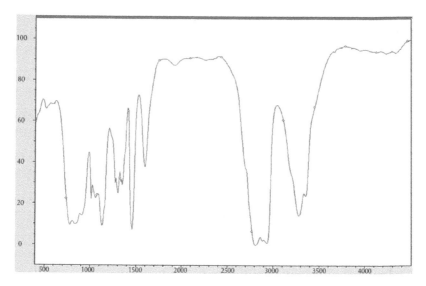

FIGURE 6.21   IR spectrum of polyethylene polyamine.

Special attention in PEPA spectrum should be paid to the peak at 1,598 cm^{-1} attributed to deformation oscillations of N–H bond, where hydrogen can participate in different coordination and exchange reactions.

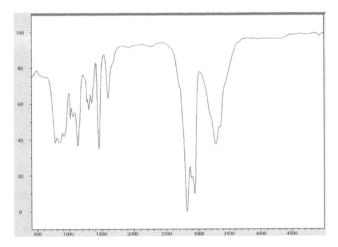

FIGURE 6.22   IR spectrum of copper or carbon nanocomposite finely dispersed suspension in polyethylene polyamine medium (ω (NC) = 1%).

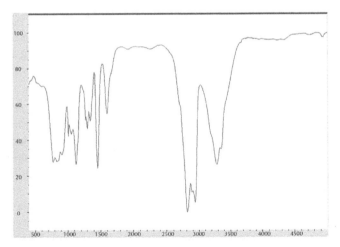

FIGURE 6.23   IR spectrum of nickel or carbon nanocomposite fine suspension in polyethylene polyamine (ω (NC) = 1%).

In the spectra wave numbers characteristic for symmetric $v_s(NH_2)$ 3,352 cm^{-1} and asymmetric $v_{as}(NH_2)$ 3,280 cm^{-1} oscillations of amine groups are present. There is a number of wave numbers attributed to symmetric $v_s(CH_2)$ 2,933cm^{-1} and asymmetric valence $v_{as}(CH_2)$ 2,803 cm^{-1}, deformation wagging oscillations vD(CH_2) 1,349 cm^{-1} of methylene groups, deformation oscillations of NHvD

(NH) 1,596 cm^{-1} and NH$_2$ νD(NH$_2$) 1,456 cm^{-1} amine groups. The oscillations of skeleton bonds at ν(CN) 1,059–1,273 cm^{-1} and ν(CC) 837 cm^{-1} are the most vivid. The analysis of intensities of IR spectra of PEPA and fine suspensions of metal or carbon nanocomposites based on it revealed a significant change in the intensities of amine groups of dispersion medium (for ν_s(NH$_2$) in 1.26 times, and for ν_{as}(NH$_2$) in approximately 50 times).

Such demonstrations are presumably connected with the distribution of the influence of nanoparticle oscillations onto the medium with further structuring and stabilizing of the system. Under the influence of nanoparticle the medium changes which is confirmed by the results of IR

spectroscopy (change in the intensity of absorption bands in IR region). Density, dielectric penetration, and viscosity of the medium are the determining parameters for obtaining fine suspension with uniform distribution of particles in the volume. Simultaneously, the structuring rate and, consequently, the stabilization of the system directly depend on the distribution by particle sizes in suspension. At the wide range of particle distribution by sizes, the oscillation frequency of particles different by size can significantly differ, in this connection; the distortion in the influence transfer of nanoparticle system onto the medium is possible (change in the medium from the part of some particles can be balanced by the other). At the narrow range of nanoparticle distribution by sizes the system structuring and stabilization are possible. With further adjustment of the components such processes will positively influence the processes of structuring and self-organization of final composite system determining physical-mechanical characteristics of hardened or hard composite system.

The effects of the influence of nanostructures at their interaction into liquid medium depends on the type of nanostructures, their content in the medium and medium nature. Depending on the material modified, fine suspensions of nanostructures based on different media are used. Water and water solutions of surface-active substances, plasticizers, and foaming agents (when modifying foam concretes) are applied as such media to modify silicate, gypsum, cement, and concrete compositions. To modify epoxy compounds and glues based on epoxy resins the media based on polyethylene polyamine, isomethyltetrahydrophthalic anhydride, toluene, and alcohol-acetone solutions are applied. To modify polycarbonates and derivatives of polymethyl methacrylate dichloroethane and dichloromethane media are used. To modify polyvinyl chloride compositions and compositions based on phenolformaldehyde and phenolrubber polymers alcohol or acetone-based media are applied. Fine suspensions of metal or carbon nanocomposites are produced using the above media for specific compositions. In IR spectra of all studied suspensions the significant change in the absorption intensity, especially in the regions of wave numbers close to the corresponding nanocomposite oscillations, is observed. Simultaneously, it is found that the effects of nanocomposite influence on liquid media (fine suspensions) decreases with time and the activity of the corresponding suspensions drops. The time period in which the appropriate activity of nanocomposites is kept, changes in the interval of 24 hr–1 month depending on the nanocomposite type and nature of the basic medium (liquid phase in which nanocomposites dispergate). For instance, IR spectroscopic investigation of fine suspension based on isomethyltetrahydrophthalic anhydride containing 0.001% of Cu/C nanocomposite indicates the decrease in the peak intensity, which sharply increased on the third day when nanocomposite was introduced (Figure 6.24).

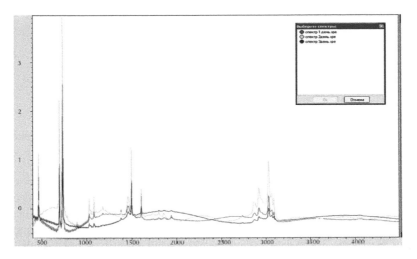

FIGURE 6.24   Changes in IR spectrum of copper or carbon nanocomposite fine suspension based on isomethyltetrahydrophthalic anhydride with time (a) IR spectrum on the first day after the nanocomposite was introduced, (b) IR spectrum on the second day, (c) IR spectrum on the third day.

Similar changes in IR spectra take place in water suspensions of metal or carbon nanocomposites based on water solutions of surface-active nanocomposites.

Figure 6.25 shows IR spectrum of iron or carbon nanocomposite based on water solution of sodium lignosulfonate in comparison with IR spectrum of water solution of surface-active substance.

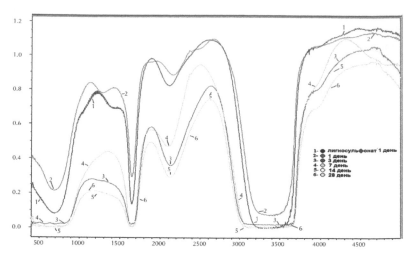

FIGURE 6.25   Comparison of IR spectra of water solution of sodium lignosulfonate (1) and fine suspension of iron or carbon nanocomposite (0.001%) based on this solution on the first day after nanocomposite introduction (2), on the third day (3), on the seventh day (4), 14th day (5) and 28th day (6).

As it is seen, when nanocomposite is introduced and undergoes ultrasound dispergation, the band intensity in the spectrum increases significantly. Also the shift of the bands in the regions 1,100–1,300 cm^{-1}, 2,100–2,200 cm^{-1} is observed, which can indicate the interaction between sodium lignosulfonate and nanocomposite. However, after 2 weeks the decrease in band intensity is seen. As the suspension stability evaluated by the optic density is 30 days, the nanocomposite activity is quite high in the period when IR spectra are taken. It can be expected that the effect of foam concrete modification with such suspension will be revealed if only 0.001% of nanocomposite is introduced.

6.3   THE COMPOSITIONS MODIFICATION PROCESSES BY METAL OR CARBON NANOCOMPOSITES

6.3.1   General Fundamentals of Polymeric Materials Modification by Metal or Carbon Nanocomposites

The material modification with the using of metal or carbon nanocomposites is usually carried out by finely dispersed suspensions containing solvents or components of polymeric compositions. It can be realized the modification of the following materials: concrete foam, dense concrete, water glass, polyvinyl acetate, polyvinyl alcohol, polyvinyl chloride, polymethyl methacrylate, polycarbonate, epoxy resins, phenol-formaldehyde resins, reinforced plastics, glues, pastes, including current conducting polymeric materials, and filled polymeric materials. Therefore, it is necessary to use the different finely dispersed suspension for the modification of enumerated materials. The series of suspensions consists of the suspensions on the basis of following liquids: water, ethanol, acetone, benzene, toluene, dichlorethane, methylene chloride, oleic acid, polyethylene polyamine, isomethyl tetra hydrophtalic anhydrite, water solutions surface-active substances, or plasticizers. In some cases, the solutions of correspondent polymers are applied for the making of the stable finely dispersed suspensions. The estimation of suspensions stability is given as the change of optical density during the definite time (Figure. 6.26).

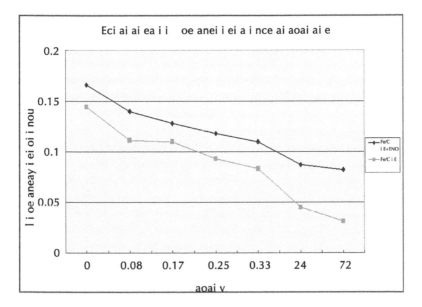

FIGURE 6.26   The change of optical density of typical suspension depending on time.

Relative change of free energy of coagulation process also may be as the estimation of suspension stability:

$$\Delta\Delta F = \lg k_{NC}/k_{NT} \tag{6.3}$$

where, k_{NC} – exp($-\Delta F_{NC}/RT$)—the constant of coagulation rate of nanocomposite, k_{NT} - exp($-\Delta F_{NT}/RT$)—the constant of coagulation rate of carbon nanotube, ΔF_{NC}, ΔF_{NT}—the changes of free energies of corresponding systems (Figure 6.27).

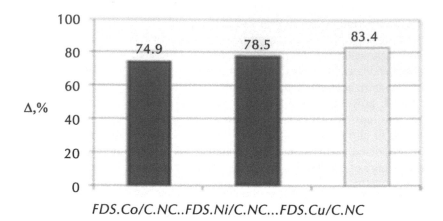

FIGURE 6.27   The comparison of finely dispersed suspensions stability.

The stability of metal or carbon nanocomposites suspension depends on the interactions of solvents with nanocomposites participation (Figure 6.28).

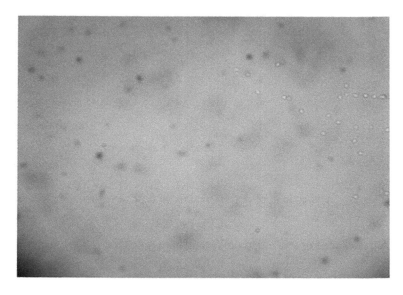

a

FIGURE 6.28   *(Continued)*

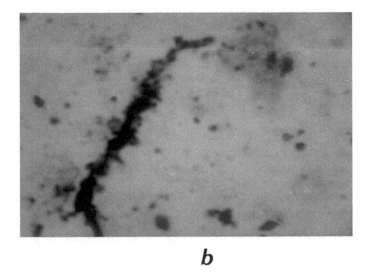

b

FIGURE 6.28 Microphotographs of Co/C nanocomposite suspension on the basis of mixture "dichlorethane–oleic acid" (a) and oleic acid (b).

The introduction of metal or carbon nanocomposites leads to the changes of kinematic and dynamic viscosity (Figure 6.29).

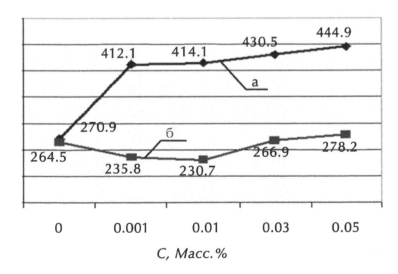

FIGURE 6.29 The dependence of dynamic and kinematic viscosity on the Cu/C nanocomposite quantity.

For the description of polymeric composition self organization, the critical parameters (critical content of nanostructures critical time of process realization, critical energetic action) may be used.

The equation of nanostructure influence on medium is proposed

$$W = n/N \exp\{an\tau^{\beta}/T\} \qquad (6.4)$$

where, n—number of active nanostructures, N—number of nanostructure interaction, a—activity of nanostructure, τ—duration of self organization process, T—temperature, β—degree of freedom (number of process direction).

6.3.2   About the Modification of Foam Concrete by Metal or Carbon Nanocomposites Suspension

The ultimate breaking stresses were compared in the process of compression of foam concretes modified with copper or carbon nanocomposites obtained in different nanoreactors of polyvinyl alcohol [21, 22]. The sizes of nanoreactors change depending on the crystallinity and correlation of acetate and hydroxyl groups in PVA which results in the change of sizes and activity of nanocomposites obtained in nanoreactors. It is observed that the sizes of nanocomposites obtained in nanoreactors of PVA matrixes 16/1(ros) (NC2), PVA 16/1 (imp) (NC1), PVA 98/10 (NC3), correlate as NC3 > NC2 > NC1. The smaller the nanoparticle size and greater its activity, the less amount of nanostructures is required for self-organization effect.

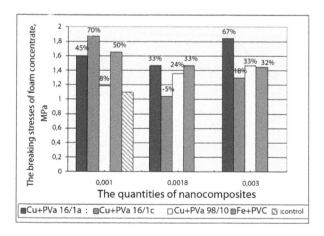

FIGURE 6.30   The dependence of breaking stresses on quantity (%) of metal or carbon nanocomposites.

Simultaneously, the oscillatory nature of the influence of these nanocomposites on the compositions of foam concretes is seen in the fact that if the amount of nanocomposite is 0.0018% from the cement mass, the significant decrease in the strength of NC1 and NC2 is observed. The increase in foam concrete strength after the modification with iron/carbon nanocomposite is a little smaller in comparison with the effects after the application of NC1 and NC2 as modifiers. The corresponding effects after the modification of cement, silicate, gypsum, and concrete compositions with nanostructures is defined by the features of components and technologies applied. These features often explain the instability of the results after the modification of the foregoing compositions with nanostructures. Besides, during the modification the changes in the activity of fine suspensions of nanostructures depending on the duration and storage conditions should be taken into account.

In this regard, it is advisable to use metal or carbon nanocomposites when modifying polymeric materials whose technology was checked on strictly controlled components.

In the paper the possibilities of developing new ideas about self-organization processes and about nanostructures and nanosystems are discussed on the example of metal or carbon nanocomposites. The obtaining of metal or carbon nanocomposites in nanoreactors of polymeric matrixes is proposed to consider as self-organization process similar to the formation of ordered phases which can be described with Avrami equation. The application of Avrami equations during the synthesis of nanofilm structures containing copper clusters has been tested. The influence of nanostructures on active media is given as the transfer of oscillation energy of the corresponding nanostructures onto the medium molecules.

The IR spectra of metal or carbon and their finely dispersed suspensions in different media (water and organic substances) have been studied for the first time. It has been found that the introduction of super small quantities of prepared nanocomposites leads to the significant change in band intensity in IR spectra of the media. The attenuation of oscillations generated by the introduction of nanocomposites after the time interval specific for the pair "nanocomposite medium" has been registered.

Thus, to modify compositions with finely dispersed suspensions it is necessary for the latter to be active enough that should be controlled with IR spectroscopy.

A number of results of material modification with finely dispersed suspensions of metal or carbon nanocomposites are given, as well as the examples of changes in the properties of modified materials based on concrete compositions, epoxy and phenol resins, polyvinyl chloride, polycarbonate, and current-conducting polymeric materials.

6.3.3   The Modification of Epoxy Resins by Metal or carbon Nanocomposites Suspensions

The IR spectra of polyethylene polyamine and metal or carbon nanocomposite fine suspensions based on it with concentrations 0.001–0.03% from polyethylene polyamine mass are analyzed and their comparative analysis is presented. The possible processes flowing in fine suspensions during the interaction of copper or carbon nanocomposite and polyethylene polyamine are described, as well as the processes influencing the increase in adhesive strength and thermal stability of metal or carbon nanocomposite or epoxy compositions.

This decade has been heralded by a large-scale replacement of conventional metal structures with structures from polymeric composite materials (PCM). Currently, we are facing the tendency of production growth of PCM with improved operational characteristics. In practice, PCM characteristics can be improved when applying modern manufacturing technologies, for example, the application of "binary" technologies of prepreg production [23], as well as the synthesis of new polymeric PCM matrixes or modification of the existing polymeric matrixes with different fillers.

The most cost-efficient way to improve operational characteristics is to modify the existing polymeric matrixes, therefore, currently the group of polymeric materials modified with nanostructures (NS) is of special interest. NS are able to influence the supermolecular structure, stimulate self-organization processes in polymeric matrixes in super small quantities, thus, contributing to the efficient formation of a new phase "medium modified nanocomposite" and qualitative improvement of the characteristics of final product—PCM. This effect is especially visible when NS activity increases which directly depends on the size of specific surface, shape of the particle and its ultimate composition [22]. Metal ions in the NS used in this work also contribute to the activity increase as they stimulate the formation of new bonds.

The increase in the attraction force of two particles is directly proportional to the growth of their elongated surfaces, therefore, the possibility of NS coagulation and decrease in their efficiency as modifiers increases together with their activity growth. This fact and the fact that the effective concentrations to modify polymeric matrixes are usually in the range below 0.01 mass % impose specific methods for introducing NS into the material modified. The most justified and widely applicable are the methods for introducing NS with the help of fine suspensions (FS).

Quantum-chemical modeling methods allow quite precisely defining the typical reaction of interaction between the system components, predict the properties of molecular systems, decrease the number of experiments due to the imitation of technological processes. The computational results are completely comparable with the experimental modeling results.

The optimal composition of nanocomposite was found with the help of software HyperChem v. 6.03. Cu/C nanocomposite is the most effective for modifying the epoxy resin. Nanosystems formed with this NC have higher interaction energy in comparison with nanosystems produced with Ni/C and Co/C nanocomposites. The effective charges and geometries of nanosystems were found with semi-empirical methods. The fact of producing stable fine suspensions of nanocomposite on PEPA basis and increasing the operational characteristics of epoxy polymer was ascertained. It was demonstrated that the introduction of Cu/C nanocomposite into PEPA facilitates the formation of coordination bonds with nitrogen of amine groups, thus resulting in PEPA activity increase in epoxy resin hardening reactions.

It was found out that the optimal time period of ultrasound processing of copper or carbon nanocomposite fine suspension is 20 min.

The dependence of Cu/C nanocomposite influence on PEPA viscosity in the concentration range 0.001–0.03% was found. The growth of specific surface of NC particles contributes to partial decrease in PEPA kinematic viscosity at concentrations 0.001%, 0.01% with its further elevation at the concentration 0.03%.

IR investigation of Cu/C nanocomposite fine suspension confirms the quantum-chemical computational experiment regarding the availability of NC interactions with PEPA amine groups. The intensity of these groups increased in several times when Cu/C nanocomposite was introduced.

The test for defining the adhesive strength and thermal stability correlate with the data of quantum-chemical calculations and indicate the formation of a new phase facilitating the growth of cross-links number in polymer grid when the concentration of Cu/C nanocomposite goes up. The optimal concentration for elevating the modified epoxy resin (ER) adhesion equals 0.003% from ER weight. At this concentration, the strength growth is 26.8%. From the concentration range studied, the concentration 0.05% from ER weight is optimal to reach a high thermal stability. At this concentration, the temperature of thermal destruction beginning increases up to 195°C.

Thus, in this work the stable fine suspensions of Cu/C nanocomposite were obtained. The modified polymers with increased adhesive strength (by 26.8%) and thermal stability (by 110°C) were produced based on epoxy resins and fine suspensions.

6.3.4 The Modification of Glues Based on Phenol-formaldehyde Resins by Metal or carbon Nanocomposites Finely Dispersed Suspension

In modern civil and industrial engineering, mechanical engineering and so on extra strong, rather light, durable metal structures, wooden structures, and light assembly structures are widely used. However, they deform, lose stability, and load-carrying capacity under the action of high temperatures, therefore, fire protection of such structures is important issue for investigation.

To decrease the flammability of polymeric coatings, it is advisable to create and forecast the properties of materials with external intumescent coating containing active structure-forming agents—regulators of foam cokes structure. At present, metal or carbon nanocomposites are such perspective modifiers. The introduction of nanocomposites into the coating can improve the material behavior during the combustion, retarding the combustion process [24, 25].

In this regard, it is advisable to modify metal or carbon nanocomposites with ammonium phosphates to increase the compatibility with the corresponding compositions. The grafting of additional functional groups is appropriate in strengthening the additive interaction with the matrix and, thus, in improving the material properties. The grafting can contribute to improving the homogeneity of nanostructure distribution in the matrix and suspension stability. Due to partial ionization, additional groups produce a small surface charge with the result of nanostructure repulsion from each other and stabilization of their dispersion. Phosphorylation leads to the increase in the influence of nanostructures on the medium and material being modified, as well as to the improvement of the quality of nanostructures due to metal reduction [26–28].

The glue BF-19 (based on phenol-formaldehyde resins) is intended for gluing metals, ceramics, glass, wood, and fabric in hot condition, as well as for assembly gluing of cardboard, plastics, leather, and fabrics in cold condition. The glue composition: organic solvent, synthetic resin (phenol-formaldehyde resins of new lacquer type), synthetic rubber.

When modifying the glue composition, at the first stage the mixture of alcohol suspension (ethyl alcohol + Me/C NC modified) and ammonium polyphosphate (APPh) was prepared. At the same time, the mixtures containing ethyl alcohol, Me/C NC and APPh, ethyl alcohol and APPh were prepared. At the second stage, the glue composition was modified by the introduction of phosphorus containing compositions prepared into the glue BF-19.

The interaction of nanocomposites with ammonium polyphosphate results in grating phosphoryl groups to them that allowed using these nanocomposites to modify intumescent fireproof coatings.

The sorption ability of functionalized nanocomposites was studied. It was found out that phospholyrated nanostructures have higher sorption ability than nanocomposites not containing phosphorus. Since, the sorption ability indicates the nanoproduct activity degree, it can be concluded that nanocomposite activity increases in the presence of active medium and modifier (ammonium polyphosphate).

IR spectra of suspensions of nanocomposites on alcohol basis were investigated. It was found that changes in IR spectrum by absorption intensity indicate the presence of excitation source in the suspension. It was assumed that nanostructures are such source of band intensity increase in suspension IR spectra.

The glue coatings were modified with metal or carbon nanocomposite. It was determined that nanocomposite introduction into the glue significantly decreases the material flammability. The samples with phospholyrated nanocomposites have better test results. When phosphorus containing nanocomposite is introduced into the glue, foam coke is formed on the sample surface during the fire exposure. The coating flaking off after flame exposure was not observed as the coating preserved good adhesive properties even after the flammability test.

The nanocomposite surface phospholyration allows improving the nanocomposite structure, increases their activity in different liquid media thus increasing their influence on the material modified. The modification of coatings with nanocomposites obtained finally results in improving their fire-resistance and physical and chemical characteristics.

6.3.5 The Modification of Polycarbonate by Metal or carbon Nanocomposites

Recently, the materials based on polycarbonate have been modified to improve their thermal-physical and optical characteristics and to apply new properties to use them for special purposes. The introduction of nanostructures into materials facilitates self-organizing processes in them. These processes depend on surface energy of nanostructures which is connected with energy of their interaction with the surroundings. It is known [21] that the surface energy and activity of nanoparticles increase when their sizes decrease.

For nanoparticles the surface and volume are defined by the defectiveness and form of conformation changes of film nanostructures depending on their crystallinity degree. However, the possibilities of changes in nanofilm shapes with the changes in medium activity are greater in comparison with nanostructures already formed. At the same time, sizes of nanofilms formed and their defectiveness, that is, tears and cracks on the surface of nanofilms play an important role [22]

When studying the influence of supersmall quantities of substances introduced into polymers and considerably changing their properties, apparently, we should consider the role which these substances play in polymers possessing highly organized supermolecular regularity both in crystalline and amorphous states. It can be assumed that the mechanism of this phenomenon is in the nanostructure energy transfer to polymer structural formations through the interface resulting in the changes in their surface energy and mobility of structural elements of the polymeric body. Such mechanism is quite realistic as polymers are structural-heterogenic (highly dispersed) systems.

Polycarbonate modification with supersmall quantities of Cu/C nanocomposite is possible using fine suspensions of this nanocomposite which contributes to uniform distribution of nanoparticles in polycarbonate solution.

Polycarbonate "Actual" was used as the modified polycarbonate.

Fine suspensions of copper or carbon nanocomposite were prepared combining 1.0, 0.1, 0.01, and 0.001% of nanocomposite in polycarbonate solution in ethylene dichloride. The suspensions underwent ultrasonic processing.

To compare the optical density of nanocomposite suspension in polycarbonate solution in ethylene dichloride as well as polycarbonate and polycarbonate samples modified with nanocomposites, spectrophotometer KFK-3-01a was used.

Samples in the form of modified and non modified films for studying IR spectra were prepared precipitating them from suspension or solution under vacuum. The obtained films about 100mcm thick were examined on Fourier-spectrometer FSM 1201.

To investigate the crystallization and structures formed the high-resolution microscope (up to 10 mcm) was used.

To examine thermal-physical characteristics the lamellar material on polycarbonate and polymethyl methacrylate basis about 10mm high was prepared. Three layers of polymethyl methacrylate and two layers of polycarbonate were used. Thermal-physical characteristics (specific thermal capacity and thermal conductivity) were investigated on calorimeters IT-c-400 and IT-λ-400.

During the investigation, the nanocomposite fine suspension in polycarbonate solution in ethylene dichloride was studied; polycarbonate films modified with different concentrations of nanocomposite were compared with the help of optical spectroscopy, microscopy, IR spectroscopy, and thermal-physical methods of investigation.

The results of investigation of optical density of Cu/C nanocomposite fine suspension (0.001%) based on polycarbonate solution in ethylene dichloride are given in Figure 6.31. As seen in the figure, the introduction of nanocomposite and polycarbonate into ethylene dichloride resulted in transmission increase in the range 640–690 nm (approximately in three times). At the same time, at 790 nm and 890nm the significant increase in optical density was observed. The comparison of optical densities of films of polycarbonate and modified materials after the introduction of different quantities of nanocomposite (1.0; 0.1; 0.01%) into polycarbonate is interesting (Figure 6.32).

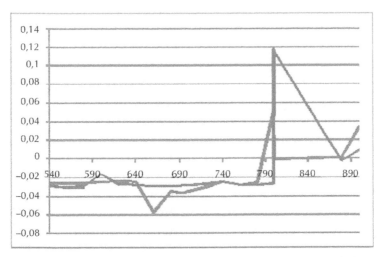

FIGURE 6.31   Curves of optical density of suspension based on ethylene dichloride diluted with polycarbonate and Cu/C nanocomposite in the concentration to polycarbonate 0.001% (1) and ethylene dichloride (2).

Comparison of optical density of suspension containing 0.001% of nanocomposite and optical density polycarbonate sample modified with 0.01% of nanocomposite indicates the proximity of curves character. Thus, the correlation of optical properties of suspensions of nanocomposites and film materials modified with the same nanocomposites is quite possible.

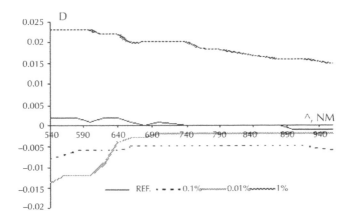

FIGURE 6.32   Curves of optical density of reference sample modified with Cu/C nanocomposites in concentration 1, 0.1, and 0.01%.

Examination of curves of samples optical density demonstrated that when the nanostructure concentration was 1% from polycarbonate mass, the visible-light spectrum was absorbed by about 4.2% more if compared with the reference sample. When the nanocomposite concentration was 0.1% the absorption decreased by 0.7%. When the concentration was 0.01%, the absorption decreased in the region 540–600 nm by 2.3% and in the region 640–960 nm by 0.5%.

During the microscopic investigation of the samples the schematic picture of the structures formed was obtained at 20-mcm magnification. The results are given in Figure. 6.33. From the schematic pictures it is seen that volumetric structures of regular shape surrounded by micellae were formed in polycarbonate modified with 0.01% Cu/C nanocomposite. When 0.1% of nanocomposite was introduced, the linear structures distorted in space and surrounded by micellae were formed. When Cu/C nanocomposite concentration was 1%, large aggregates were not observed in polycarbonate.

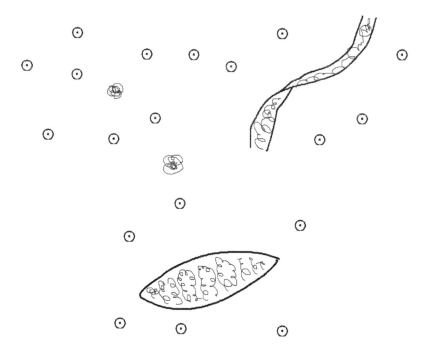

FIGURE 6.33   Schematic picture of structure formation in polycarbonate modified with Cu/C nanocomposites in concentrations 1, 0.1, and 0.01% (from left to the right), 20-mcm magnification.

Apparently, the decrease in nanocomposite concentration in polycarbonate can result in the formation of self-organizing structures of bigger size. In [3] there is the hypothesis on the transfer of nanocomposite oscillations onto the molecules of polymeric composition, the intensity of bands in IR spectra of which sharply increases even after the introduction of super small quantities of nanocomposites. In this case, hypothesis was checked on modified and non-modified samples of polycarbonate films.

In Figure 6.34, the IR spectra of polycarbonate and polycarbonate modified with 0.001% of Cu/C nanocomposite.

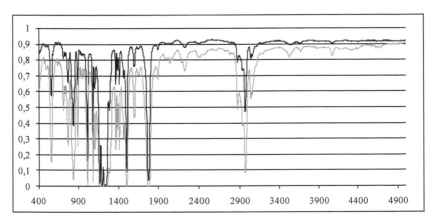

FIGURE 6.34 IR spectra of reference sample (upper) and polycarbonate modified with Cu/C nanocomposites in concentration 0.001% (lower).

As seen in IR spectra, the intensity increases in practically all regions, indicating the influence of oscillations of Cu/C nanocomposite on all the system. The most vivid changes in the band intensity are observed at 557cm^{-1}, in the region 760–930 cm^{-1}, at 1,400 cm^{-1}, 1,600 cm^{-1}, and 2,970 cm^{-1}.

Thus, polycarbonate self-organization under the influence of Cu/C nanocomposite takes place with the participation of certain bonds, for which the intensity of absorption bands goes up. Simultaneously, the formation of new phases is possible which usually results in thermal capacity increase. But thermal conductivity can decrease due to the formation of defective regions between the aggregates formed.

The table contains the results of thermal-physical characteristics studied.

TABLE 6.11   Thermal-physical characteristics of polycarbonate and its modified analogs.

Parameters	Nanocomposite content in polycarbonate, %			
	0%, ref.	1%	0.1%	0.01%
m 10^{-3}, kg	1.9	1.955	1.982	1.913
h 10^{-3}, m	9.285	9.563	9.76	9.432
Csp, J/kg K	1440	1028	1400	1510
□	0.517	0.503	0.487	0.448

It is demonstrated that when the nanostructure concentration in the material decreases, thermal capacity goes up that is confirmed by the results of previous investigations. Thermal conductivity decline, when the nanostructure concentration decreases, is apparently caused by the material defectiveness.

When Cu/C nanocomposites are introduced into the material modified, the nanostructures can be considered as the generator of molecules excitation, which results in wave process in the material.

It is found that polycarbonate modification with metal or carbon containing nanocomposites results in the changes in polycarbonate structure influencing its optical and thermal-physical properties.

KEYWORDS

- Hartree–Fock method
- Hydroxyfullerene molecule
- Metal-carbon nanocomposites
- PC-GAMESS
- Quantum-chemical investigations

REFERENCES

1. Schmidt, M. W., Baldridge, K. K., & Boatz, J. A., et al. (1993). *Journal of Computational Chemistry, 14,* 1347–1363.
2. Copley, J. R. D., Neumann, D. A., Cappelletti, R. L., & Kamitakahara, W. A. (1992). *Journal of Physical Chemistry of Solids, 53*(11), 1353–1371.
3. Lo, Sh., & Li, V.(1999). *Russian Chemical Journal, 5,* 40–48.
4. Ponomarev, A. N. (2007). *Constructional Materials, 6,* 69–71.
5. Krutikov,V. A., Didik, A. A., Yakovlev, G. I., Kodolov, V. I., & Bondar, A.Yu. (2005). *Alternative Power Engineering and Ecology, 4,* 36–41.
6. Granovsky, A. A. PC GAMESS version 7.1.5 (http://classic.chem.msu.su/gran/gamess/index.html).
7. Schmidt, M. W., Baldridge, K. K., Boatz, J. A., et al. (1993). *Journal Computational Chemistry, 14,* 1347–1363.
8. Bürgi, T., Droz, T., & Leutwyler, S. (1995). *Chemical Physics Letters, 246,* 291–299.
9. Matsumoto, Y., Ebata, T., & Mikami, N. (1998). *Journal of Chemical Physics, 109,* 6303–6311.
10. Schemmel, D., & Schütz, M. (2008). *Journal of Chemical Physics, 129,* 034301.
11. Famulari, A., Raimondi, M., Sironi, M., & Gianinetti, E. (1998). *Chemical Physics, 232,* 275–287.
12. Hedberg, K., Hedberg, L., & Bethune, D. S., et al. (1991). *Science, 254,* 410–412.
13. Copley, J. R. D., Neumann, D. A., Cappelletti, R. L., & Kamitakahara, W. A. (1992). *Journal of Physics and Chemistry Solids, 53,* 1353–1371.
14. Lipanov, A. M., Kodolov, V. I., Khokhriakov, N. V., Didik, A. A., Kodolova, V. V., & Semakina, N. V. (2005). *Alternative Power Engineering and Ecology, 2,* 58–63
15. Valiev, M., Bylaska, E. J., Govind, N., Kowalski, K., Straatsma, T. P., van Dam, H. J. J., Wang, D., Nieplocha, J., Apra, E., Windus, T. L., & de Jong, W. A. (2010)., *Computer Physics Communication, 181(9),* 1477–1489.
16. Famulari, A., Raimondi, M., Sironi, M., & Gianinetti, E. (1998). *Chemical Physics, 232,* 289–298
17. Spectral database for organic compounds AIST. http://riodb01.ibase.aist.go.jp/sdbs/cgi-bin/cre_index.cgi?lang=eng.
18. HyperChem 8 manual. (2002). *Computational Chemistry, 149,* 369.
19. Seeger, D. M., Korzeniewski, C., & Kowalchyk, W. (1991). *Jounal of Physical Chemistry, 95,* 68–71.
20. Kazicina, L. A., & Kupletskaya, N. B. (1971). *Application of UV-, IR- and NMR spectroscopy in organic chemistry* (p. 264). Moscow: Vyshaya Shkola.
21. Nakamoto, K. (1966) *Infrared spectra of inorganic and coordination compounds* (p. 412). New York: Wiley.
22. Kodolov, V. I., & Khokhriakov, N. V. (2009). *Chemical physics of formation and transformation processes of nanostrcutures and nanosystems* (Vol.1, p. 365; Vol. 2, p. 4153). Izhevsk: Izhevsk State Agricultural Academy.
23. Kodolov, V. I., Khokhriakov, N. V., Trineeva, V. V., & Blagodatskikh, I. I. (2008). Activity of nanostructures and its expression in nanoreactors of polymeric matrixes and active media. *Chemical Physics and Mesoscopy, 10*(4), 448–460.

24. Panfilov, B. F. (2010). Composite materials: production, application, market tendencies. *Polymeric Materials,* (2–3), 40–43.
25. Bulgakov, V. K., Kodolov, V. I., & Lipatov, A. M. (1990). *Modeling of polymeric materials combustion* (p. 238). Moscow: Khimia.
26. Shuklin, S. G., Kodolov, V. I., Larionov, K. I., & Tyurin, S. A. (1995). Physical and chemical processes in modified two-layer fire- and heat-resistant epoxy-polymers under the action of fire sources. *Physics of Combustion and Explosion, 31*(2), 73–79.
27. Glebova, N. V., & Nechitalov, A. A. (2010). Surface functionalization of multi-wall carbon nanotubes. *Journal of Technical Physics, 36*(19), 12–15.
28. Patent 2393110 Russia Technique of obtaining carbon metal containing nanostructures/Kodolov V. I., Vasilchenko Yu. M., Akhmetshina L. F., Shklyaeva D. A., Trineeva V. V., Sharipova A. G., Volkova E. G., Ulyanov A. L., Kovyazina O. A.; declared on 17.10.2008, published on 27.06.2010.
29. Patent 2337062 Russia Technique of obtaining carbon nanostructures from organic compounds and metal containing substances/Kodolov V. I., Kodolova V. V. (Trineeva), Semakina N. V., Yakovlev G. I., Volkova E. G. et al; declared on 28.08.2006, published on 27.10.2008.
30. Kodolov, V. I., Trineeva, V. V., Kovyazina, O. A., & Vasilchenko, Yu. M. (2012). Production and application of metal or carbon nanocomposites. *The problems of nanochemistry for the creation of new materials* (pp. 17–22). Torun Poland: IEPMD.
31. Kodolov, V. I., & Trineeva, V. V. (2012). Perspectives of idea development about nanosystems self-organization in polymeric matrixes. The problems of nanochemistry for the creation of new materials (pp. 75–100). Torun Poland: IEPMD.

CHAPTER 7

THE INVESTIGATION OF NANOCOMPOSITES ELECTRON STRUCTURE INFLUENCE CHANGES ON DIFFERENT MEDIA AND COMPOSITIONS: PART I

I. N. SHABANOVA, G. V. SAPOZHNIKOV, and N. S. TEREBOVA

CONTENTS

7.1   THE INVESTIGATION OF POLYMERIC AND BIOLOGICAL MATE-RIALS MODIFIED BY METAL OR CARBON NANOCOMPOSITES

7.1.1   Modification of Polymer Coating with D-metals and Carbon Copper-containing Nanostructures

To improve the service properties of materials, the modification of materials with nanostructures is used.

In the present paper, the formation of chemical bond between the atoms of polymer coating components and carbon metal-containing nanostructures with addition of silver or zinc is studied for improving electrical conduction in coatings.

The analysis has been conducted by the X-ray photoelectron spectroscopy method (XPS). The XPS allows to investigate an electronic structure, chemical bond, and nearest surrounding of an atom.

Experiment

The investigation was conducted on a unique automated X-ray electron magnetic spectrometer with double focusing allowing the investigation of samples in both solid and liquid state, which has the following performance specifications: the resolution is 10^{-4} and luminosity is 0.185% [1].

The electron magnetic spectrometer has a number of advantages in comparison with electrostatic spectrometers, which are the constancy of luminosity and resolution independent of the electron energy, and high contrast of spectra. The XPS method is a non destructive method of investigation which is especially important for studying metastable systems.

Two samples of polymer coatings were studied:

- Silver-containing coating modified with carbon metal-containing nanostructures (70% Ag and 1% nanostructures);
- zinc-containing coating modified with carbon metal-containing nanostructures (60% Zn, and 1% nanostructures).

The carbon metal-containing nanostructure samples were prepared by low-temperature synthesis. Carbon copper-containing nanostructures were mixed with polyethylene-polyamine $H_2N[CH_2-CH_2-NH]_mH$, where m = 1–8, through mechanic activation.

For improving the interaction of the polymer coating and carbon metal-containing nanostructures, the interaction of nanostructures with d-metals, that is, Ag and Zn, was used.

The spectra of the C1s, Ag3d, Ag3p, and Zn2p core levels and the spectra of valence bands of the composites prepared were studied. The composites were carbon metal-containing nanostructures containing silver, a polymer coating with addition of carbon copper-containing nanostructures functionalized with silver. The spectra of reference samples, that is, Ag, Zn, and carbon copper-containing nanostructures were studied as well. The cleanness of the sample surface was controlled by the O1s-spectrum. At heating to 300°C, oxygen on the surface is absent.

The decomposition of the spectra into components was performed with the help of a program based on the least squares method. For the spectrum decomposition, energy position, width of spectrum components and components' intensity based on the data obtained from the reference samples spectra were entered into the program. The accuracy in the determination of the peak positions was 0.1 eV. The error in the determination of the contrast of the electronic spectra was less than 5%.

Results and Discussion

The C1s-spectrum (Figure 7.1a) of carbon copper-containing nanostructures consists of three components: C-Cu (283 eV), C–H (285 eV) and C–O (287 eV).

In the C1s-spectrum (Figure 7.1b) of the studied composite consisting of carbon copper-containing nanostructures functionalized with silver, the intensity of the first component C–Me significantly increases and the binding energy changes (283.8 eV), which corresponds to the data from [2] and is associated with a larger localization of d-electrons in silver than that in copper.

In [3], a model is offered, which describes the connection of the parameters of multiplet splitting of the XPS 3s-spectra with a spin state, namely, the relative intensity of the maxima of the multiplets in 3s-spectra correlates with the value of the magnetic moment of atoms in d-metal systems; distance between the maxima of the multiplets (Δ) provide the information about the exchange interaction of the 3s-3d shells. The change of their overlapping is associated with changes in the distance between atoms and average atomic volume. The presence of changes in the shape of 3s-spectra provides information about the changes in chemical bond of atoms in nearest surrounding in the composite.

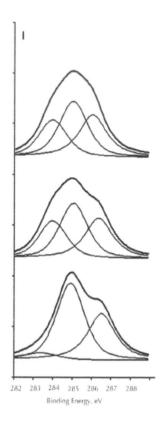

FIGURE 7.1 C1s spectrum (a C1s spectrum of the copper or carbon nanocomposite, b C1s spectrum of the copper or carbon nanocomposite with silver, c C1s spectrum coating filled by silver and modified by copper or carbon nanocomposite.

The analysis of the parameters of the multiplet splitting in the Cu3s- and Ag4s-spectra shows the presence of the atomic magnetic moments on copper (1.6мB, Δ = 2.5eV) and argentum atoms (2.2мB, Δ = 2.0eV). In contrast to carbon copper-containing nanostructures, decrease in distance between the multiplets indicates the enhancement of the chemical bond between copper d-electrons and carbon p-electrons and the formation of a strong covalent bond. The appearance of magnetic moment on silver atoms is associated with the formation of non compensated d-electrons on silver atoms and the involvement of d-electrons of silver atoms into the covalent bond with carbon atoms.

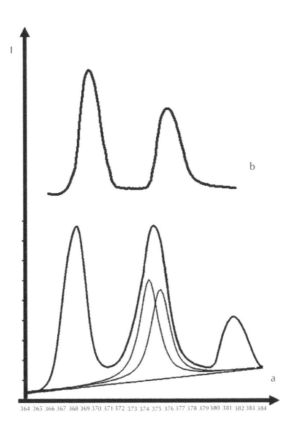

Binding Energy, eV

FIGURE 7.2 Ag3d spectrum (a Ag3d spectrum of the carbon copper-containing nanostructures during interaction with silver, b Ag3d spectrum of coating).

The pure silver Ag3d$_{5/2, 3/2}$ spectra are formed by two components of the spin-orbital splitting (Figure 7.2a). Similar situation is observed in carbon copper-containing nanostructures functionalized with silver (Figure 7.2b).

The complex composite Ag3d-spectrum (Figure 7.2) consists of the following maxima: First—Ag or Ag–C–Cu, and at the distance of 7eV from it, there is a less intensive maximum corresponding to an ionic component of the chemical bond between silver and carbon.

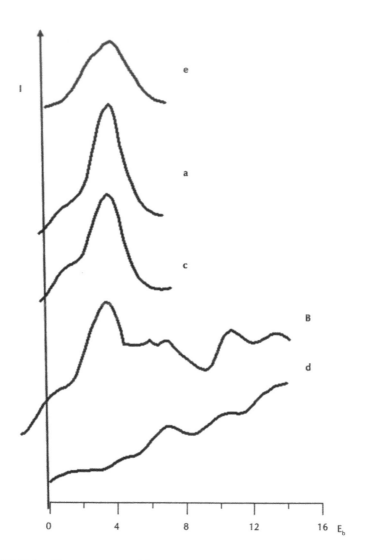

FIGURE 7.3   The spectrum of the valence band a spectrum of valence band nanostructures interacted with silver, b spectrum of valence band coating, c spectrum of valence band Ag$_2$O, d spectrum of valence band C, e spectrum of valence band pure Ag.

To clarify the nature of the first Ag or Ag–C component, the spectrum of valence band of the studied composite has been taken (Figure 7.3). In contrast to pure silver, the growth of an additional maximum at E_f and a similar maximum in the Ag_2O valence band indicates the hybridization of Ag4d5s-electrons with 2p-electrons of the second component—carbon. The shape and intensity of the valence band at a distance of 7 eV is associated with the presence of Ag–C–O component, which is confirmed by the coincidence of them with the maxima in valence band of pure graphite [4] and calculations of the oxygen density of states.

The increase of the density of states at E_f leads to the growth of the electron density of the polymer composite, and, correspondingly, of the electrical conduction in it, similar to that observed for Ag_2O which is used for increasing conduction and is an obligatory element in fabricating conductive glass [5].

The data on the measurement of electrical resistance shows that when 1% carbon metal-containing nanostructures are added in polymer, the electrical resistance decreases from 10^{-4} to 10^{-5} $\Omega \cdot cm$.

CONCLUSIONS

It is shown that nano-modification of the argentum-containing polymer coating improves the electrical conduction in it by one order of magnitude.

The analysis of the X-ray photoelectron spectra of the nano-modified composite shows the presence of the covalent bond of the Ag–C–Cu atoms and the growth of the electron density at E_f due to the hybridization of the C2p- and Ag4d5sp-electrons.

KEYWORDS

- **Binding energy**
- **C1s-spectrum**
- **Hybridization**
- **Polymer Coating**
- **Spectrum of valence band**

REFERENCES

1. Trapeznikov, V. A., Shabanova, I. N., Varganov, D. V., & Dobysheva, L. V., et al. (1986). New automated X-ray electron spectrometers: Spectrometers with technologi-

cal adapters and manipulators, spectrometer for studying melts. *Izvestiya Akademii Nauk SSSR. Seriya Matematicheskaya, 50*(9), 1677–1682.

2. Nefedov, V. I. (1984). *X-ray photoelectron spectroscopy of chemical compounds* (p. 255). Moscow: Khimia.

3. Lomova, N. V., & Shabanova, I. N. (2004). The study of the electronic structure and magnetic properties of invar alloys based on transition metals. *Journal of Electron Spectroscopy and Related Phenomena, 137–140,* 511–517.

4. Klobova, K. M., Shabanova, I. N., Kulyabina, O. A., et al. (1981). Investigation of the electronic structure of globular graphite in aluminum cast iron. *FMM, 51*(4), 890–893.

5. Gordienko, A. B., Zhuravlev, Yu. N., & Fedorov, D. G. (2007). Band structure and chemical bonding in Cu_2O and Ag_2O oxides. *Physics of Solid State, 49*(2). 223–228.

CHAPTER 8

THE INVESTIGATION OF NANOCOMPOSITES ELECTRON STRUCTURE INFLUENCE CHANGES ON DIFFERENT MEDIA AND COMPOSITIONS: PART II

I. N. SHABANOVA, N. S. TEREBOVA, and E. A. NAIMUSHINA

CONTENTS

8.1 DETERMINATION OF AN OPTIMAL MODIFIER FOR ALBUMIN AND MEDICOBIOLOGICAL TECHNOLOGY BY THE X-RAY PHOTO-ELECTRON SPECTROSCOPY METHOD

8.1.1 Introduction

The modification of functional groups in the composition of a protein macromolecule is one of the approaches in the development of biotechnology for pharmaceutics. The virus protection of pharmaceutical preparations of plasma is an acute international task. It is necessary to find albumin modifiers which do not negatively influence human organism.

The main task of the present investigation is the establishment of the regularities of the formation of the energy spectra of electrons and the determination of the chemical bond between the atoms of protein and a modifier by X-ray photoelectron spectroscopy, which allows to define the direction in the investigation of an increase in the protein stability and to choose an optimal modifier for albumin.

The XPS method has been chosen for the above purpose because in contrast to methods using ion and electron beams, it is a non destructive method. The choice of an X-ray electron magnetic spectrometer is conditioned by a number of advantages over electrostatic spectrometers, that is, constant luminosity and resolution independent of the electron energy and high contrast of spectra. In addition, the constructional separation of the magnetic-type energy-analyzer from the spectrometer vacuum chamber allows different effect in a sample in vacuum directly during spectra taking [1].

The present investigation tasks are as follows:

- The development of the method for the decomposition of X-ray photoelectron spectra into components for finding the spectra parameters indicating the transition of protein atoms into the state of stabilization.
- The development of the method for the determination of temperature of protein change and the establishment of a criterion for structural transitions.
- Investigation of the spectra of ordinary and compound amino acids with the purpose of interpretation of protein C1s-, O1s- and N1s-spectra.
- The study of formation of chemical bond between the atoms of protein and modifiers, namely, copolymers, super-dispersed particles of d-metals, and carbon metal-containing nanoforms.
- The study of influence of protein modification degree (protein with polymer—vinyl pyrrolidone acrolein diacetal copolymer) on the protein thermal stability.

- The study of influence of protein modification on the protein thermal stability.
- The selection of an optimal modifier for protein, which would provide the best protein thermal stability.

The objects of investigation are the native form of protein and modified protein. Protein was modified with carbon metal-containing nanostructures and additions of functional sp-groups for increasing the activity of the interaction of nanostructures with the environment, NiO super-dispersed powder and vinyl pyrrolidone acrolein diacetal copolymer at temperature changing from room temperature to 623K.

Carbon metal-containing nanostructures are multilayer nanotubes growing on a metal particle due to the penetration of carbon atoms and carbon adsorption on the particle surface. Samples were prepared by the low-energy synthesis method from polymers in the presence of metal systems. Cu and Ni (3-d metals) were used in the form of oxides and super-dispersed particles [2]. To increase the activity of the synthesis, the functional sp-elements of ammonium polyphosphate were used [3].

The investigation of the formation of bio-nanostructures of a certain form and their properties is based on the concept of studying the interatomic interaction of initial components, and the formation of the hybridized chemical bonds of d-electrons of metal atoms with p-electrons of atoms of sp-elements.

8.1.2 Experiment and Discussion of Results

The C1s-, O1s- and N1s-spectra of the core levels of the samples of native albumin and modified albumin were studied at temperatures from room temperature to 573 K. For investigating the state of atoms of carbon, oxygen, and nitrogen, the investigation results of reference samples of amino acids (glycine and histidine), copolymer, carbon metal-containing nanostructures, and super-dispersed d-metal particles were studied and the data on the electronic structure of graphite and hydrocarbons were used [4].

Figure 8.4 shows the X-ray photoelectron 1s-spectra of carbon and nitrogen taken from glycine and albumin samples at room temperature.

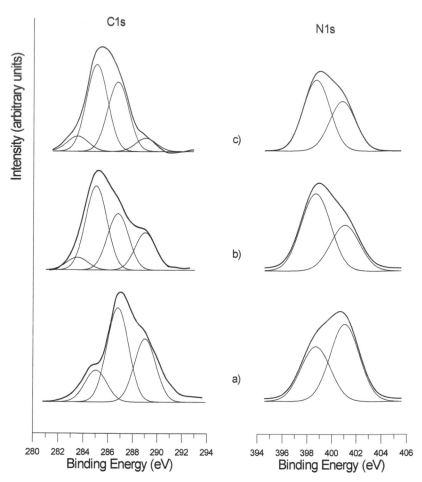

FIGURE 8.1   The X-ray photoelectron C1s and N1s-spectra of glycine (a), histidine (b), and albumin (c) at 300 K.

At room temperature the C1s-spectrum of glycine (Figure 8.4a) consists of three components, bound with different surroundings of carbon atoms, namely, C–H (285 eV), CH–NH (286.5 eV), and COOH (290.1 eV). At room temperature, the N1s-spectrum of glycine (Figure 8.4b) consists of two components bound with different surroundings of nitrogen atoms, namely, CH–NH (398.5 eV), and N–O (401 eV).

The histidine C1s-spectrum (Figure 8.4b) consists of four components bound with different surrounding of carbon atoms at room temperature: C–C (283.5 eV), C–H (285 eV), CH–NH (286.5 eV), and COOH (289.1 eV). The histidine N1s-spectrum (Figure 8.4b) consists of two components bound with different surrounding of nitrogen atoms: CH–NH (398.5 eV) and N–O (401 eV).

At room temperature, the albumin C1s-spectrum (Figure 8.4c) also consists of four components bound with different surrounding of carbon atoms: C–C (283.5 eV), C–H (285 eV), CH–NH (286.5 eV), and COOH (289.1 eV). The presence of the C–C bond is also indicated by a satellite at 306 eV [4]. The albumin N1s-spectrum (Figure 8.4c) consists of two components reflecting the bonds of nitrogen with hydrogen (N–H) and oxygen (N–O). The appearance of oxidized nitrogen can be explained by the oxidation of protein on the sample surface (several tens of angstroms) or by the formation of the N–O bonds in the protein structure.

At temperature growing above 350 K, the spectra shapes of the studied samples change significantly. In the C1s-spectrum, the contributions from the components C–C (283.5 eV) and C=O (287.1 eV) appear; and the components CH–NH (286.5 eV) and COOH (290.1 eV) disappear, which indicates the decomposition of the samples at heating, and the spectrum consists of three components C–C, C–H, and C=O. In N1s-spectrum, the contribution from CH–NH component disappears at heating, and the spectrum consists of one component reflecting the bond N–O (401 eV).

Thus, the oxidation of protein leads to its destruction. The appearance of the carbonyl group C=O indicates the protein destruction (Figure 8.4a, b, c). The concentration of the carbonyl groups shows the degree of the protein destruction [5]. The growth of the C–C bonds indicates the partial breakage of the C–H bonds. Thus, we have determined the parameters of X-ray photoelectron spectra characterizing the state of protein. The NH components in N1s-spectrum and COOH in C1s-spectrum indicate the presence of protein. The absence of these components in N1s- and C1s-spectra and growth of N–O and C=O bonds indicate the oxidative destruction of protein. With change of temperature, amino groups of glycine, histidine, and albumin have similar behavior.

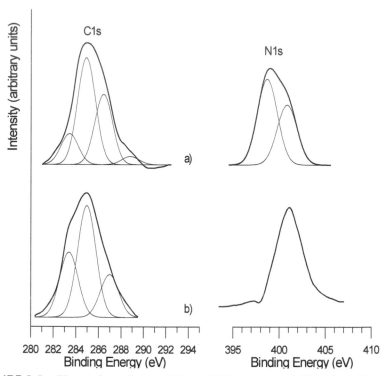

FIGURE 8.2   X-ray photoelectron C1s and N1s-spectra of albumin: a T = 300K, b T = 450K.

The samples of vinyl pyrrolidone acrolein diacetal copolymer have been studied. The copolymer contains 84% of vinyl pyrrolidone and 16 mol% of acrolein diacetal.

The X-ray photoelectron C1s- and N1s-spectra of vinyl pyrrolidone acrolein diacetal copolymer were taken at room temperature and at heating to 473 K. At room temperature, the C1s-spectrum of vinyl pyrrolidoneacrolein diacetal copolymer consists of three components bound with different surrounding of carbon atoms: C–H (285 eV), N–C (N–CH) (287.3 eV), and COOH (290.1 eV). With the increase in temperature to 373 K, the contribution of the COOH component decreases in the carbon spectra and the components C–H (285 eV) and N–C (N–CH) (287.3 eV) remain. At temperature above 473 K, the component C=O (287.1 eV) appears in the C1s-spectrum. At room temperature and temperatures up to 373 K, the N1s-spectrum consists of two components N–C (397 eV) and N–O (401 eV) and at temperature above 373 K, there is one component N–O (401 eV) in the N1s-spectrum.

The comparison of the C1s-spectra (Figure 8.6) of albumin, vinyl pyrrolidone acrolein diacetal copolymer and albumin modified with vinyl pyrrolidone acrolein diacetal copolymer (conjugate) taken at room temperature shows that in the conjugate prepared from albumin and vinyl pyrrolidone acrolein diacetal copolymer in the ratio 1:5, there are three component characteristics of albumin, that is, C–C (283.5 eV), C–H (285 eV) and COOH (290.1 eV), and the fourth component has the binding energy 286.8eV characteristic of the CH–NH bond and vinyl pyrrolidone acrolein diacetal copolymer which has a component with the binding energy 287.3 eV (CH–N). Figure 8.6 shows the N1s-spectra of albumin, vinyl pyrrolidone acrolein diacetal copolymer, and conjugate. In the conjugate N1s-spectrum, there are components characteristic of vinyl pyrrolidone acrolein diacetal copolymer and albumin. In contrast to albumin, an increase in the thermal stability of conjugate can be explained by the formation of stronger bonds C–N in it.

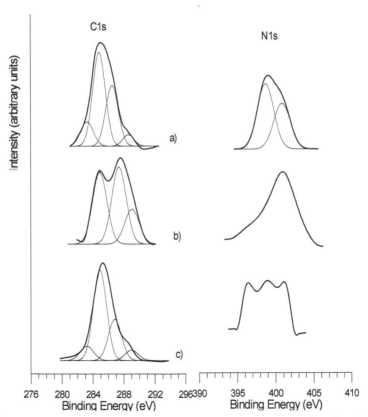

FIGURE 8.3   The X-ray photoelectron C1s and N1s-spectra of a albumin, b vinyl pyrrolidone acrolein diacetal copolymer, and c conjugate.

The dependence of the conjugate thermal stability on the degree of the albumin modification with vinyl pyrrolidone acrolein diacetal copolymer (1:1, 1:3, 1:5) has been studied.

The samples of albumin modified with vinyl pyrrolidone acrolein diacetal copolymer in the ratios 1:1, 1:3, and 1:5 were investigated by the XPS method. The C1s- and N1s-spectra of all the samples (Figure 8.7) were studied.

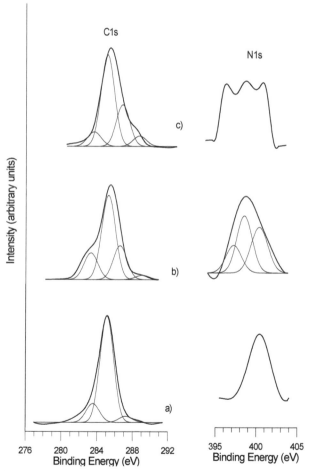

FIGURE 8.4　The X-ray photoelectron C1s-spectra of protein modified with vinyl pyrrolidone acrolein diacetal copolymer in the ratios: a) 1:1; b) 1:3; c) 1:5.

In the C1s-spectrum (Figure 8.7a) of the sample with equal content (1:1) of the modifier (vinyl pyrrolidone acrolein diacetal copolymer) and albumin, there are three components C–C, C–H, and C=O. In the N1s-spectrum of the above sample, there is only one component N–O and the contributions from N–H and COOH characteristic of protein are absent at room temperature.

For the sample containing protein and vinyl pyrrolidone acrolein diacetal co-polymer in the ratio 1:3, the C1s-spectrum (Figure 8.7b) shows maxima corresponding to the bonds C–C, C–H, C–NH and, COOH and in the N1s-spectrum, maxima characteristic of the C–NH, N–H, and N–O bonds are observed. Consequently, there are components N–H and COOH characteristic of protein.

The amino acid group NH is bound with carbon (C–NH), which stabilizes the N–H group and prevents its decomposition. With further growth of the vinyl pyrrolidone acrolein diacetal copolymer content in protein (5:1), the C–NH and COOH components' contributions grow relative to that of C–H (Figure 8.7c). The thermal stability grows with the increase of the relative content of vinyl pyrrolidone acrolein diacetal copolymer in the conjugate and it reaches 473K for the 5:1 modification.

For studying the consequences of the interaction of nano-biosystems with artificial nanostructures, the modification of protein with nanostructures was conducted. At the protein modification with multilayer carbon copper-containing nanostructures, the position and relative content of the components in the C1s- and N1s spectra (Figure 8.8a) differ from those of the components observed for unmodified protein. For modified protein, in the C1s- and N1s-spectra, in addition to C–C, C–H and COOH, the C–N (H$_2$) components are observed. Heating up to 523 K changes insignificantly the shape of spectra (Figure 8.8a, b). Apparently, the bond between protein atoms (N–H) and carbon atoms become stronger due to the formation of double or triple bond of C with NH$_2$ or, which is more likely, the formation of the hybridized sp^3 bond on the C–N(H$_2$) atoms. In contrast to the weak bond of the NH$_2$ groups, the presence of the strong covalent bond C–N(H$_2$) leads to an increase in the thermal stability of the protein modified with carbon copper-containing nanostructures to 523 K. With the temperature increase above 523K, the intensity of the C–N(H$_2$) and COOH components decreases in the C1s-spectrum and at T = 623K, these components completely disappear. In the C1s-spectrum, the components similar to destructed albumin, that is, C–C, C–H, and C=O remain and in the N1s-spectrum, the component N–O remains which is characteristic of destructed protein. Sharp worsening of vacuum and the appearance of the Au4f-spectrum from the substrate indicate the protein coagulation. Similar to the case of albumin, the growth of C–C bonds is observed.

Nanoparticles can influence protein molecules by penetrating into cells and often destructing them [6]. There is a minimal radius of a nanoparticle at which it can be captured inside the cell. Consequently, the biocompatibility of protein and nanoparticles depends on the nanoparticle size, that is, the surface area can be used as a measure of oxidation (toxicity).

Then, the protein modified by multilayer carbon nickel-containing nanotubes prepared by low-temperature synthesis of polymers and nickel oxides was investigated. The multilayer carbon nickel-containing nanotubes have external dimensions much smaller (20–40 nm) than those of the above multilayer copper-containing nanotubes (50 nm and larger). The protein modified by multilayer carbon nickel-containing nanotubes destruction is observed at room temperature. In the C1s-spectrum, the COOH carboxyl group is absent, and C=O carbonyl group, C–C, and C–H appear. In the N1s-spectrum the N–H group disappears and the component N–O grows.

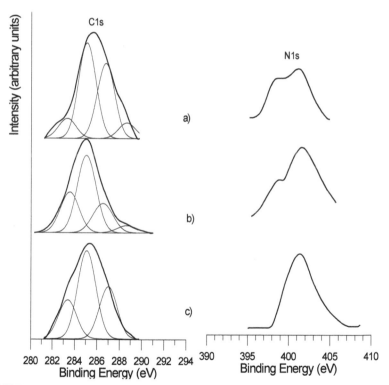

FIGURE 8.5 The X-ray photoelectron C1s and N1s-spectra of albumin with multilayer carbon copper-containing nanotubes: a T = 373 K; b T = 523 K; and c T = 623 K.

The same result is observed when protein is modified with NiO nanoparticles which are oxidation catalysts. The destruction of albumin modified with multilayer carbon nickel-containing nanotubes is observed at room temperature. The decomposition of the spectra into the components shows that in the C1s- and N1s spectra, the components COOH and N–H are absent and the components C–C,

C–H, C=O, and N–O are present similar to the case of modification with carbon nickel-containing nanostructures, that is, the protein destruction takes place at room temperature. Thus, the growth of carbonyl groups, the breakage of the N–H bonds, and the formation of the N–O bonds occur at an elevated temperature or in the presence of the interaction with a metal catalyst.

8.1.3   Conclusion

Based on the conducted fundamental investigations, the regularities are obtained, which can be used for the choice of the direction in the purposeful increase of protein stability. Investigations of protein modification show that

- Low thermal stability of protein (albumin) is due to the weak bond of the $NH_3(NH_2)$ group and the decrease of the content of NH (NH_3, NH_2) groups in the protein composition and nitrogen oxidation at heating.
- One of the consequences of the formation of the strong covalent bond between the atoms of the NH_2 amino group and the carbon atoms (C–NH) at the protein modification is an increase in the thermal stability of the modified protein. The destruction of carboxyl groups (COOH) and the formation of carbonyl groups (C=O) as it follows from the change of the C1s-spectrum at heating indicate the destruction of protein. At the protein destruction, the bond between hydrogen and nitrogen is broken, and nitrogen forms the bond with oxygen.
- The presence of metal catalysts (Ni, NiO) in protein leads to the oxidation and destruction of protein due to the formation of carbonyl groups and the destruction of the N–H groups.

Also:

- A method for the determination of the temperature of the protein destruction has been developed and the criteria of the protein destruction have been established.
- During the modification of protein with carbon copper-containing nanotubes, a strong covalent bond (C–NH) of the atoms of the N–H amino group with the carbon atoms is formed, which results in the increase of the modified protein thermal stability up to 523–573K.
- The compatibility of bio- and nanostructures depends on the size of the nanostructures, which can serve as a measure of oxidation.
- In addition to the influence of the temperature on the oxidative destruction of protein, the presence of metal catalysts (Ni, NiO) in protein leads

to the oxidation and destruction of protein due to the formation of carbonyl groups and the decomposition of the N–H groups at room temperature.

- In contrast to the modification of protein with carbon copper-containing nanotubes, the modification of protein with carbon nickel-containing nanotubes leads to the protein destruction at room temperature due to the presence of Ni atoms.
- One of the consequences of the formation of the covalent bond C–N(–H) between nitrogen atoms and carbon atoms, is an improvement in the thermal stability of the modified proteins.
- The dependence of the thermal stability of proteins modified with vinyl pyrrolidone acrolein diacetal copolymer on its content in the mixture is found. The growth of the temperature of stabilization with the increasing content of vinyl pyrrolidone acrolein diacetal copolymer is shown.
- At the vinyl pyrrolidone acrolein diacetal copolymer concentration three times smaller than that of protein, the protein destruction is observed at room temperature. With the threefold and more increase of the vinyl pyrrolidone acrolein diacetal copolymer concentration, the temperature of the protein stability grows and is maximal at the ratio 1:5.

Based on the results obtained, some recommendations on the modification of protein for increasing its stability are given, which can be used in pharmaceutical biotechnology.

A model for the protein stabilization is offered on the basis of the formation of the strong hybridized chemical bond of the atoms of protein and a modifier.

The results of the work on choosing an optimal modifier for protein for medico-biological and pharmaceutical technologies show that the highest thermal stability of protein (523–573 K) is achieved at the modification of protein with carbon copper-containing nanotubes having functional groups of sp-elements in the ratio 1:0.01, which is more rational in comparison with the protein modification with vinyl pyrrolidone acrolein diacetal copolymer because in this case the required amount of the modifier is significantly larger 3–5 times) than the amount of protein.

KEYWORDS

- **Magnetic-type energy-analyze**
- **Protein thermal stability**
- **Super-dispersed particles**
- **Vinyl pyrrolidone acrolein diacetal copolymer**
- **X-ray photoelectron spectroscopy**

REFERENCES

1. Shabanova, I. N., Varganov, D. V., Dobysheva, L. V., et al. (1986). New automated X-ray electron magnetic spectrometers: Spectrometer with technological adaptors and manipulators, a spectrometer for studying melts. *Izvestiya Akademii Nauk SSSR. Seriya Matematicheskaya, 50*(9), 1677.
2. Kodolov, V. I., Khohryakov, N. V. (2009). Chemical physics of the processes of the formation and transformations of nanostructures and nanosystems. *Izvestiya Akademii Nauk SSSR. Seriya Matematicheskaya, 1–2,* 728.
3. Shabanova, I. N. & Terebova, N. S. (2011). Dependence of the value of the atomic magnetic moment of d-metals on the chemical structure of nanoforms. *Polymers Research Journal, 5*(2), 7.
4. Makarova, L. G., Shabanova, I. N., Kodolov, et al. (2004). X-ray photoelectron spectroscopy as a method to control the received metal-carbon nanostructures. *Journal of Electron Spectroscopy and Related Phenomena, 137–140,* 239.
5. Shabanova, I. N., & Kodolov, V. I. (2011) X-ray photoelectron spectroscopy investigation of thermal stability of protein modified with metal-containing ultra- or nanostructures. *Polymers Research Journal, 5*(2), 15.
6. Muravleva, L. Ye., Molotov-Luchansky, V. B., Kluyev, D. A., et al. (2010). Oxidative modification of proteins: Problems and prospects of investigation. *Fundamentalnie Issledovaniya, 1,* 74.

CHAPTER 9

THE INVESTIGATION OF NANOCOMPOSITES ELECTRON STRUCTURE INFLUENCE CHANGES ON DIFFERENT MEDIA AND COMPOSITIONS: PART III

L. G. MAKAROVA, N. S. TEREBOVA, I. N. SHABANOVA, V. I. LADYANOV, and R. M. NIKONOVA

CONTENTS

9.1 INVESTIGATION OF NANOSTRUCTURES IN METALS AND AL-LOYS

9.1.1 Investigation of Carbon Nanostructures in Iron Matrix

For many years, the improvement of the mechanical properties of structural materials was mainly performed by the development of new alloys with new chemical and phase compositions. Lately, new ways have appeared for improving properties of structural materials, namely by the well-directed formation of micro- and nano-crystalline structure.

The set of the experimental methods that are used for studying the chemical structure of carbon cluster nanostructures is limited. Therefore, one of the main tasks is the development of diagnostic methods, which will allow controlling intermediate and final results in the creation of new materials.

At present, the analysis of numeral works shows that classical methods for determining shapes, sizes and compositions of carbon nanostructures are transmission electron microscopy, methods based on electron diffraction, and Raman spectroscopy. However, more and more publications appear referring to the investigation of nanostructures with the use of the X-ray photoelectron spectroscopy (XPS) method. Further development of the XPS method and related methods for the surface (from 1 to 10nm) investigation will lead to an increase in the number of methods for studying compositions, electronic properties and structures of nanostructures.

The XPS has been used for the determination of the type of carbon structures. The XPS method allows investigating the electron structure, chemical bond and nearest environment of atoms. One of the important specific features of the method is its non destructive character of action since the X-ray radiation used for photoelectron excitation does not practically cause any damages in most materials. This cannot be said about surface analysis methods that involve ion or electron bombardment of a surface. In most cases, a sample can further be used for some other investigations after it has been studied by the XPS method.

In addition, the method provides the possibility to analyze thin layers and films, which is very important for the case of formation of fullerenes, nanotubes, and nanoparticles, and to obtain the information on sample chemical compositions based on spectra, which provides the control over chemical purity of materials. The XPS method allows investigating electron structure, chemical bond, the nearest environment of atoms with the use of an X-ray photoelectron magnetic spectrometer. Russian X-ray photoelectron magnetic spectrometers with automated control system [1] are not inferior to the best foreign spectrometers in their main parameters. The preference is given to the X-ray electron magnetic spectrometer because of a number of advantages compared with electrostatic spectrometers [2], which are the high spectrum contrast, the permanency of optical efficiency and resolution capacity that are not influenced by electron energy. Moreover, the XPS method is a non destructive investigation method.

Experimental

In this work, the samples prepared from iron powder modified with fullerenes or graphite were studied with the use of the XPS method. The samples were prepared in two ways: fusion or pressing. Modification was carried out for obtaining nano-carbon structures in metal matrices in order to improve strength properties of a material. The samples were verified with the use of X-ray diffraction, which showed that the sample structures were mainly fcc iron. The description of the samples studied is given in Table 9.1.

TABLE 9.1   The investigated samples.

Sample no.	Sample composition	Sample form
1	Fe + 0.5% $C_{60/70}$	Ingot (T = 1410°C)
2	Fe + 0.5% graphite	Ingot (T = 1410°C)
3	Fe + 2% $C_{60/70}$	Powder
4	Fe + 2% graphite	Powder
5	Fe + 2% $C_{60/70}$	Pellet (P = 800MPa)
6	Fe + 2% graphite	Pellet (P = 800MPa)

The X-ray photoelectron investigations were carried out for studying the changes in the nearest environment of the carbon atoms in the samples prepared in different ways.

The investigations were carried out using the X-ray photoelectron magnetic spectrometer with double focusing and instrumental resolution of 0.1eV at the excitation of AlKα-lines (1486.6eV).

For the XPS investigations of carbon-metal cluster nano-materials, the method of the C1s spectra identification by the satellite structure was employed. To do this, reference samples were studied, the carbon components of which could give the C1s spectrum: C–H—hydrocarbons [3], C–C (sp^2)—graphite [4], C–C (sp^3)–diamond [5]. The spectra parameters are presented in Table 9.2, where E_b is binding energy, E_{sat} is the energy characterizing the satellite position, I_{sat} is the satellite intensity, I_0 is the intensity of the main maximum.

TABLE 9.2　The C1s spectra parameters for reference samples.

	E_b (eV)	E_{sat} (eV)	$\Delta E = E_{sat} - E_b$ (eV)	FWHM (eV)	I_{sat}/I_0, $\Delta = 10\%$
C–H [3] Hydrocarbons	285.0 ± 0.1	292.0 ± 1.0	~6-7	2.0 ± 0.1	0.10
C–C (sp^2) [4] Graphite	284.3 ± 0.1	306.0 ± 1.0	~22	1.8 ± 0.1	0.10
C–C (sp^3) [5] Diamond	286.1 ± 0.1	313.0 ± 1.0	~27	1.8 ± 0.1	0.15

To identify the structures studied in [6], the C1s spectra of carbon nanostructures were studied. The nanostructures were obtained in the electric arc during graphite electrode sputtering. The carbon nanostructures obtained were fullerenes C_{60}, single-walled and multi-walled carbon nanotubes and amorphous carbon.

It is shown that in all the C1s spectra, there is a satellite structure related to different effects (a shake-up process in which characteristic losses—plasmons [7]), which allows to create a calibration technique and to determine not only the energy position of the components but their intensities as well. In the C1s spectrum of fullerenes C_{60}, there is a satellite with binding energy of 313 eV [8] and the relative intensity of 15% from the main peak. This satellite is characteristic of the sp^2-hybridization of the valence electrons of the carbon atoms.

In the C1s-spectrum of single-walled carbon nanotubes, in addition to a gradually rising spectrum in the high-energy region, two satellites are observed with characteristics of the C–C bonds with the sp^2- and sp^3-hybridization of the valance electrons of the carbon atoms. Consequently, in the C1s-spectrum of the one-layer nanotubes, there are these two components with the binding energies of 284.3 and 286.1 eV and the intensities of 1:0.1 and 1:0.15 relative to their satellites and the width of 1.8 eV. The ratio between the C–C bonds with the sp^2- and those with sp^3-hybridization of valence electrons is 2.

The similar situation is observed for the C1s-spectrum of multi-walled carbon nanotubes.

In the region of 313 eV of the amorphous carbon spectrum, there is a satellite characteristic of sp^3-hybridization of the valence electrons with the relative intensity of 15%. Consequently, in the C1s spectrum, there is a component characteristic of C–C bond with sp^3 hybridization of the valence electrons at a distance of 27 eV from the satellite. Thus, the amorphous carbon presents carbon inclusions that are like globe-shaped form of graphite.

The development of a calibration method for spectra in the X-ray photoelectron investigations of reference samples allows to realize the decomposition of the C1s spectrum to the components that determine the chemical bond, the hybridization type of s-p valence electrons of the carbon atoms and the nearest environment of the carbon atoms. Studying the nanoparticles with a known structure gives the possibility of the identification of studied carbon structures by examining the C1s spectra shapes.

The C1s-spectra identification method developed by us was successfully used for the investigation of nanostructures in the iron matrix.

Results and Discussion

The XPS method was used to obtain the C1s, O1s and Fe2p spectra. The O1s spectra show that large amounts of adsorbed oxygen and iron oxides are present on the sample surfaces. The Fe2p spectra are also indicative of the presence of the oxidized iron layer on the surfaces of the samples. During the experiment, the shift of the spectra was not observed.

The experimental C1s-line spectra for sample no. 1 and 2 are shown in Figure 9.1a, b.

The mathematical treatment of the C1s spectra was performed, that is, background subtraction and procedure of the spectra smoothing and decomposition. The results are given in Figure 9.2.

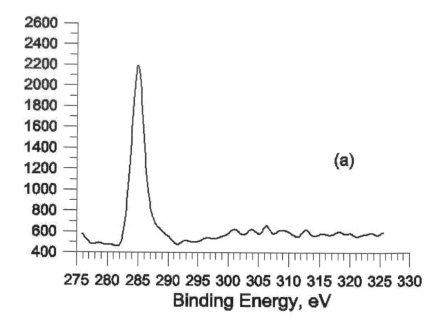

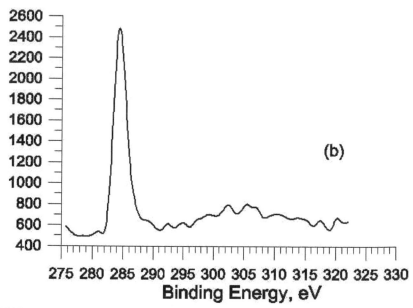

FIGURE 9.1   The experimental C1s spectra of sample no. 1 (a) and 2 (b).

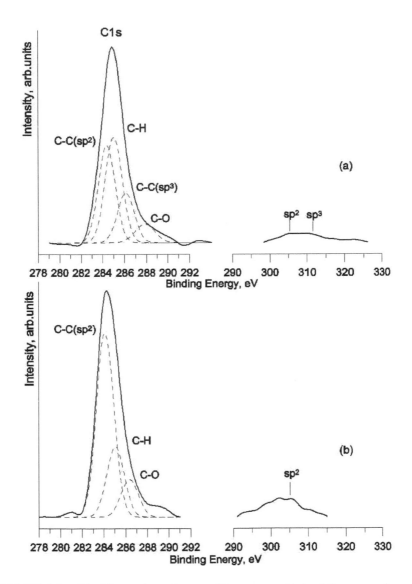

FIGURE 9.2   The X-ray photoelectron C1s spectra obtained from sample no. 1 (a) and 2 (b) after the spectra were decomposed into their components.

In Figure 9.2*a*, the X-ray photoelectron spectrum of the C1s-line is displayed, which was obtained from sample no. 1 without heating in the spectrometer chamber. In the high-energy region two satellites are observed, which are characteristics of sp^2 and sp^3 hybridization of the valence electrons of the carbon atoms judged by their binding energies (306.0 eV and 313.0 eV, respectively). According to the data in Table II in the C1s spectrum, there are component characteristics of the C–C (sp^2) and C–C (sp^3) bonds at distances of ~22 and ~27eV. The relation of the C–C (sp^2) bond intensity to the C–C (sp^3) bond intensity is ~2, which is characteristic of carbon nanostructures. In the spectrum, there are also components of C–H and C–O bonds, which characterize surface contaminations.

When the sample is heated, the breakup of the C–C bonds with sp^3-type of hybridization of the valence electrons of the carbon atoms is taking place. At further heating of the sample, in the C1s spectrum there is a component characteristic of Fe–C bonds. In addition, at heating, the Fe2p spectra are observed on the sample surface, which is indicative of the fact that the sample surface is cleaned during heating.

In Figure 9.2b, the C1s spectrum is displayed, which is obtained from sample no. 2 without heating in the spectrometer chamber. In the high-energy region, a satellite with the binding energy of ~306.0 eV is observed, which is characteristic of the sp^2 hybridization of the valence electrons of the carbon atoms. Consequently, in the C1s spectrum, at the distance of ~22 there is a component characteristic of C–C bonds with sp^2 hybridization. There are also components characteristic of C–H and C–O bonds in the spectrum.

The X-ray photoelectron investigations of samples 3, 4, 5, and 6 show that, the C1s spectra do not have a satellite and they have low intensity. Thus, only hydrocarbon contaminations are present on the surfaces of these samples.

Conclusion

The investigations conducted have demonstrated that when iron powder is mixed with fullerene mixture C_{60}/C_{70} or with graphite powder and then the mixtures obtained are treated differently (fusion and pressing), there are strong differences in the X-ray photoelectron spectra.

On the surfaces of the samples prepared from the mixture of the iron powder and the mix of fullerenes C_{60}/C_{70} and subjected to fusion, carbon nanostructures are present. On the surface of the samples prepared from the mixture of the iron powder and graphite powder, which were also subjected to fusion, graphite-like structures are present.

KEYWORDS

- **Chemical bond**
- **Fullerenes**
- **Raman spectroscopy**
- **Spectra smoothing**
- **Transmission electron microscopy**

REFERENCES

1. Trapeznikov, V. A., Yefimenko, A. I., Yevstafyev, A. V., et al (1975). Automated electron magnetic spectrometer (B430326, p. 176). Moscow: VNTITsentr.
2. Siegbahn, K., Nordling, C., Fahlman, A., et al. (1967). ESCA: Atomic, molecular and solid state structure studied by means of electron spectroscopy (p. 493). Uppsala.
3. Briggs, D., & Sih, M. P. (Eds.). (1987). The analysis of surface with the use of the auger- and X-ray photoelectron spectroscopy (p. 600). Moscow: Mir.
4. Shabanova, I. N. (1990). Dissert. na soisk. stepeni doktora fiz.-mat. nauk; Izhevsk, p. 502.
5. Kolobova, K. M., Shabanova, I. N., Kulyabina, O. A., et al (1981). The investigation of the electron structure of globe-like graphite in aluminum iron. FMM, 54(4), 890–893.
6. Makarova, L. G., Shabanova, I. N., & Terebova, N. S. (2005). The Development of the X-ray electron spectroscopy method for studying the chemical structure of carbon cluster nanostructures. Nanotekhnika, 4, 55–57.
7. Nefedov, V. I. (1984). X-ray electron spectroscopy of chemical compounds (p. 256). Moscow: Khimia.
8. Khodorkovskiy, M. A., Shakhmin, A. L., & Leonov, N. B. (1994). The investigation of the C60 coating with different thickness by X-ray photoelectron spectroscopy. FTT, 36(3), 626–630.

CHAPTER 10

THE INVESTIGATION OF NANOCOMPOSITES ELECTRON STRUCTURE INFLUENCE CHANGES ON DIFFERENT MEDIA AND COMPOSITIONS: PART IV

V. I. RYABOVA, G. V. SAPOZHNIKOV, I. N. SHABANOVA, and
N. S. TEREBOVA

CONTENTS

10.1 INVESTIGATION OF NANOSTRUCTURES IN MODIFIED CAST IRONS AND STEELS

10.1.1 Introduction

The question of the mechanisms of modified cast irons and steels and the reasons for the formation of ordering phases in them is one of the most important in obtaining new materials with enhanced properties of use.

Inspite of the large number of theoretical and experimental investigations devoted to the nanotypes, there is no single model that allows us to explain the structure and properties of steel and cast iron samples we obtain. The number of experimental techniques used to investigate the chemical structure of nanostructures at the atomic level is limited.

10.1.2 Experiment

In this work, industrial samples of nanomodified cast irons and steels were examined by the means of X-ray photoelectron spectroscopy (XPS) on an EMS-100 photoelectron magnetic spectrometer with double focusing [1]. The XPS allows us to investigate electronic structure, chemical bonds, and local atomic environments. Photoelectron magnetic spectrometer was choosen because of its numerous advantages over electrostatic spectrometers.

These include constant light power and accuracy independent of electron energies and high spectral sharpness. In addition, XPS is a non destructive technique, which is particularly important in investigations of metastable systems.

The objects of our investigation were non legated cast iron samples obtained by means of rotary casting, modified aluminum doping agents, samples of 08X18N10T steel and stainless non magnetic modified nanostructures of 08X21G11AN6 steel containing *sp* elements (carbon and nitrogen).

The chemical composition of the samples is shown in Table 10.1

TABLE 10.1   Results of chemical analysis.

Sample	Chemical composition, weight %							
	C	Mn	Si	Cr	Ni	Cu	Al	N
Cast iron	3.03	0.54	1.92	0.28	0.13	0.31	0.19	0.01
Modified cast iron	3.06	0.55	2.07	0.27	0.12	0.30	1.40	0.01
Stainless steel	0.08	1.20	0.50	18.10	9.85	0.25	0.10	0.01
Modified stainless steel	0.05	10.84	0.36	20.98	5.39	0.09	0.01	0.61

Experiments were performed on a unique EMS 100 automatic X-ray photo-electron magnetic spectrometer [1] with a resolution of 10^{-4} and a luminosity of 0.185%.

The spectra of inner energy levels Fe2p, Al2p, C1s, and O1s were investigated. The time of signal accumulation at each point of a spectrum was from 3 to 30s. The purity of the surface was monitored using the C1s and O1s spectra. Spectral decomposition was performed using software based on the least square method. Energy status, spectral width, and intensity data were considered in the software and compared to the spectra of reference samples. Decomposition was performed using a Gaussian function with maximum approximation of the envelope curve to the experimental curve. The accuracy of determining peak positions was 0.1eV. The error in determining the sharpness of the electron spectra was no more than 5%.

Decomposition of the C1s spectrum was performed to examine chemical bonds, *sp*-hybridization of valence electrons, and the local atomic environment of carbon atoms. To accomplish this, reference spectra (the carbon components of the C1s spectrum) were investigated: C–C (sp^2) graphite, C–C (sp^3) fullerene, and C–H hydrocarbons [2].

Figure 10.1 shows the C1s spectra of carbon in the ordinary and modified cast irons.

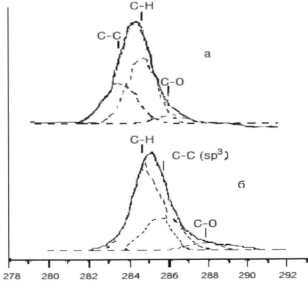

Bond energy, eV

FIGURE 10.1   C1s spectra of carbon in a ordinary and b modified cast iron.

The spectra are shifted toward each other by 1.7eV. The energy value of the C1s level in the ordinary cast iron is close to the one corresponding to the line in the Fe_3C spectrum and the C–C bond in graphite. Consequently, the bonds between carbon and iron in these samples are close in nature. Hydrocarbon and adsorbed oxygen impurities were present in all of the investigated samples, since all of the samples were placed into the chamber from the air. The energy values for the C1s level in modified cast iron coincide with the energy of the C1s level for diamond. Diamond like structures with sp3electron configuration thus form in the carbon atoms of modified cast iron.

The C1s spectra of C_{60} carbon nanostructures obtained in an electric arc during the diffusion of the carbon electrode were examined to identify the objects under investigation. A satellite with bond energy 313 eV and an intensity of 15% relative to the main peak intensity is observed in the C1s fullerene spectrum. This particular satellite peak is characteristic of the sp^3 hybridization of valence electrons in a carbon atom. Satellite peaks are observed in graphite and diamond spectra at 22 and 27 eV, respectively, from the main maximum with relative intensities of 10 and 15%, respectively [3].

The values of the bond energies of the electrons of the Fe2p line of the cast iron samples are shown in Table 10.2.

TABLE 10.2 Values of electron bond energies of inner bonds.

Substance	Energy level (eV)	C1s energy level
Fe		
Fe_3C	708, 7 ± 0.2	284, 1 ± 0.2
Fe_3Al	708, 1 ± 0.2	
Fe_3Si	708, 5 ± 0.3	
Graphite	-	284, 4 ± 0.2
Cast iron	708, 8 ± 0.3	284, 2 ± 0.2
Modified cast iron	708, 1 ± 0.3	286, 2 ± 0.2

In contrast to the spectra of pure iron and like Fe_3C in highly carbon phase, the Fe2p spectra have no clear features.

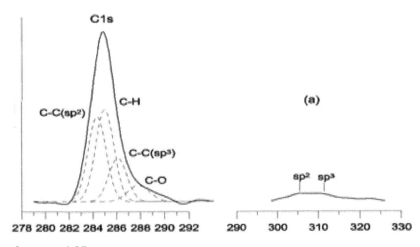

Bond energy (eV)

FIGURE 10.2   C1s spectra of carbon.

Figure 10.2 presents the XPS spectrum of a C1s line obtained from the surface of the modified steel sample.

Two satellites are observed in the high energy area that in terms of energy bonds are characteristics of the sp^2 (306.0eV) and sp^3 (313.0eV) hybridizations of the valence electrons of carbon atoms, respectively. Components characteristics of C–C (sp^2) and C–C (sp^3) bonds are observed in the C1s spectrum at distances of around 22 and 27eV, respectively. The ratio of the intensities of the main maximum of C–C (sp^2) and the main maximum of C–C (sp^3) bonds is ~2, a feature of carbon structures. The spectral components of C–H and C–O bonds also indicate surface contamination. Our investigations reveal that the obtained steel is 1.5–3 times more durable, with identical or increased plasticity (Table 10.3).

TABLE 10.3   Mechanical properties of samples.

Sample	Yield limit, $\sigma_{0.2}$ (kgf/mm²)	Breakdown limit, σ_t (kgf/mm²)	Relative elongation, δ (%)
08X18H10T	25	55	35
08X21Г11AH6	75	99	55

10.1.3 Results and Discussion

Iron and aluminum atoms in cast irons and steels have different properties. Aluminum, distinguished by its high chemical activity, is an energetic reductant in steels and cast irons, and partially interacts with oxygen, drawing the oxygen to itself.

The high carbon phase plays an important role in the formation of structures of cast iron and steel. In triple Fe–Al–C alloys, their physical nature undergoes qualitative change. Aside from two known phases of the high carbon type (graphite and cement carbide) special types of carbon appear in alloys in the Fe–Al–C system (spherical graphite and Fe–Al phase). Aluminum atoms interact with carbon atoms and are adsorbed primarily at crystal faces where free bonds are present. As a result, the spread of graphite along the basis planes is suppressed, graphitization is slowed, and conditions are created for transverse crystal growth and the formation of compact (spherical) graphite inclusions. Our conclusions are also confirmed by optical microscopy data. Figure 10.3 presents photographs of the microstructure of our samples.

Direct Fe–Al and C–C bonds with sp^3 hybridization of valence electrons begin to form during the production of modified iron by means of rotary casting and aluminum alloying. This leads to the appearance of new structural elements in electron spectra, relative to ordinary cast irons in which Fe–Si and C–C bonds with sp^2 hybridization are present. The durability of aluminum cast irons is enhanced due to the stronger interatomic interaction of Fe–Al relative to Fe–Si, since hybrid $3d$ (Fe) and $3p$ (Al) bonds are formed [4].

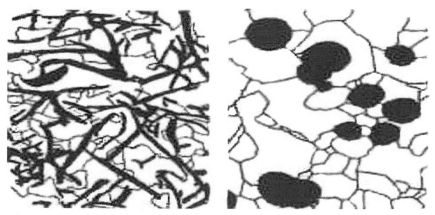

FIGURE 10.3 Graphite in modified and non modified cast iron.

The modification of cast iron with aluminum leads to the transfer of some aluminum valence electrons to graphite atoms, accompanied by the formation of a certain proportion of atoms with more energetically stable and diamond like sp^3 hybridization of electrons.

Investigations of iron modified by fullerene have revealed the presence of carbon nanostructures on the surface of a sample, leading to its increased durability. Easily formed carbide nanostructures do not appear in steels alloyed with metals (V, Cr, Mn). Such steels are alloyed with nitrogen. Alloying steel with nitrogen demonstrates that of all incorporated elements, nitrogen is the most effective hardener of solid alloys, and nitrides are more easily separated than carbides. Nitrogen's solubility in solid alloys of the Fe–Cr system is especially high due to the strong austenite forming properties of this element.

10.1.4 Conclusions

Using an X-ray photoelectron magnetic spectrometer, the XPS technique was applied for the first time to the investigation of metallic carbon nanomodified cast irons obtained by means of rotary casting.

The results from X-ray photoelectron spectroscopy allowed us to determine the character of the ineratomic interaction of cast irons alloyed with aluminum and the mechanisms of the formation of spherical graphite in them.

The satellite structure of C1s spectra was used to identify C1s spectra and determine the sp hybridization type of valence electrons in the investigated samples.

Scientifically valuable results were thus obtained:

- The role of the electronic structure in obtaining nanomodified cast irons was identified.
- The dependence of structure formation on the composition of initial substances was revealed.
- The high durability of modified cast irons is explained by the formation of strong hybrid $3d$ (Fe) and $3p$ (Al) bonds, and by the presence of C–C complexes with diamond like $sp3$ hybridization of electrons.
- Carbon nanostructures were present on the surface of the sample prepared from a mixture of iron powders and a C_{60}/C_{70} fullerene mixture and exposed to melting.
- The high effectiveness of nitrogen's influence on the properties of steel use was demonstrated.

Results from our X-ray photoelectron investigations of nanomodified cast irons and steels point the way for subsequent development of the synthesis of new cast irons and steels. XPS could be used to control the synthesis of new cast irons and steels with enhanced characteristics of use.

KEYWORDS

- **Aluminum doping agents**
- **Carbon nanostructures**
- **Gaussian function**
- **Graphite inclusions**
- **Photoelectron magnetic spectrometer**

REFERENCES

1. Trapeznikov, V. A. (1998). *Uspekhi Fizicheskikh Nauk*, 7, 793.
2. Shabanova, I. N., Makarova, L. G., Kodolov, V. I., et al (2002). *Anal Poverkhn Mezhfazn Gran*, 34, 80.
3. Makarova, L. G., Shabanova, I. N., & Terebova, N. S. (2005). *Zavod Laboratory for Diagnostic Materials*, 71(5), 26.
4. Shabanova, I. N., Kormilets, V. I., & Terebova, N. S. (2001). *Journal of Electron Spectroscopy and Related Phenomena*, 114–116, 609.

CHAPTER 11

THE CHANGES OF PROPERTIES OF MATERIALS MODIFIED BY METAL OR CARBON NANOCOMPOSITES

V. I. KODOLOV, A. M. LIPANOV, V. V. TRINEEVA,
YU. M. VASIL'CHENKO, L. F. AKHMETSHINA, M. A. CHASHKIN,
YU. V. PERSHIN, and YA. A. POLYOTOV

CONTENTS

11.1 GENERAL INFORMATION ABOUT THE MATERIALS PROPERTY CHANGES AT THE MODIFICATION WITH METAL /CARBON NANO- COMPOSITES USING

For the production of materials with the improved characteristics, the modification occurs with the use of metal/carbon nanocomposites finely dispersed suspensions. The content of active nanocomposites usually makes up super small quantities (0.01–0.0001%). For even distribution of metal/carbon nanocomposites super small quantities, numbers of suspensions are applied in correspondent compositions:

- The water suspensions containing nanocomposite and surfactant are used for the modification of foam concrete and dense concrete.
- The suspension based on the solution of polyvinyl chloride (PVC) in acetone (or chlorinated paraffins) are used for the modification of PVC.
- The suspensions based on the solution of polymethyl methacrylate (PMMA) in dichloromethane are used for the modification of PMMA.
- The suspension based on the solution of polycarbonate (PC) in methylene chloride (or dichloroethane) are used for the modification of PC.
- The suspensions based on the solution of phenol-formaldehyde resins in alcohol (or in mixture of alcohol and toluene) are used for the modification of glue BF-19.
- The suspensions based on the polyethylene polyamine are used for modification of epoxy resins (cold hardening).
- The suspensions based on the isomethyl tetrahydrophtalate anhydrate are used for modification of epoxy resins (hot hardening).

Before the application of the suspensions, they are usually diluted to the necessary quantity of nanocomposite.

To select the components of fine suspensions with the help of quantum-chemical modelling by the scheme described before [1], first, the interaction possibility of the material component being modified (or its solvent or surfactant) with metal/carbon nanocomposite is defined. The suspensions are prepared by the dispersion of the nanopowder in ultrasound dispersator. The stability of fine suspension is controlled with the help of nephelometer (photocolorimeter) and laser analyzer. The action on the corresponding regions participating in the formation of fine suspension or solution is determined with the help of IR spectroscopy. As an example, below you can see the brief technique for obtaining fine suspension based on polyethylene polyamine.

IR spectra of metal/carbon and their fine suspensions in different (water and organic) media have been studied for the first time. It has been found that the introduction of super small quantities of prepared nanocomposites lead to the significant change in band intensity in IR spectra of the media. The attenuation of oscillations generated by the introduction of nanocomposites after the time interval specific for the pair "nanocomposite–medium" has been registered.

Thus to modify compositions with fine suspensions, it is necessary for the latter to be active enough that should be controlled with IR spectroscopy.

Up to now, we have carried out the laboratory and industrial experiments together with research and manufacturing divisions to modify inorganic and organic composite materials (e.g., concretes, foam concretes, epoxy binders, glue compositions, PVC, PC, polyvinyl acetate, and fireproof coatings). A number of results of material modification with finely dispersed suspensions of metal or carbon nanocomposites are given, as well as the examples of changes in the properties of modified materials based on concrete compositions, epoxy and phenol resins, PVC, PC, and current-conducting polymeric materials:

- The introduction of metal/carbon nanostructures (0.005%) in the form of fine suspension into polyethylene polyamine or the mixture of amines into epoxy compositions allows increasing the thermal stability of the compositions by 75–100°C and also improving adhesive characteristics of glues and lacquers.

- Hot vulcanization glue was modified by toluene-alcohol finely dispersed suspensions of copper/carbon or nickel/carbon nanostructures. On the test results of samples of four different schemes, the tear strength σ_t increased up to 50% and shear strength τ_s up to 80%, concentration of metal / carbon nanocomposite introduced was 0.0001–0.0003%.

- The introduction of fine suspension of copper/carbon nanostructures (0.01%) leads to the significant decrease in temperature conductivity of the material (by 1.5 times). The increase in the transmission of visible light in the range 400–500nm and decrease in the transmission in the range 560–760nm were observed.

- The PVC film modified containing 0.0008% of NC does not accumulate the electrostatic charge on its surface.

- The introduction of nickel/carbon nanocomposite (0.01% of the mass of polymer filled on 65% of silver micro particles) into the epoxy polymer, hardened with polyethylene polyamine, leads to the decrease in electric resistance to 10^{-5}Ohm·cm (10^{-4}Ohm·cm without nanocomposite).

Thus, Metal/Carbon Nanocomposites, obtained in nanoreactors of polymeric matrixes are effective for the different polymeric compositions.

Recently the unique properties of nanostructured composites formed in the process of nanoparticle introduction into composite materials have been attracting close attention of researchers. These properties are determined by nanometric sizes of the particles introduced and are not observed for bigger ones. Researches in this field demonstrate that nanostructured materials possess much better characteristics than similar compositions with chaotic particle layout. Nanoparticles are able to stimulate self-organization processes [1]. This feature of nanoparticles makes it possible to define the part of nanotechnology linked with their application as the technology that allows using the capability of nanoparticles to stimulate self-organization of systems for directed production of materials with the required properties.

It is rather perspective to use the nanocomposites obtained to improve the properties of composite materials. The introduction of such nanocomposites in super small quantities (0.0001–0.001% by mass) into materials has a positive effect on their structure and properties.

A more complex task during the modification of composite materials is the introduction of nanoadditives into the composite with uniform distribution of the additive in the material volume. Currently the obtaining of fine suspensions and colloid solutions of nanoparticles in various media is a widely spread and standard method for uniform distribution of nanoparticles.

Up to now we have carried out the laboratory and industrial experiments together with research and manufacturing divisions to modify inorganic and organic composite materials (e.g., concretes, foam concretes, epoxy binders, glue compositions, PVC, PC, polyvinyl acetate, and fireproof coatings).

11.2 PROPERTY CHANGES OF CONCRETE AND ANALOGOUS MATERIALS, MODIFIED BY METAL/CARBON NANOCOMPOSITES

The modification of foam concrete and dense concrete is usually carried out for the increase of compressive strength. Therefore, the investigations of foam concrete strength changes depending on content and nature of metal/carbon nanocomposites are accomplished. Recently the results of these investigations were given in Part 7 on Figure 7.30. The dependence of foam concrete strength on Cu/C nanocomposite content (in %) is given below (Figure 11.1).

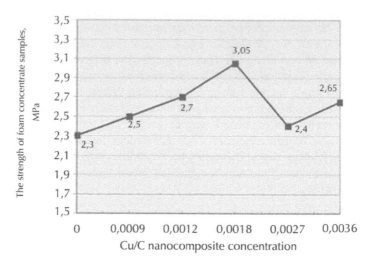

FIGURE 11.1   The dependence of foam concrete strength on Copper/Carbon nanocomposite content (%).

The introduction of super small quantities of nanocomposites leads to the increasing of foam concrete surface layer density (Figure 11.2).

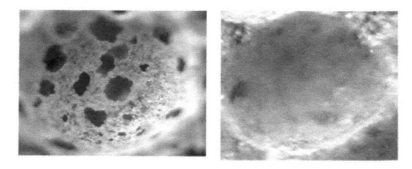

FIGURE 11.2   The foam concrete structure without nanocomposite and with the addition of nanocomposite (0.002%)(×90).

At the same time, the kinetics of these material's strength growth is studied (Figure 11.3). In this case, the modification of foam concrete is realized with the use of nanocomposite's finely dispersed water suspension. According to the results, adducted on Figure 11.3, the strength increasing of foam concrete modified by nanocomposite is more than 80% in comparison with foam concrete without nanocomposite. After 56 days, this increasing is decreased to 30%. It is possible that the re coordination of nanocomposite supermolecule with composition molecules takes place during the period of composition hardening. Analogous effect is observed during the modification of dense concrete (Figure 11.4).

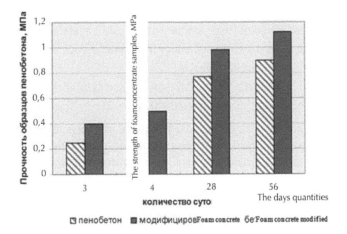

FIGURE 11.3   The kinetics of the growth of foam concrete strength.

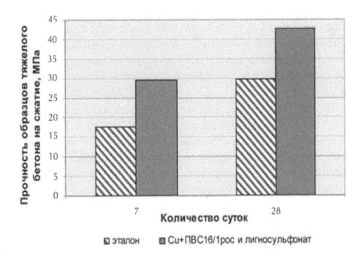

FIGURE 11.4   The kinetics of the dense concrete growth.

The investigation results of kinetics of dense concrete strength growth during 7 and 28 days are given in Figure 11.5. After 7 days, the growth of modified dense concrete strength more than the increasing of strength of non modified dense concrete by 66% and after 28 days, the increasing of strength was about 43%. Owing to large density of the dense concrete composition, the nanocomposite recoordination process became difficult.

According to investigation results, the following conclusion may be made:

- The production of foam concrete modified by metal or carbon nanostructures leads to the improvement of its characteristics by 1.5–2 times and to the increase of its durability.
- The strength characteristics of dense concrete modified by metal / carbon nanocomposites increase by 30–50%.

The introduction of a modifier based on metal/carbon nanocomposites into the composition, results in medium structuring and decrease in the number of defects, thus, improving the material's physical and mechanical characteristics.

11.3   RESULTS OF SILICATES MODIFICATION WITH THE USE OF METAL/CARBON NANOCOMPOSITES

11.3.1   Influence of Metal/Carbon Nanocomposites on Silicate Composition Properties, Including Properties of Liquid Glass

The modification of silicates and liquid glass with nanocomposites containing metal clusters is perspective. The interest to liquid glass is conditioned, first of all, by its ecological friendliness and production and application simplicity, inflammability and non toxicity, biological stability, and raw material availability. Nanostructures can be used to improve characteristics of paints such as elasticity, adhesion to the base, hydrophobic behavior and also helps solving the problems connected with coating flaking-off from the base, discoloration, limited color range, and so on, Due to the unique properties of nanostructures we can apply new properties to silicate paints, for example, to produce the coating protecting from electromagnetic action. The application of nanostructures in silicate materials to improve their stability to external action is known [1]. The introduction of nanosize structures into the liquid glass allows qualitative change of the material's behavior in a positive way. This possibly occurs in the process of binder super molecular structure change.

Based on the aforesaid, it is important to improve operational properties of compositions on liquid glass basis, modifying them with metal/carbon nanocomposites, providing a wider field of their application.

11.3.2   Equipment and Investigation Technique

Spectrophotometry

To define the interaction between liquid glass and nanocomposite, the optical density of liquid glass sample films with nanostructures was investigated. The films were produced applying a thin layer of silicate modified with nanostructures onto transparent film with further drying to remove the moisture. Films without nanocomposites were used as reference. The investigations were carried out with spectrophotometer KFK-3-01 in the wave range 350–950nm.

Ultrasound Processing of Suspensions

Spectrophotometer KFK-3-01 was also used to define the optimal action time interval on nanocomposite fine suspension processed with ultrasound. In the process of selecting the work wavelength (λ) on spectrophotometer KFK-3-01 to define the optimal time interval of ultrasound processing of fine suspension on liquid glass basis, the optimal wavelength of 430nm was found, as with this wavelength the optical density of liquid glass with NC before the US processing was 0.818, thus corresponding to the midpoint of the work range of this instrument (work range: optical density (D) = 0.001–1.5). At $\lambda = 430$nm, the optical densities of all fine suspensions investigated were found. The D was measured for the samples of nanocomposite fine suspension and liquid glass with the concentration 0.003% processed in ultrasound bath Sapfir UZV 28 with US power 0.5 kW and frequency 35 kHz. Liquid glass not modified with nanostructures was used as a reference sample. The processing time with maximum D was selected as optimal.

Investigation of Heat-Physical Characteristics

Heat capacity of the samples was investigated with dynamic technique measuring the specific heat capacity of solids with c-calorimeter IT-c-400 [4]. The measuring device was based on comparative technique of c-calorimeter with heat meter and adiabatic shell. Temperature delay time on heat meter was measured with stopwatch at room temperature and the specific heat capacity value was found by the following formula:

$$C_p = K_T / m_0 (\tau_T - \tau_T^0) \tag{11.1},$$

where C_p—specific heat capacity, J/kg·K; m_0—sample mass, g; K_T—instrument constant which depends on the temperature and is found by the instrument calibration; τ_T^0—temperature delay time with empty ampoule, s; τ_T—temperature delay time on heat meter, s.

λ-calorimeter IT-λ-400 based on monotonous heating mode was used to measure heat conductivity. In the experiment, the readings of n_0—temperature difference by plate thickness was measured with a stopwatch by the following formula to define the heat resistance:

$$R_s = ((1+\sigma_c)S_{n0}/K_T*n_T) - R_K \tag{11.2},$$

where, R_s—sample heat rtesistance, $m^2 \cdot K/W$; S—sample area, m^2; $\sigma_c = C_0/2(C_0+C_c)$—allowance taking into account the sample heat capacity C_0 (C_0—total heat capacity of the sample tested, J/(kg*K); found before with the instrument IT-s-400; C_c, K_T, R_K—constant of the measuring instrument. The time was measured with a stopwatch with the accuracy 0.01s.

Determination of Relative and Absolute Viscosity of Suspensions

Relative viscosity of the modified liquid glass was found with viscometer VZ-246 [5]. The viscometer was placed in the rack and fixed horizontally with the balance. The vessel was placed under the viscometer nozzle. The nozzle hole was closed with a finger, the material to be tested was poured into the viscometer to excess in order to have convex meniscus above the upper edge of the viscometer. The viscometer was filled slowly to avoid air bubbles. Then the nozzle hole was opened and as soon as the material being tested appeared, the stopwatch was started, then stopped and the discharge time was calculated. The values of relative viscosity were translated into the kinematic based on Russian Standard 8420-74.

The absolute (dynamic) viscosity of liquid glass was measured with a falling sphere viscometer (Geppler viscometer) intended for precise measurement of transparent Newton liquids and gasses. In accordance with Geppler principle, the liquid viscosity is proportional to the falling time being measured. The falling time of a sphere rolling down inside the inclined pipe filled with the liquid being tested was measured. The time required for the sphere to cover the distance between the upper and lower ring marks on the pipe with the sample was measured with a stopwatch with the resolution 0.01s. Further the values of dynamic viscosity were calculated.

IR Spectrometry Technique

IR Fourier-spectrometer FSM 1201 was used to obtain IR spectra of suspensions based on liquid glass and nanocomposites. The spectra were taken in the range of wave numbers 399–4500 cm^{-1}.

11.3.3   Preparation of Fine Suspension

According to reference sources, the most advantageous and urgent technique to introduce nanostructures into material is the application of fine suspensions of nanoproducts. They provide the fine modifier uniform distribution through the volume of the material modified.

The necessary concentration of nanostructures in suspension was chosen based on the basic binder mass and varied from 0.003 up to 0.3%. The suspensions were prepared in mechanical mortar, mixing nanocomposites with liquid glass following the pre-selected mode with further processing in ultrasound bath (Figure 11.5) to prevent the nanostructure coagulation. The metal in nanostructures will possibly interact with liquid glass forming insoluble silicates. Due to this the hydrosilicic acid is isolated from the liquid glass as an insoluble gel that results in compressing the liquid glass structure and increasing its moisture stability.

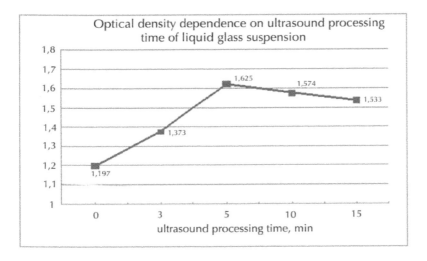

FIGURE 11.5   Diagram of suspension optical density on liquid glass.

11.3.4   Analysis of Compositions Based on Liquid Glass

Ultrasound processing: According to reference sources, the ultrasound technique is an effective disperstion method. Such techniques are widely applied today. However, the lengthy soaking in ultrasound baths can result in partial or complete decomposition of nanostructures and their re-coagulation. The decomposition of nanostructures due to lengthy ultrasound action is conditioned by high temperature and pressure during the cavitation. The coagulation can be caused with the decomposition of solvate shell on disperse phase particles. It is necessary to define the processing time interval during which the optical density will be at the maximum, that is, it will correspond to the maximal saturation of nanocomposite suspension.

To define the time interval of ultrasound processing, four suspensions were prepared for each soaking period—3, 5, 10, 15 min, respectively, and also the reference solution without ultrasound—for comparison. The concentration in all solutions was 0.003%.

Thus the optimal time interval for ultrasound processing is 5min. Further processing of suspensions for investigation will be carried out within this interval.

Spectrophotometric investigations: In accordance with the results of spectrophotometric investigations of modified films in the glue sample with Fe and glue with Ni, the shift of optical density is observed at some wavelengths indicating the changes taking place when nanostructures are introduced (Figure 11.6).

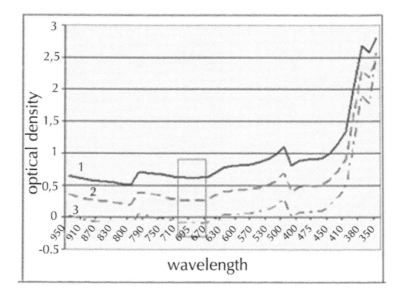

FIGURE 11.6   *(Continued)*

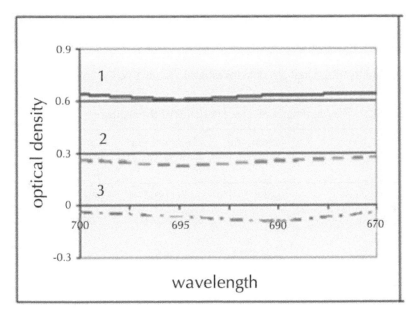

FIGURE 11.6   Changes of liquid glass (3) optical density in comparison with optical densities of liquid glasses, modified by Fe/C (1) or Ni/C (2) nanocomposites.

At other wavelength values, the optical density only increases indicating the possibility to apply nanostructures as coloring pigments in paints. At the same time, the increase in the optical density of films with nanostructures in comparison with liquid glass films possibly indicates the increase in the density of compositions and formation of new structural elements in them.

Investigation of heat-physical properties: To obtain the details of changes in heat-physical characteristics, the heat capacity and thermal conductivity of the samples based on cardboard and modified liquid glass were investigated. The samples were prepared gluing several layers of cardboard with liquid glass modified with nanostructures. The sample dimensions were found by the technique for measuring heat capacity and thermal conductivity. The sample being tested was 15mm in diameter and 10 mm in height. The sample dimensions were measured with micrometer with 0.01 mm accuracy. The sample mass was measured with the allowance not exceeding 0.001 g. Heat capacity of the samples was investigated on c-calorimeter IT-c-400. To find thermal conductivity the calorimeter IT-λ-400 was used. Further the specific thermal conductivity of the sample was calculated as follows:

$$\lambda = h/R_s \qquad (11.3),$$

where, λ—specific thermal conductivity, W/m*K; h—sample thickness, m. The results of thermal physical investigations are given in Table 11.1.

TABLE 11.1   Thermal physical characteristics of the samples.

Samples	Cardboard/glue	Cardboard/glue with Fe (change in %)	Cardboard/glue with Ni (change in %)
Density (kg/m³⁾	624.5	744 (↑19%)	669 (↑7%)
Heat capacity, C_{spec} (J/kg*K)	1,790	2,156 (↑20%)	2,972 (↑66%)
Thermal conductivity, λ (W/m*K)	0.083	0.061 (↓27%)	0.064 (↓23%)

Heat capacity of the sample, containing Fe/C nanocomposite, increased by 20% in comparison with the standard sample, that is, sample with Ni by 66%. At the same time, thermal conductivity decreased by 27 and 23% for the samples with Fe and Ni, respectively. So, when nanostructures are introduced, in the average the characteristics change as follows: density increases by 13%, heat capacity by 40%, and thermal conductivity decreases by 25%. Further, using the experimental results, the temperature conductivity is calculated by the following formula:

$$a = \lambda/c\rho \qquad (11.4),$$

where, a—temperature conductivity coefficient, λ—thermal conductivity coefficient, c—heat capacity, ρ—density.

Inserting the experimental data into the formula, we can define the temperature conductivity values (in percent):

$$a = \lambda/c\rho = (0.75\lambda_0/1.4c_0) \times 1.13\rho_0 = 0.47a_0 \qquad (11.5)$$

or calculate separately:

$$a_1/a_0 = \lambda_1\rho_0 C_0/\lambda_0\rho_1 C_1 = 0.51 \qquad (11.6)$$

$$a_2/a_0 = \lambda_2\rho_0 C_0/\lambda_0\rho_2 C_2 = 0.43 \qquad (11.7)$$

where a_1, λ_1, ρ_1, C1—characteristics of the sample with Fe/C NC, a_2, λ_2, ρ_2, C_2—characteristics of the sample with Ni/C NC, a_0, λ_0, ρ_0, C_0—characteristics of non modified sample.

Thus, temperature conductivity decreased by nearly 50% in comparison with the initial values (a_0).

When nanostructures are introduced, self-organization takes place. Nanoparticles structure the silicate matrix leading to the formation of new elements in the structure, thus increasing the material density and influencing its heat-physical characteristics. When additional structural elements and new bonds are formed, the system internal energy increases leading to heat capacity elevation and, consequently, temperature conductivity decrease. Thermal conductivity decrease of silicate paints when applied as a coating allows improving heat-physical characteristics of the whole protective structure of a building. In turn, temperature conductivity decrease results in decreasing the amount of heat passing through the coating, thus preserving adhesive characteristics of the coating for a long time.

IR spectroscopy: To determine the interactions between nanostructures and liquid glass, the IR spectroscopy investigations were carried out as well. The spectra were taken in relation to water as the suspensions contained nanocomposite solution with liquid glass and water (Figure 11.7).

The spectra were read with the reference tables. Water spectra demonstrate the region 2,750–3,750 cm^{-1} connected with O–H bond oscillations. They were observed in other regions as indicated by the values of wave numbers: 2,380, 3,850, 3,840 cm^{-1}. At the same time, the values 3,200, 3,500 cm^{-1} appropriate for valence bound OH appeared in the spectra of nanostructures on Fe basis. In the region 1,600–1,650 cm^{-1}, the bands of deformation oscillations of OH-groups were observed. Bands appropriate for the oscillations of Si–O–Me and Si–O–bonds (1,100–400 cm^{-1}) [5] were seen in low frequency region.

IR spectra contained wave numbers reflecting the interaction between metal / carbon nanocomposites and liquid glass, for example, peaks with wave numbers 407–420 cm^{-1} (Si-Me). The spectra of suspensions vividly demonstrated peaks at 1,020 and 1,100 cm^{-1}, appropriate for the oscillations of Si–O–Si and Si–O–C. However, in the spectrum of liquid glass containing Fe/C nanocomposite those values were shifted from 1,104 to 1,122 cm^{-1}, there was also the shift from 600 to 660cm^{-1} caused by the change in oscillations of some bonds under the nanocomposite action. The spectrum obtained resembles the liquid glass spectrum from reference sources [5].

The spectra demonstrate the absence of strong interactions, mostly the intensity changes, thus indicating the formation of a large number of definite bonds, for example, Si–O showing the structuring of liquid medium.

For suspensions on liquid glass basis, the relative and absolute viscosity were found. Taking into account the aforementioned data for defining the optimal ultrasound processing time, the suspensions for viscosity test were prepared with ultrasound pre-processing within 5min.

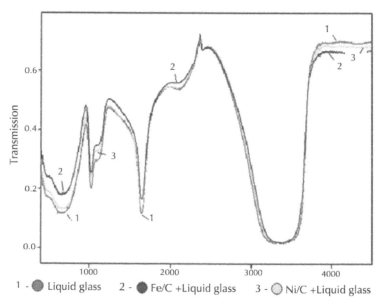

1 - ● Liquid glass   2 - ● Fe/C +Liquid glass   3 - ◯ Ni/C +Liquid glass

FIGURE 11.7   IR spectrum of suspensions on the basis of liquid glass and various types of nanocomposites (1—liquid glass, 2–Fe/C NC and liquid glass, 3–Ni/C NC and liquid glass).

Determination of relative and absolute viscosity of suspensions: Dynamic viscosity is the ratio of force unit required to shift the liquid layer for distance unit to the layer area unit. In metric system, it is given as dyne-second per square centimeter called "poise".

$$\eta = \tau \,.(\rho_1 - \rho_2)\,.\,\kappa\,.\,F \tag{11.8}$$

where, η—dynamic viscosity (mPa·s), τ—sphere movement time (s), ρ_1—sphere density (g/cm³) according to the test certificate, ρ_2—sample density (g/

cm^3), κ—sphere constant according to the test certificate (mPa·cm^3/g), F—work angle constant.

The experiment was carried out with pipe inclination of 80° for the sample.

$$=> F=1.0; \rho_1 = 15.2 \text{ g/cm}^3;$$

$$\rho_2 = 1.45 \text{ g/cm}^3; \kappa = 0.7.$$

Liquid glass viscosity: $\tau_{cp} = 9.60$s; $\eta = 9.6 \cdot (15.2 - 1.45) \cdot 0.7 \cdot 1 = 92.4$ Pa·s

Viscosity of liquid glass modified with iron containing nanocomposite: $\eta = 6.97 \cdot (15.2 - 1.45) \cdot 0.7 \cdot 1 = 67.09$ Pa·s

Viscosity of liquid glass modified with nickel containing nanocomposite: $\eta = 8.76 \cdot (15.2 - 1.45) \cdot 0.7 \cdot 1 = 84.32$ Pa·s

Kinematic viscosity is the ratio between dynamic viscosity and liquid density. Based on the data obtained when measuring the viscosity of liquid glass and fine suspension with Fe/C and Ni/C NC, diagrams were constructed (Figure 11.8 and 11.9)

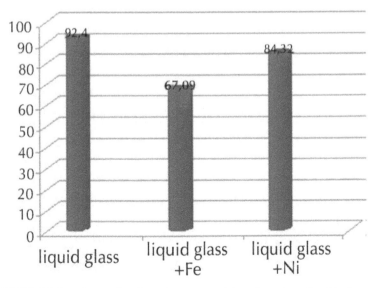

FIGURE 11.8 Dynamic viscosity compositions based on liquid glass and liquid glasses modified by Fe/C or Ni/C nanocomposites.

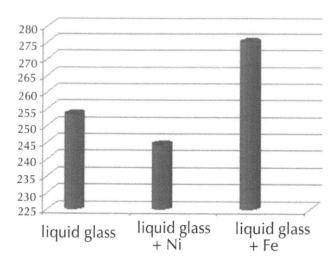

FIGURE 11.9   Kinematic viscosity compositions based on liquid glass and liquid glasses modified by Fe/C or Ni/C nanocomposites.

TABLE 11.2   Characteristics of kinematic and dynamic viscosity of liquid glass and its analogs modified with Ni/C and Fe/C nanocomposites.

Sample	Kinematic viscosity $(mm^2 \cdot s^{-1})$	Dynamic viscosity (Pa·s) (change in %)
Liquid glass	253.3	92.4
Liquid glass with Ni/C nanocomposite	244.3 (↓4%)	84.32 (↓9%)
Liquid glass with Fe/C nanocomposite	275.1 (↑9%)	67.09 (↓27%)

After analyzing this diagram, it can be concluded that when a nanocomposite is introduced, the viscosity decreases. Such phenomenon will have a positive effect in the process of silicate material production and application. For example, better application in silicate paint and much easier foaming process in foam glass.

In this diagram, it can be seen that nanostructures work differently. The viscosity of fine suspension with Ni/C NC is less than kinematic viscosity with Fe/C NC. This can be caused by coagulation and aggregation of Fe/C NC.

The viscosity of compositions on liquid glass basis depends on nanocomposite composition: in some cases the viscosity increase is observed, in other cases the decrease. The viscosity increase is connected with the elevation of intermolecular friction, which can be caused by the influence of nanoparticles on liquid glass structuring and density increase. The decrease in suspension viscosity can be explained by the fact that NC is surrounded by liquid glass molecules resulting in the decrease in their interactions and intermolecular friction.

Based on the results obtained, it can be concluded that nanostructures have a positive effect on liquid glass properties. The optimal time interval of suspension ultrasound processing was found experimentally. It equals 5min and corresponds to the maximum solution saturation with nanocomposites.

The data of optical density investigation of liquid glass films and IR spectra of liquid glass solutions indicate the changes in material structure when introducing metal / carbon nanocomposites.

The changes in heat-physical and viscous properties of suspensions on liquid glass indicate the self-organization processes when introducing nanostructures. At the same time, the results differ when introducing nanocomposites of different compositions.

The application of nanocomposites will allow increase in the life of paints, and consequently, their storage and covering ability. Also not only potassium but sodium liquid glass can be used as well, since potassium liquid glass is significantly more expensive than sodium one and its manufacturing is limited. To improve the characteristics only small concentrations of nanoproduct are required, which will positively affect the material production cost.

11.4 PROPERTIES OF EPOXY RESINS, GLUES AND PLASTICS BASED ON EPOXY RESINS MODIFIED BY METAL/CARBON NANOCOMPOSITES

The compositions based on epoxy resins of cold and hot hardening for different purpose (binder of plastics, epoxy compounds, glues and coating) are modified by Metal/Carbon Nanocomposites. It is shown that the introduction of Cu/C nanocomposite (0.02%) into binder based on epoxy resins for the reinforced glass plastics production lead to their strength growth on 32.2% (Figure 11.10).

FIGURE 11.10   The image of glass reinforced plastic armature.

The samples of epoxy polymer modified with Cu/C NC were produced by mixing the epoxy resin heated up to 60°C and fine suspension of Cu/C NC on PEPA basis in proportion 10:1. The mixing took 5–10 min.

It is noted [6] that, copper/carbon nanocomposite promotes the formation of well regulated polymeric net with epoxy groups conversion nearly 100% in the interval two times less than analogous interval of hardening of the composition without nanocomposite.

The formation of well regulated netted structure promotes the substantial growth of heat capacity modified materials in comparison with epoxy materials without nanocomposite (Figure 11.11).

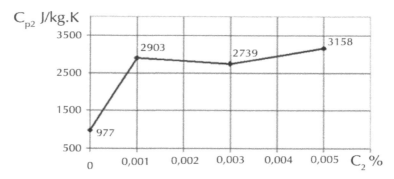

FIGURE 11.11   The modified epoxy resin heat capacity dependence on the copper /carbon nanocomposite quantities. According to the results obtained of the heat capacity determination, the heat capacity of epoxy materials modified by 0.005% Cu/C nanocomposite is 223% more than the heat capacity of non modified analog.

The formation of coordination bonds of nanocomposites with macromolecules of polymer and also the formation of additional net, the increasing of degree of epoxy groups conversion and the formation of well regulated structure stipulates the increasing of destruction beginning temperature to 110°C for epoxy polymeric materials modified by 0.005% copper / carbon nanocomposite (Figure 11.12).

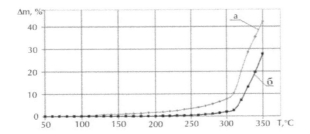

FIGURE 11.12 The epoxy polymers (non modified (a), modified (b)) mass losses dependence on temperature.

Adhesion

The adhesive strength of modified material in comparison with non modified analog to metals such as copper and steel is studied. Two methods used for the determination of adhesive strength are: original and standard.

In the first case, the following method is proposed:

A part of the polymer modified with NC was poured into the mold with copper wire to test the adhesive strength (Figure 11.13) and was further hardened in the mold. Another part was hardened in the form of plates to grind the samples for thermal gravimetric investigation.

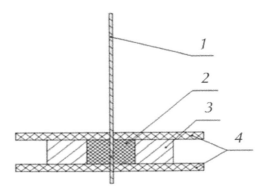

FIGURE 11.13 Sample for determining the adhesive strength of cold-hardened epoxy composition (CHEC): (1) copper wire; (2) hardened composition; (3) metal mold, and (4) anti-adhesive liner.

Two sets of samples were produced to determine the adhesive strength including the reference and modified FS of Cu/C NC with concentrations 0.001, 0.01, and 0.03%. For oxide film removal and degreasing, the copper wire was treated with 0.1M solution of hydrochloric acid and acetone.

Determination of adhesive strength: The adhesive strength of NC/EC was found with the technique described above. The tests were carried out on the tensile testing machine. The strength was found by comparing the breaking stresses of CHEC and NC/EC samples.

The tests for defining the adhesion of modified epoxy resin to copper wire were carried out on tensile testing machine, the values of destruction load were found. The adhesive strength was calculated by the following formula:

$$\sigma = F/A \ [\text{MPa}] \qquad\qquad (11.9)$$

where, F—average load values at which the breaking-off took place, (kgs); A—area of wire interaction with the hardened composition (9cm²)

$A = 2 \cdot \pi \cdot r \cdot h = 2 \times 3.14 \times 0.05 \times 1 = 0.314 \text{ cm}^2$, where r—wire radius = 0.05cm; h—height of metal mold = 1.00 cm.

The following diagram was prepared based on tests (Figure 11.14)

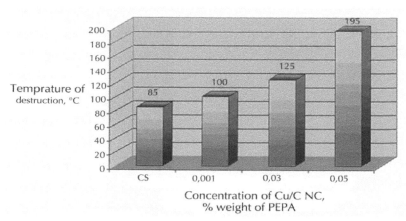

FIGURE 11.14   Dependence of adhesive strength of NC/EC on Cu/C NC composition.

The data from the diagram of adhesive strength (Figure 11.14) indicates that the strength maximum of the modified epoxy resin is reached when Cu/C nanocomposite concentration is 0.03%. Probably the maximum strength of the modified epoxy resin at this concentration is conditioned by the optimal number of a new phase growth center. The adhesion decrease with the concentration increase indicates that the number of Cu/C nanocomposite particles exceeded the critical value which depends on their activity [6]. Therefore, probably the number of cross-links in the polymer grid increased and the material became brittle.

The modification of industrial epoxy materials EZC–11 and ferric epoxy material with the using of 0.005% of Cu/C nanocomposite increases their adhesive strength more than 60%. These materials contain fillers, the particles of which are organized within well regulated net formed under the influence of nanocomposite. In this case the formation of filler interacted particles chains is possible [7–10].

Below the diagram of relative data on adhesive strength of named above industrial polymeric materials is given (Figure 11.15).

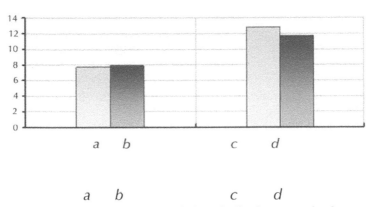

a   b   c   d

FIGURE 11.15   The relative characteristics of adhesive strength of epoxy material EZC-11 (a) and ferric epoxy material (b) in comparison with their analogs (c, d) modified by 0.005% Cu/C nanocomposite.

11.4.1 Thermal Gravimetric Investigations

For thermal gravimetric investigations also, two sets of samples were produced including the reference and modified FS of Cu/C NC with concentrations 0.001, 0.01, 0.03, and 0.05%.

Thermal gravimetric technique: Thermal stability was found by the destruction temperatures of modified epoxy resins. The temperatures of destruction beginning were determined by thermal gravimetric (TG) curves. Thermal balance DIAMOND TG/DTA was applied to obtain TG curves. TG curves of modified epoxy resins with NC concentrations 0.001, 0.03, 0.05% from PEPA weight were studied. The sample heating rate was 5°C/min.

To define the influence of nanocomposite on thermal stability of epoxy composition, a number of thermogravimetric investigations were carried out on reference and modified samples. The concentrations 0.001, 0.03, and 0.05% from PEPA weight were used. Based on the results of thermogravimetric investigations, the following diagram was prepared (Figure 11.16).

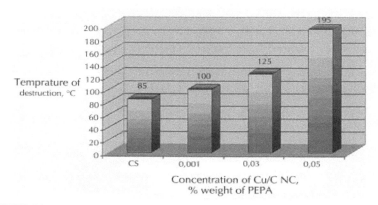

FIGURE 11.16   Dependence of thermal stability of cold-hardened epoxy composition on the concentration of copper or carbon nanocomposite.

When the concentration of nanocomposite elevates, the growth of polymer thermal stability is observed (Figure 11.16). The thermal stability growth is apparently connected with the increase in the number of coordination bonds in epoxy polymer.

According to the investigation on the modification results of cold hardened epoxy resins, the following conclusion may be made:

The test for defining the adhesive strength and thermal stability correlate with the data of quantum-chemical calculations and indicate the formation of a new phase facilitating the growth of cross-links number in polymer grid when the concentration of Cu/C nanocomposite goes up. The optimal concentration for elevating the modified epoxy resin (ER) adhesion equals 0.003% from ER weight. At this concentration, the strength growth is 26.8%. At the same time, the optimal quantity of Cu/C nanocomposite for elevating the modified industrial epoxy materials adhesion equals 0.005% that leads to the strength growth equals to 60.7%. From the concentration range studied, the concentration 0.05% from ER weight is optimal to reach a high thermal stability. At this concentration, the temperature of thermal destruction beginning increases up to 195°C.

The modification of hot hardened epoxy resins by mans of metal / carbon nanocomposites is carried out with the application of the finely dispersed suspension based on isomethyl tetraphthalic anhydrate.

The modification of hot vulcanization glue with copper/carbon and nickel/carbon nanostructures to increase the level of adhesive characteristics was carried out with the help of fine suspensions produced based on toluene. After testing the samples of four different schemes, the increase in the strength at detachment σ_{det} up to 50% and shear τ_{sh} up to 80% was observed, the concentration of metal /carbon nanocomposite introduced was 0.0001–0.0003% (Figure 11.17).

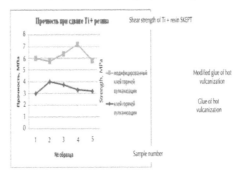

FIGURE 11.17   Results of glue compounds modification after introducing 0.0001% of copper or carbon nanocomposite.

The modification of conventional recipes of hot vulcanization glues with metal/carbon nanocomposites (51-K-45) leads to significant increase in adhesions characteristics on all glue boundaries investigated (Table 11.3).

TABLE 11.3   Results of samples tear and shear tests.

Tear strength σ_{tear} (kgs/cm²)			Shear strength τ_{shear} (Mpa)		
Conventional recipe of the glue 51-K-45	Modified 51-K-45		Conventional recipe of the glue 51-K-45	Modified 51-K-45	
	Ni/C	Cu/C		Ni/C	Cu/C
38.3	48.6	56.4	3.5	6.3	6.3

The application of these materials as adhesives for the gluing of metals and vulcanite is realized on the schemes "metal$_1$–adhesive$_1$–vulcanite–adhesive$_2$ metal$_2$". To define the adhesive tear and shear strengths, the above proposed scheme was used (Figure 11.18 and 11.19). The investigations carried out revealed that the modification of the conventional recipe of the glue 51-K-45 results not only in increasing the glue adhesive characteristics but also in changing the decomposition character from adhesive–cohesive to cohesive one.

The availability of metal compounds in nanocomposites can provide the final material with additional characteristics such as magnetic susceptibility and electric conductivity.

The modification of different materials with super small quantities of metal/carbon nanocomposite allows improving their technical characteristics, decreasing material consumption, and extending their application.

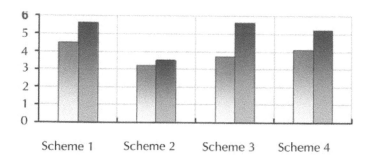

Scheme 1   Scheme 2   Scheme 3   Scheme 4

FIGURE 11.18   Relative changes of adhesive tear strengths of epoxy glues modified by metal/carbon nanocomposites (content of NC—0.0001%).

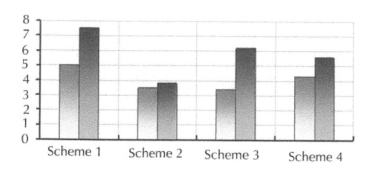

Scheme 1   Scheme 2   Scheme 3   Scheme 4

FIGURE 11.19   Relative changes of adhesive shear strengths of epoxy glues modified by metal/carbon nanocomposites (content of NC–0.0001%).

Thus the application of metal/carbon nanostructures is effective for the modification of epoxy compounds and binders on the basis of epoxy resins for the reinforced glass plastics and different epoxy glues.

11.5   RESULTS OF MODIFICATION OF FIREPROOF MATERIALS, FIRE RESISTANT INTUMESCENT COATINGS AND GLUES MODIFIED BY NANOSTRUCTURES

11.5.1   The Creation Problems of Fireproof Materials: Theory and Practice of Modification by Nanostructures

Previously [11], a number of criteria for providing the formation of fireproof surface layers were proposed including the carbon ones, the formation of which is connected with several reactions proceeding during the action of heat flows upon the materials. At the same time, the approaches for assessing the possibility of such layers formation and methodology for selecting components of fireproof polymeric materials and coatings were developed. The assessments and methodology proposed are quite applicable and can be adapted to fireproof nanocomposites. It is of interest to consider generalized results based on experimental data and theoretical models of fireproof materials and coatings formation taking into account the formation of nanostructures of certain shape and composition in them.

Comparative Evaluation to Protect Materials from Fire

Based on the experience gained on protecting combustible materials from fire, the main variations are in decreasing the possibility of exothermic redox reaction, this can be expressed in oxidizer and/or reducer concentration decrease below a certain limit, temperature drop below critical. This implies considerable changes during ignition and combustion. The ways of protecting polymeric materials from high-temperature flows or flame when using fireproof coatings or introducing fire-retardants into materials are known [11]. However, this and other ways for reducing combustibility have certain disadvantages. First, this can cause coating peeling during exploitation or combustion source action. Second, the changes in material basic characteristics sometimes considerable can sharply reduce its application fields. Therefore, when fireproof materials are applied, the changes in thermal-physic and adhesive characteristics during fire action are very important, but when fire-retardants are introduced into polymeric materials, it is necessary to decrease the content of additives introduced into the material to an effective minimum increasing their activity to reduce combustibility. In both cases nanophases of certain extension are formed, thus stimulating the formation of organic and inorganic cokes, preventing the destruction of basic mass of polymeric materials. During the combustion of fireproof polymeric materials the mixed combustion takes place [12], that is, in coking process the regularities of heterogeneous combustion appear when unstable combustion takes place [13] if a coked surface occupies over 80% of the surface subject to combustion. When introducing active additives representing nanostructures that can have the properties of nanocatalysts into polymeric matrixes, nanophases stimulating self-organization of matrix macromolecules into the "blank" of future coked surface (S_c) can be formed. In this case, S_c can be written down as a formula resembling Silberberg equation [14]:

$$S_c = 1 \cdot E(\%) \cdot k_s \cdot k_e \qquad (11.10),$$

where, 1—coordination number of fire retarder active element participating in the interaction or maximum possible number of active groups in fire retarder or fire-retardant system; E (%)—active element optimum content in fire-retardant; k_s—number of macromolecules of first layer adsorbed on fire-retardant active surface; k_e—number of macromolecules of second layer included into the nanophase being formed.

Product $k_s \cdot k_e$ depends on adsorption variants and interactions between fire-retardants and macromolecules of matrix polymers. The decrease of fire-retardant particles and their uniform distribution in a material contributes to the increase of nanophase and "blank" of coked surface, and consequently, to the decrease of active element content. The combustion values of phosphorus-containing polyesters and semi-empirical calculations that arise from the following aspects are given as an example:

- Increase in active group or active element content increases their intermolecular interaction, thus leading to the association and formation of particle aggregates; the surface of which is depleted by the centers actively interacting with macromolecules.
- Decrease in active element content during the dispersion degree growth contrary to the aforesaid, intensifies the chemisorption and structuring of polymer macromolecules by analogy with structure-forming agents [13].
- Since the coked surface cannot exceed 100%, the number of active phosphorous-active groups is less than 4 and experimental effective phosphorous content is 5.9% [11], the product of $k_s \cdot k_e$ approaches 4. At the same time, with the decrease of adsorbed molecules number on active centers, the amount of changes in the structure of macromolecules of second layer increases, that is, coefficient k_s characterizing the chemisorption of macromolecules on active surfaces of fire-retardant decreases together with the growth of active element content and coefficient k_e characterizing the aggregation degree increases.

From the above theses, it follows that the dependences of combustibility and other characteristics, for instance, physic-mechanical ones and density properties upon fire-retardant content will be described with curves having extremes (maximums and minimums). The combustibility assessment during weight losses can be inversely proportional to coked surface percentage in combustion area (Figure 11.20).

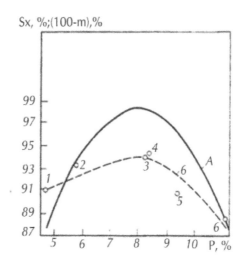

FIGURE 11.20 Dependence of Sc upon phosphorus content.

Block curve is calculated on the basis of the aforesaid theses, and dashed-line curve represents experimental data for phosphorus-containing polyesters of similar structure (Table 11.4), when Sc ~ (100-m)%.

TABLE 11.4 Changes in weight losses during the combustion with phosphorus content changes

Phosphorus-contain-ing polyesters	Phosphorus content (%)	Weight losses during combustion (%)	100-m (%)
PPE 1	4.7	10	90
PPE 2	5.7	6.5	93.5
PPE 3	8.0	5.5	94.5
PPE 4	8.2	5.4	93
PPE 5	9.0	7.0	93
PPE 6	11.2	11.5	88.5

Surely there is no complete coincidence of curves, but their character and location of maximums confirms to assessments made.

Analogous dependences were obtained experimentally for phosphorus–vanadium-containing polyamides (Figure 11.21).

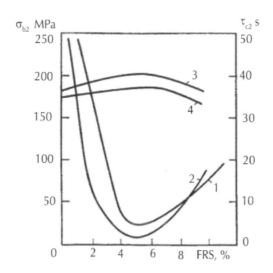

FIGURE 11.21 Time changes in independent combustion τcr (1, 2) and bending breaking stress (3, 4) depending upon the content in polyamides (kapron 1, 3 and nylon 2, 4). Fire-retardant systems (FRS) $[P\text{-}Ba(VO_3)_2]$.

In this case, the combustibility is determined by the duration of independent burning, and physic-mechanical properties are represented by bending breaking stress.

The fire-retardant content and the proportion of components in complex fire-retarding systems are suggested to be assessed taking space-energy parameters into consideration. If protective carbon layer is formed on material or coating surface, the fire-retardants applied must be transformed into a complex structure with interface distance close or slightly over the interface distances of graphite-like substances.

Energy characteristics must correspond to the transition of chemical bonds of one type into another. If the half-sum of destructing and forming bonds is less or equal to the adsorption potential, the process is possible. Such an approach is based on concepts of multiple catalysis theory. For polyesters containing phosphorus-oxide groups, the schemes of bond transformation in polyester with couples P–O (Figure 11.22) can be recorded.

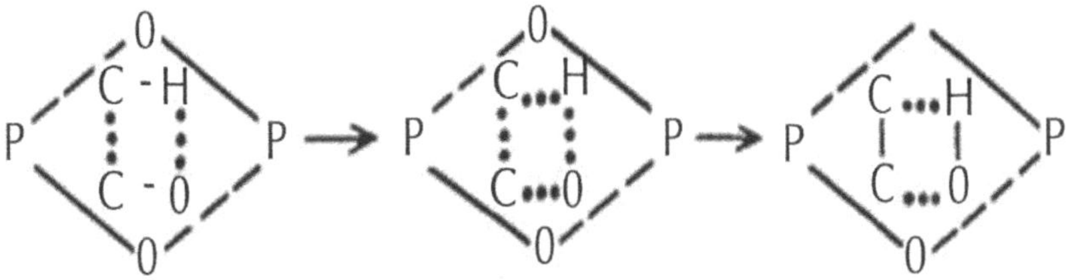

FIGURE 11.22   Scheme of the formation of multiplets in phosphorus-containing polyester.

Interface layers d (polyphosphates or polyphosphoric acid) are taken as equal to the sum of Van der Waals radii $d = R_p + R_0 = 3.3$pm. Adsorption potential q is expressed through half-sums of electronegativities of atoms participating in the interaction of active centers of fire-retardants and chemisorbed atoms of macro-molecule fragments and equals to q = 2.72. Half-sum of energies of destructing and forming bonds is calculated by the analogy through the half-sums of electronegativities of atoms participating in the bonds. For polyesters—$\Sigma E/2 \approx 2.65$. Since $q \geq \Sigma E/2$, the reaction is directed at the primary water and carbon substances formation. The application of phosphorus–vanadium-oxide fire-retarding systems, when assessing carbonization products is more favorable in comparison with phosphorus-oxide systems or vanadium-oxide systems, since the percentage of fire-retarding phosphorus–vanadium-oxide system appears to be less due to the increase of coordination number 1. Depending upon vanadium coordination saturation, the content of coked surface changes, the thickness of boundary layers and "embryo" of coked surface increase together with coordination number growth. Thus, the registration of coordination processes during the formation of compositions and fire-retardant systems gives the possibility to predict and assess the efficiency of fireproofed material or coating. At the same time the assessment is made on the level of molecular and atomic interactions with nanophase formation, the appearance of which is conditioned by the formation of boundary layers.

Efficiency of Nanostructure Application in Polymeric Materials

The formation of nanophases and nanostructures of different composition and structure is possible when fire-retardant systems in the form of powders or liquids are introduced into the composition.

In case of liquid components mixing, their thermodynamic interaction with the formation of extensive phases is possible. However, it is known [15, 16] that even polymerhomologs restrictedly dissolve in each other. Therefore in multi-component composites, isotropy and density of materials are ensured by the distribution of components and phases from the surface to the volume of samples. At the same time, boundary layers are formed between phases, which have properties slightly different from the properties of contacting phases and which can be identified with nanophases. The density of nanophases can be intermediate between the planes of interacting phases (Figure 11.23a) or can be below the density of both phases if there is no interaction between the phases (Figure 11.23b).

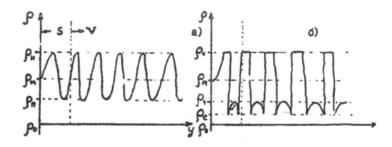

FIGURE 11.23 Diagram of density with the formation of nanophase with high density (a) and nanophase with low density (b).

When the sizes of phases decrease, their activity and interaction between them increase, thus leading to the growth of nanophase density. If the sizes of component phases are in the range of nanophases, then the composite density as a whole increases together with heat capacity due to the appearance of dense boundary layers (nanophases).

Therefore, the formation of nanostructures in composites can results in the decrease of temperature conductivity of the material. This is confirmed by experimental data based on thermalphysic characteristics of elastic material, in which, when introducing fire-retarding phosphorus-vanadium-containing system nanophases are formed and when the temperature grows from 20 to 200°C the heat capacity changes to a greater extent than thermal conductivity:

	Non modified material	Modified material
C_p (kJ/kg·°C)	1.47 → 1.89	1.66 → 2.54
	30→ 200°C	20→ 200°C
λ (Wt/m²·°C)	0.26 → 0.39	0.27 → 0,33
	30 → 200C	30 → 200°C

In case fire-retardant systems of lamellar structure or lamellar-forming structure ones are introduced, when distributing in polymeric matrix the conditions for the formation of extended nanophases can arise, in which one of the dimensions exceeds the limits of nanometer formations. These can be fibrillar, lamellar or tubular structures with the length over 1mcm. In such formations as nanocontainers the transformation of chemical particles with the formation of new nanophases increasing the total density and stability of materials to the action of heat and flame sources are possible.

At the same time, as follows from literature, for instance [16], there are interactions, and on the primary structure the secondary one "directed" is formed. After fire-retardants are introduced into the compositions, the formation of nanophases contributing to heterogeneous coke-formation catalysis leads, in some cases, to their "memorizing" and repetition of corresponding reflexes in difractograms of pyrolysis or coke residues (Figure 11.24).

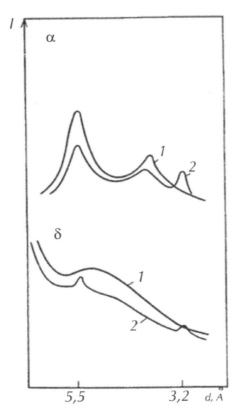

FIGURE 11.24   Broad-angle difractograms of polycarbonates not containing (1) and containing (2) phosphorus–vanadium-containing system (a) and corresponding pyrolysis residues (b).

As shown in the Figure 11.24, the nanophase appeared during the introduction of phosphorus–vanadium-containing fire-retarding system is preserved on broad-angle difractograms of the residues of fire-protected polycarbonate pyrolysis.

Thus, nanophases appearing in polycarbonate are stable to high thermal loads. Therefore, it can be expected that the formation of nanocomposites containing a lot of nanophases can lead to the increase of their stability to mechanical and thermal loads.

At the same time, the presence of directed processes along the layers formed in the material can lead to material reinforcement and anisotropy appearance. For fire-retardant coatings, this property contributes to heat flow along the surface in a certain direction. The appearance of nanostructures of certain directivity in the material contributes to the formation of double electric layers and catalytic processes of fire-retardant coating formation.

Perspective materials are fire-retardant intumescent coatings, the introduction of nanostructures into which allows increasing the coating adhesion to polymeric material being protected and the strength of foamcoke being formed. What is the action mechanism of nanostructures being formed or introduced? When fire-protected polymeric materials are formed by adding combustion fire-retardants to polymeric compositions, which are structure-formers and nanophase formation stimulators in materials, the increase of "embryos" of coked surface is observed when the temperature goes up. In turn, the quick formation of coked surface layer having good adhesion to the bulk of material prevents the development of combustion process under the action of one and the same fire source. More often the heat flow along the surface increases, and in some cases this leads to the significant decrease of mass combustion rate at comparatively high flame spreading velocity. In these cases it is necessary to carry out additional surface treatment of polymeric materials with thermal shocks. At the same time, additional nanostructures contributing to the decrease of amount of volatile and highly inflammable products under the action of fire sources onto the materials appear in interface spaces, nanocontainers in a way. The density of surface layers and thermal capacity considerably increase, and thermal conductivity decreases along the normal to the surface, therefore sharp decrease of temperature conductivity is observed.

The introduction or formation of nanostructures in fireproof intumescent coatings leads to the formation of foamcokes with regularly located closed pores and sufficiently strong bubble walls. Moreover, depending upon the location and shape of nanostructures, the gas bubbles from gas-forming agent are distributed in foamcoke bulk in a certain order. If fine powders of ammonium polyphosphate (APP) or its derivatives are used as gas-forming agents and dispersed with ultrasonic field in the composition, it is possible to approach the model system, in which the distribution of fire-retardant and gas-forming agent particles, as APP, is practically uniform (Figure 11.25)

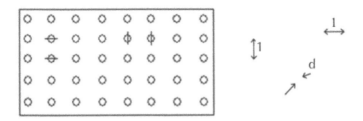

FIGURE 11.25   APP distribution in epoxy composition.

The Figure 11.25 shows a simplified scheme of particle location. The particles have a ball shape, their size is about 10 mcm and the distance between them is about 1mm. The foamcoke formed represents carbonaceous material with bubbles filled with the mixture of ammonium and water. At the same time a carbonaceous barrier containing the layer of polyphosphoric acid on the inner surface is formed between the bubbles. In contrast to the previous coatings, in which, according to X-ray photoelectron spectroscopy data the surface is carbonized, in intumescent coatings first the carbonization of inner wall of bubbles being formed takes place [17]. The introduction of carbon-metal-containing nanostructures obtained from polyvinyl alcohol and metal salts into intumescent fireproof compositions [18, 19] contributes to the increase of location regularity of bubbles being formed from PVA in foamcokes. At the same time the strength of foamcoke formed increases [20]. Such a result is conditioned by the decreased velocity of ammonium bubble formation due to the coordination of ammonium molecules on metal clusters, for instance, copper, nickel or cobalt, which are placed in carbon shells open from the ends. Such carbon-metal-containing tubules are filled with metal phases up to 60–70%. They mainly represent twisted branched structures 40–60 nm in diameter and 300–1000 nm long. The bubbles formed from APP partly sorb on the surface of these structures. At the same time, foamcokes during their formation become more stable to mechanical loads due to the formation of new carbon nanophases on the "frame" of nanostructures introduced. Thus, when the intumescence degree slightly goes down, the foamcoke strength and stability under various loads increase.

Possible Models of Carbonization Processes and Foamcokes Formation

Modeling of carbonization process at the temperatures of polymeric materials decomposition can proceed with and without formation of liquid phase [21]. The first variant can be represented in one or two stages. In two-stage process intermediate formation of liquid stage is possible, which speedily "structures" into solid amorphous and crystalline products or transforms into gaseous products.

When putting a task to model processes in two-stage process, which in practice is used more often, it is necessary to take into consideration possible reactions
responsible for structuring and reactions resulting in coking. In general form reactions proceeding during structuring can be given as:

$$k_{is} = (1-\beta)^{n_{is}} k_{is}(T) \qquad (11.11),$$

where, β—share of decomposed active component, n_{is}—order of 1st structuring reaction. The main difficulties when building up the models given in details
in monogram [21], arise when determining the proceeding possibilities of this or
another reaction its way, velocity and activation energy constant. When building
up the models the application of computation schemes of quantum chemistry,
molecular mechanics, and thermal-dynamic concepts based on activated complex
theory is possible. It is advisable to use adapted programs—Hyperchem, Gamess,
and Gaussian. Fragments of molecular structures are picked out, based on which
interactions can proceed and which can be identified with reaction centers. In
reaction centers the transformation processes of one bond into another take place.
The calculations of energies of molecules and molecular fragments participating
in the interaction are carried out with the help of Hartry–Fock method in minimum basis 3-21G. In this process program complex GAMESS is used [22]. In the
frameworks of energy model mentioned, the balanced atomic geometries of fragments and interaction energy of reaction centers are determined. When kinetic parameters such as activation energy and pre-exponential factor k_0 are determined,
ab initio calculations using activated complex theory are carried out [23]. In case
of reactions connected with intermolecular rearrangement of chemical bonds, the
system transition state was looked for. Afterwards, the system movement along
the reaction coordinate via transition state obtained (activated complex) was investigated. The reaction activation energy is calculated as the difference between
the energies of activated complex and initial state taking into account the energies
of zero oscillations. This value is a criterion of reaction proceeding possibility.
The pre-exponential factor is calculated based on the analysis of statistical sums
of transition state and initial molecular fragments or reaction centers in accordance with the following equation:

$$k_0 = \frac{k_{\ddot{a}}T}{h} \frac{z_{act}}{z_0} \qquad (11.12),$$

Where, $k_{\scriptsize e}$—Boltzmann constant, h—Plank constant, z_{act}—statistical sum of
transition state, z_0—statistical sum of initial molecular fragments.

If reactions were connected with molecular fragment breaking-off, the activation energy was assessed as the difference between the total energy of isolated molecular fragments formed during the reaction and initial state energy. At the same time the transition state energy is usually higher than the total energy of fragments formed. Therefore, both approximations partly compensate each other. To assess errors, calculations were made in the frameworks of semi-empirical method PM3 taking configuration interaction C1 into account [22]. The surroundings influence upon the process was estimated with the help of molecular-dynamic modeling in the frameworks of the model of molecular mechanics MM+ [24]. The proposed approach was used in all the cases when the directivity of fast-proceeding reactions including combustion, carbonization, and heterogeneous catalysis processes was predicted. The process can proceed with liquid phase formation, in which convection processes can occur. The process gets complicated in case of radiation from soot particles to the surface. Therefore, it is necessary to take these circumstances into account when carrying out a computation experiment for reactions proceeding in corresponding areas and phases considering nanophases and interactions between them. Computing experiment of pore-formation acquires extreme importance when modeling carbonization or mineralization processes. If volume shares of crystalline and amorphous cokes increase in the process of liquid decomposition with the formation of gaseous phase, the porosity of material condensed layers being formed decreases. Bubbles appearing as a result of the gas-formation and evaporation processes exceeding over condensation, form a summation distributed in accordance with sizes and satisfying continuous Zeldovich equation [25]. The bubbles are supposed to grow further until their volume share reaches the porosity value. Afterwards the bubbles can merge into one big bubble and gaseous substances will flow out from it as a result of bubble integrity breakage due to the decrease of bubble wall adhesion strength with basic partly pyrolyzed material. The problem of obtaining a stable strong foamcoke is solved in the models of fireproof intumescent coatings [26], on the example of epoxy-polymer hardened with polyethylene-polyamine with polyammonium-phosphate (PAP) added. At the same time, minimum eight gross reactions are recorded that proceed in temperature-time and spatial zones. These zones are limited by gas-forming agent surface, inner surface of bubble being formed (polymer-gas boundary) and bubble cavity, in which the mixture ammonium-water is being formed. At high destruction temperatures and oscillations by bond with low frequency,

statistic oscillating sums can be used— $z_k = \dfrac{kT}{h\nu}$, where ν—oscillation frequency of breaking chemical bond. If the bond oscillations correspond to the transfer along the reaction coordinate, it can be said that $k_0 = \frac{kT}{h} \frac{z^{\dagger}}{z_A}; \partial$, and since the oscillations of chemical bonds are characterized by wave numbers, $k_0; \omega c$, where ω—wave number, c—light speed. The calculations of pre-exponential factor for monomolecular reactions carried out with the help of quantum-chemical methods are correlated with the values obtained experimentally, and without experimental data with numeric values of frequencies of breaking bonds. For bimolecular reactions the calculations of k_0 were carried out using semi-empirical method by statistic rotational sums:

$$k_0 = \frac{h^2 Z_{RT}}{(2\pi)^{3/2} k_B^{1/2} (m^*)^{3/2} T^{1/2} (Z_A)_{RT} (Z_B)_{RT}} \qquad (11.13)$$

for reaction A + B → C + D, where $m^* = \dfrac{m_A \cdot m_B}{m_A + m_B}$ reduced mass of fragments A and B, $(Z_A)_{RT}$ and $(Z_B)_{RT}$—statistic rotational sums A and B, Z_{RT}—statistic rotational sum of transition state. Computing of statistic rotational sums is carried out based on the following formula:

$$Z_{RT} = \frac{8\pi^2 (k_B T)^{3/2} (8\pi^3 I_A I_B I_C)^{1/2}}{\sigma^* h^3} \qquad (11.14)$$

where, σ^* corresponds to the section of impacts and is found in accordance with the following formula:

$$\sigma^* = \frac{2^{1/3}}{8N^{2/3}} \left[\left(\frac{M_A}{\rho_A} \right)^{1/3} + \left(\frac{M_B}{\rho_B} \right)^{1/3} \right]^2 \qquad (11.15)$$

M_A and M_B—molar masses of interacting fragments, ρ_A and ρ_B—densities of corresponding reagents, I_A, I_B, I_C—inertia moments of fragments.

To find inertia moments it is necessary to pick up the coordinate system and rule to select distances in activated complex. It is decided that reacting atoms in activated complex are located at the distance equal to the sum of Van der Waals radii, and changes in valence angles in activated complex are insufficient. Inertia moments are found in relation to the selected center in coordinate system based on the following scheme:

$$I_x^{(A)} = \sum M_i\left(z_i^2 + y_i^2\right) \quad I_{xy} = \sum M_i x_i y_i$$

$$I_y^{(A)} = \sum M_i\left(x_i^2 + z_i^2\right) \quad I_{xz} = \sum M_i x_i z_i \quad I_A I_B I_C = \begin{vmatrix} I_x - I_{xy} - I_{xz} \\ -I_{xy} I_y - I_{yz} \\ -I_{xz} - I_{yz} I_z \end{vmatrix} \text{(11.16)}$$

$$I_z^{(A)} = \sum M_i\left(x_i^2 + y_i^2\right) \quad I_{yz} = \sum M_i y_i z_i$$

It should be noted that I_x, I_y, I_z_axial inertia moments in relation to any three mutually perpendicular axes going through the gravity center of interacting functional groups, and I_{xy}, I_{xz}, I_{yz}_products of inertia.

At the same time the coordinates of system gravity center are found from the following relations:

$$x^* = \frac{\sum M_i x_i}{\sum M_i}; \quad y^* = \frac{\sum M_i y_i}{\sum M_i}; \quad z^* = \frac{\sum M_i z_i}{\sum M_i}; \qquad \text{(11.17)}$$

Calculated value of k_0 by the equation

$$k_0 = \frac{h^2 \sigma^* \left(I_A^{\neq} I_B^{\neq} I_C^{\neq}\right)}{(2\pi)^{3/2} k_B^{3/2} \left(m^*\right)^{3/2} T^{1/2} I_A^A I_B^A I_C^A} \qquad \text{(11.18)}$$

Calculated k_0 for the reaction

-P(O)OH + HO(O)P → P(O)–O–(O)P- + H$_2$O

            ONH$_4$ ONH$_4$ ONH$_4$ ONH$_4$

equals to 10^{-14} cm^3/s that approximately corresponds to experimental values and values obtained with the help of quantum-chemical calculations.

Spatial-energy parameters can be applied to determine nanophase distribution in fire-protected composites and coatings, as well as to determine the interactions between [27, 28].

11.5.2   Modification of Phenol–Formaldehyde Glues for Obtaining Intumescent Fire Resistant Glues with the Use of Cu/C Nanocomposites

The glues based on phenol-formaldehyde resins (BF-19) were modified with copper/carbon nanocomposite and with phosphorylated analog. It was determined that nanocomposite introduction into the glue significantly decreases the material flammability. The samples with phospholyrated nanocomposites have better test results. When phosphorus containing nanocomposite is introduced into the glue, foam coke is formed on the sample surface during the fire exposure. The coating flaking off after flame exposure was not observed as the coating preserved good adhesive properties even after the flammability test.

The nanocomposite surface phospholyration allows improving the nanocomposite structure, increases their activity in different liquid media thus increasing their influence on the material modified. The modification of coatings with nanocomposites obtained finally results in improving their fire-resistance and physical and chemical characteristics.

Modification of the glue BF-19 with fine suspensions of metal or carbon nanocomposites including phosphorus containing ones

The glue BF-19 is intended for gluing metals, ceramics, glass, wood and fabric in hot condition, as well as for assembly gluing of cardboard, plastics, leather and fabrics in cold condition. The glue composition: organic solvent, synthetic resin (phenol–formaldehyde resins of new lacquer type), and synthetic rubber.

When modifying the glue composition at the first stage, the mixture of alcohol suspension (ethyl alcohol + Cu/C NC_{mod}) and ammonium polyphosphate (APPh) was prepared. At the same time, the mixtures containing ethyl alcohol, Cu/C NC and APPh, ethyl alcohol and APPh were prepared. At the second stage the glue composition was modified by the introduction of phosphorus containing compositions prepared into the glue BF-19.

Samples preparation for flammability determination: The samples to be tested are the plates with the dimensions 150 x 15 x 3mm. The plates consist of foam polyethylene and paper glued together with phosphorus containing glue modified with metal/ carbon nanocomposites with and without phosphorus. At the same time, check samples are prepared. These are the plates of foam polyethylene and paper glued together with the glue BF-19 filled with ammonium polyphosphate with phosphorus content in the glue 3, 4, 5% from its mass.

Technique of sample flammability testing: When studying the influence of Cu/C nanocomposites on the flammability of polymeric coatings on the basis of phenol-formaldehyde resins (PFR) to select the optimal composition of nanostructures, the lengths of carbonized parts of the samples with 1.5 minute flame exposure were determined.

Composition of sample coating: Glue BF-19 + APPh

To compare the results of coating flammability, three samples were selected and tested. The test results revealed that the length of carbonized part of the samples containing APPh and exposed to burner flame for 1.5min can be about 8.5 cm with 3% phosphorus content in the sample (Table 11.5).

TABLE 11.5 Results of testing samples containing Ammonium Polyphosphate.

Sample No	Sample composition	Phosphorus content (%)	Flame exposure time (min)	Length of carbonized part of samples (mm)
1	APPh	5	1.5	65.33
2	APPh	4	1.5	82
3	APPh	3	1.5	84.67
	Average			77.33

The tests of check samples confirmed that with the phosphorus content increase in composition the length of carbonized parts of samples goes down.

The composition of sample coating: glue BF-19 + APPh + Cu/C NC $_{pure}$.

The next step was to test samples containing nanocomposites without phosphorus content. The average value of the carbonized part of the samples was 21.81 mm. The test results (Table 11.6) allow making the conclusion that nanocomposite inclusion significantly decreases the material flammability (in 3.5 times).

Table 11.6 Results of testing samples of modified compositions.

Sample No	Sample composition	Phosphorus content (%)	NC content (%)	Flame exposure time (min)	Length of carbonized part of samples(mm)
4	Cu/C NC $_{pure}$ + APPh	5	0.00025	1.5	15.33

TABLE 11.5   *(Continued)*

5	Cu/C NC $_{pure}$ + APPh	4	0.0002	1.5	23.43
6	Cu/C NC $_{pure}$ + APPh	3	0.00015	1.5	26.66
	Average				21.81

Composition of sample coating: Glue BF-19 + APPh + Cu/C NC$_{mod}$

Phosphorus containing samples of Cu/C NC had better flammability test results than samples with APPh and samples containing APPh and Cu/C NC$_{pure}$. The length of the carbonized part of the samples was less by 3 mm in the average. The average value of the carbonized part of the samples was 18.89 mm (Table 11.7). Thus it can be concluded that the inclusion of phospholyrated nanocomposite decreases the material flammability to a greater extent than non phospholyrated nanocomposite.

TABLE 11.7   Results of testing samples of compositions modified by phosphorylated nanocomposites.

Sample No	Sample composition	Phos-phorus content (%)	NC content (%)	Flame exposure time (min)	Length of carbon-ized part of samples (mm)
7	Cu/C NC $_{mod}$ + APPh	5	0.00025	1.5	14.67
8	Cu/C NC $_{mod}$ + APPh	4	0.0002	1.5	16.67
9	Cu/C NC $_{mod}$ + APPh	3	0.00015	1.5	25.33
	Average				18.89

From the data demonstrated based on the test results, it can be concluded that nanocomposite inclusion into the glue composition significantly decreases the material flammability. The length of the carbonized part of the samples modified with nanocomposites was in 4.1 times in the average less in comparison with similar parameters of the samples not containing nanocomposites. The samples with phosphoryl groups in nanocomposites have better test results (Figure 11.7).

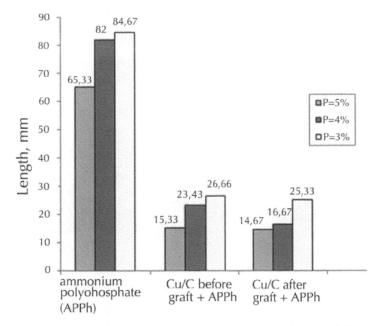

FIGURE 11.26   Diagram of the lengths of carbonized parts of the samples depending on phosphorus content in the composition.

The coating flaking off after flame exposure is not observed, that is, the coating preserves good adhesive properties even after the flammability test.

When the intumescent glue composition is modified with nanostructures, the material is structured with the formation of crystalline regions. In turn, such structuring under the influence of nanosystems results in the increased physical and mechanical characteristics including their stability against high and low temperatures.

The application of metal/carbon nanocomposites is perspective for the modification of polymeric materials on a large scale as this is described in [29–36].

11.6   PROPERTIES OF PVC AND PC MODIFIED BY METAL/CARBON NANOCOMPOSITES. ELECTRO CONDUCTED GLUES AND PASTES MODIFIED BY METAL/CARBON NANOCOMPOSITES

The modification of polymeric films based on PC or PVC with the use of metal or carbon nanocomposites decreases the antistatic quantity essential for the substantial decrease of electrostatic charge on their surfaces. Especially, this is necessary for the PVC films.

The PVC film modified by the finely dispersed suspension based on chloroparaffins contains 0.0008% Fe/C nanocomposite and does not accumulate the electrostatic charge on the surface.

At the same time, the crystalline phase in this material is increased. Image of final production of PVC films modified by Fe/C nanocomposite (0.0008%) is given in Figure 11.27.

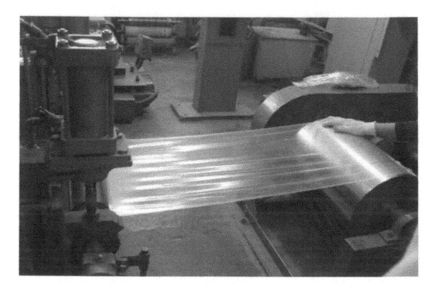

FIGURE 11.27   Final production stage of PVC film modified with Fe/C NC (0.0008%).

The material obtained completely satisfies the requirements applied to PVC films for stretch ceilings.

To modify PC based compositions, the Cu/C nanocomposite finely dispersed suspension based on mixture of methylene chloride and dichloroethane are produced. The introduction of 0.01% of copper or carbon nanostructures leads to the significant decrease in temperature conductivity of the material (in 1.5 times). The increase in the transmission of visible light in the range 400–500 nm and decrease in the transmission in the range 560–760 nm were observed.

The current conducted glues and pastes are obtained with the improved characteristics when the nickel/carbon nanocomposite is applied.

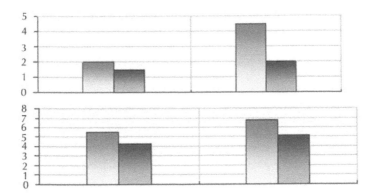

FIGURE 11.28   The tear and shear adhesive strength of current conducted epoxy material modified by nickel or carbon nanocomposite.

Table 11.8   The relative results of glues (pastes) current conductivity determination modified by the nickel or carbon nanocomposite.

Parameters (volume resistivity, Ohm·cm)	Current conducted paste	Current conducted glue
Non modified	$2.4 \cdot 10^{-4}$	$3.6 \cdot 10^{-4}$
Modified	$2.2 \cdot 10^{-5}$	$3.3 \cdot 10^{-5}$

According to Figure 11.28 and Table 11.8, the modification of polymeric materials by nickel or carbon nanocomposite leads to the increasing of their current conductivity as well as to the improve their adhesive properties.

KEYWORDS

- **IR spectrometry technique**
- **Metal or carbon nanocomposites**
- **Polymeric matrixes**
- **Thermal gravimetric investigations**
- **Viscometer**

REFERENCES

1. Kodolov, V. I. & Khokhriakov, N. V. (2009). *Chemical physics of formation and transformation processes of nanostrcutures and nanosystems* (Vol. 1, pp. 365, vol. 2, pp. 415). Izhevsk: Izhevsk State Agricultural Academy.
2. Kodolov, V. I., Vasilchenko, Yu. M., Akhmetshina, L. F., Shklyaeva, D. A., Trineeva, V. V., Sharipova, A. G., Volkova, E. G., Ulyanov, A. L., & Kovyazina, O. A. Patent 2393110. Russia. *Technique of obtaining carbon metal containing nanostructures*. Declared on 17.10.2008, published on 27.06.2010.
3. Kodolov, V. I., Kodolova, V. V. (Trineeva), Semakina, N. V., Yakovlev, G. I., Volkova, E. G. et al. Patent 2337062. Russia *Technique of obtaining carbon nanostructures from organic compounds and metal containing substances*. Declared on 28.08.2006, published on 27.10.2008.
4. Platunov, E. S., Buravoy, S. E., Kurepin, V. V., & Petrov, G. S. (1986). In E. S. Platunov (Ed.), *Heat-physical measurements and instruments*. L.: Mashinostroenie.
5. Plyusnina, I. I. (1967). IR *spectra of silicates* (pp. 190). Moscow: Moscow State University.
6. Panfilov, B. F. (2010). Composite materials: production, application, market tendencies. *Polymeric materials, 2–3,* 40–43.
7. Kodolov, V. I., Khokhriakov, N. V., Trineeva, V. V., & Blagodatskikh, I. I. (2008). Activity of nanostructures and its expression in nanoreactors of polymeric matrixes and active media. *Chemical Physics and Mesoscopy, 10*(4), 448–460.
8. Bobylev, V. A., & Ivanov, A. V. (2008). New epoxy systems for glues and sealers produced by CJSC "Chimex Ltd". *Polymer Science, Series D, Glues and Sealing Materials, 1*(3), 167–170.
9. Kodolov, V. I., Kovyazina, O. A., Trineeva, V. V., Vasilchenko, Yu. M., Vakhrushina, M. A., & Chmutin, I. A. (2010). *On the production of metal/carbon nanocomposites, water and organic suspensions on their basis*. VII International Scientific-Technical Conference "Nanotechnologies to the production" (pp. 52–53). Fryazino.
10. Chashkin, M. A. (2012). Peculiarities of modification by metal/carbon nanocomposites for cold hardened epoxy compositions and the investigation of properties of polymeric compositions obtained. Thesis of candidiasis Diseases (pp. 17). Perm: PNSPU.
11. Bulgakov, V. K., Kodolov, V. I., & Lipanov, A. M. (1990). Modeling of polymeric materials combustion (pp. 238). M.: Khimiya.
12. Maharinsky, L. E. et al. (1983). *Physics of combustion and explosion, 5,* 83.
13. Sheluhin, G. G., Buldakov, V. D., & Belov, V. P. (1969). *Physics of combustion and explosion, 1,* 42.
14. Silberberg A. (1971). *Pure and Applied Chemistry, 20*(3), 585–591.
15. Lipatov, Yu. S. (1984). *Colloid chemistry of polymers* (p. 343). Kiev: Naukova dumka.
16. Lipatov, Yu. S. (1980). *Interface phenomena in polymers* (p. 259). Kiev: Naukova dumka.
17. Shuklin, S. G., Kodolov, V. I., & Kuznetsov, A. P. (2000). *Chemical Physics and Mesoscopy, 2*(1), 5–11.

18. Didik A. A., Kodolov V. I., & Volkov A. Yu. et al. (2003). *Inorganic Materials, 39*(5), 693–697.
19. Didik, A. A., Kodolov, V. I., Volkov, A. Yu., & Volkova, E. G. (2002). *Chemical Physics and Mesoscopy, 4*(2), 214–223.
20. Schmidt, M. W. et al. (1993). *Journal of Computational Chemistry, 14,* 1347–1363.
21. Emmanuel, N. M., & Knorre, L. G. (1969). *Course in chemical kinetics* (p. 431). Moscow: Vysshaya shkola.
22. Stewart, J. J. P. (1991). *Journal of Computational Chemistry, 12,* 320–333.
23. Skripov, V. P. (1972). *Metastable liquid* (p. 312). Moscow: Nauka.
24. Shuklin, S. G., Klimenko, E. N., & Kodolov, V. I. (2001). *Chemical Physics and Mesoscopy, 3*(2), 113–125.
25. Korablev, G. A. (1999). *Application of spatial-energy concepts in prognostic assessment of phase-formation of solid solutions of refractory and relative systems* (p. 290). Izhevsk: IzhSAA.
26. Korablev, G. A., & Kodolov,V. I. (2001). *Chemical Physics and Mesoscopy, 3*(2), 243–254.
27. Shuklin, S. G., Kodolov, V. I., Larionov, K. I., & Tyurin, S. A. (1995). Physical and chemical processes in modified two-layer fire- and heat-resistant epoxy-polymers under the action of fire sources. *Physics of Combustion and Explosion, 31*(2), 73–79.
28. Glebova, N. V., & Nechitalov A. A. (2010). Surface functionalization of multi-wall carbon nanotubes. *Journal of Technical Physics, 36*(19), 12–15.
29. Rakov, E. G. (2006). *Nanotubes and fullerenes: Students' book for higher educational institutions* (pp. 376). Moscow: University book, Logos.
30. Kodolov, V. I., Vasilchenko, Yu. M., Akhmetshina, L. F., Shklyaeva, D. A., Trineeva, V. V., Sharipova, A. G., Volkova, E. G., Ulyanov, A. L., & Kovyazina, O. A. Patent 2393110. *Russia Technique of obtaining carbon metal containing nanostructures*. Declared on 17.10.2008, published on 27.06.2010.
31. Kodolov, V. I., Kodolova, V. V. (Trineeva), Semakina, N. V., Yakovlev, G. I., & Volkova, E. G. et al. Patent 2337062. *Russia Technique of obtaining carbon nanostructures from organic compounds and metal containing substances*. Declared on 28.08.2006, published on 27.10.2008.
32. Kodolov, V. I., & Trineeva, V. V. (2012). Perspectives of idea development about nanosystems self-organization in polymeric matrixes. In *The problems of nanochemistry for the creation of new materials* (pp. 75–100). Torun: IEPMD.
33. Kodolov, V. I., Trineeva, V. V., Kovyazina, O. A., & Vasilchenko, Yu. M. (2012). Production and application of metal/carbon nanocomposites. In *The problems of nanochemistry for the creation of new materials* (pp. 23–36). Torun: IEPMD.
34. Akhmetshina, L. F., Kodolov, V. I. (2012). Modification of silicates with metal/carbon nanocomposites. In *The problems of nanochemistry for the creation of new materials* (pp. 17–22). Torun: IEPMD.
35. Akhmetshina, L. F., Koreneva, E. Yu., Kodolov, V. I. et al. (2010). Nanostructures interaction with silicate compositions. *Nanotechnics, 3,* 13–16.
36. Akhmetshina, L. F., Kodolov, V. I., Tereshkin, I. P., Korotin, A. I. (2010). The influence of carbon metal containing nanostructures on strength properties of concrete composites. *Nanotechnologies in Construction, 6,* 35–46.

CHAPTER 12

PRODUCTION OF NANO-ALUMINUM AND ITS CHARACTERIZATION

JAYARAMAN KANDASAMY

CONTENTS

Dr. Jayaraman kandasamy has completed Master of Engineering (Aeronautical), at Department of Aeronautical Engineering, Madras Institute of Technology, Anna University, Chennai, India. He obtained his doctorate from Indian Institute of Technology, Madras, India, in the field of nano-particles production and characterization and its application in aerospace propulsion. He was selected for Sandwich Postdoctoral Fellowship from Science Technology and Serivce, Embassy of France, New Delhi, India. He performed the postdoctoral fellowship research at Institut de Combustion Aérothermique Réactivité et Environnement, *Centre National de la Recherche Scientifique*, France. He has published five international journal papers in the field of nano-particle production, characterization, combustion, and propulsion applications. He has presented nine international conference papers, which includes USA (AIAA) and China (International Symposium on Combustion). Now, he is the senior postdoctoral researcher at ICARE, France, for working the "Optimizing gasification of high-ash content coals for electricity generation" an Indo European collaborative project under seventh framework programme of European Union.

His doctorate research has involved different facets such as production of nano-aluminium particles by electrical wire explosion and their characterization by several different physicochemical analytical methods, processing of solid propellants and model propellants containing nano-aluminium and comparison formulations without aluminium or micron-sized aluminium, and a battery of combustion tests to study the performance and burning rate characteristics of these propellants and to understand their combustion mechanisms, and high heating rate tests of combinations of propellant ingredients including nano-aluminium and spreading rate studies of particle laden polymer slurries to simulate the burning surface behaviour of propellants. He has generating hydrogen gas using clean technologies relevant for on demand applications, avoiding storage costs and risks is of great importance for fuel cell applications for various uses and transport applications using both fuel cells or internal combustion engines and micro gas turbines for both propulsion and energy generation uses. He has started intensive research activities at ICARE on the energetics of metal combustion (Al and Mg) for space propulsion applications and for future Mars Sample Return Missions using the in-situ propellant production concept (in this case the use of Martian CO_2 to burn Al and Mg present in the Martian soil).

He has designed, fabricated and demonstrated the Pyro igniter and Pyro starter Development for Gas Turbine Engine Applications, the project sponsored by GTRE (Gas Turbine Research and Establishment), DRDO (Defence Research and Development Organisation), Govt of India. He is the principal investigator for the project titled "Development of Design Methodology for Optimizing Squeeze Film Damper (SFD) for Aero Gas Turbine Engine and Experimental Verification Studies" under scheme of GATET (Gas Turbine Enabling Technology Initiative), sponsored by ARDB (Aeronautical Research and Development Board), DRDO, Govt of India. He is the co-investigator for the project titled "Enhancement of flame stabilization using porous media stabilizers for gas turbine engine applications", sponsored by ARDB, DRDO, Govt of India.

12.1 INTRODUCTION

Nowadays, the possibility of production of nano-sized powders has gained attention in the interest of producing materials with new properties and to develop new technologies. Nano-sized materials are supposed to display highly desirable features such as increased catalytic activity, higher reactivity, and lower melting temperature.

Nano-particles are usually defined as particles of diameter <100nm. They have different and often superior properties compared with those of bulk material. A single nano-particle constitutes only a few hundreds of atoms, so the surface energy of these particles is quite different from their bulk counterpart. These properties make the nano-particles as excellent materials typically for use in catalytic applications.

Nano-aluminum particles can be produced by various physical or chemical methods. Two types of method are mostly adopted for producing the nano-sized particles, namely, (i) top-down approach (ii) bottom-up approach. In top-down approach, the nano-sized particles are produced from bulk-sized particles by different techniques, such as mechanical attrition, inert gas condensation, laser ablation, and so on. These are typically physical methods. Chemical methods, on the other hand, are commonly referred to as belonging to the bottom-up approach, wherein the nano-sized particles are formed by appropriate chemical reaction between molecules. In physical methods, vapor is produced by pyrolytic effect. The physical methods avoid unwanted products encountered in a chemical reaction, resulting in relatively high purity of the produced powder. The electrical wire explosion technique is a physical method adopted in this study.

12.2 METHODS OF PRODUCING NANO PARTICLES

Nano-particles can essentially be produced in three different ways: (i) mechanical attrition, (ii) wet phase methods, (iii) gas phase methods. Mostly, the properties of the final product may differ depending upon the fabrication process.

In mechanical attrition, the bulk material is reduced in size by milling. This is a simple technique and economically advantageous; however, it produces particles of a broad size distribution, and contamination from the milling parts is often a problem. Metal oxides and inter-metallic compounds are produced by this method.

A well-defined quantity of different ionic solutions is mixed in the wet phase methods. The conditions (temperature and pressure) are precisely controlled to promote the formation of insoluble compounds that precipitate out of the solution. Then, the precipitate solution is dried and filtered out to obtain the powder. This process is also economically feasible, but the yield is typically quite low.

In the gas phase methods, the energy applied on the material is of the order of the sublimation energy. The energy supply comes from various sources such as resistive heating, radio-frequency heating, electron-beam heating and laser or plasma heating. This is a simple top-down approach. The shape, size, and purity of the material can be maintained at a considerable level in this process.

12.3 LITERATURE SURVEY

12.3.1 Nano-Aluminium Production

Nano-particles can be produced using different methods, as outlined earlier. Out of these, the electrical wire explosion process is one of the promising methods. Among many methods of producing nano-sized metal powders, the electrical wire explosion technique consumes relatively low levels of energy at ~2kWh/kg [3]. Sabari Giri et al. [16] generated the nano-Al_2O_3 particles using aluminium wire by wire explosion method at air ambience. Sarathi et al. [17] reported the material characterization of nano-Al_2O_3 and high-speed imaging of the explosion event to understand the mechanism of nano-particle formation by the wire explosion technique. Sindhu et al. [21] studied the production of nano-aluminum nitride by explosion of aluminum wire in a nitrogen ambience. Sarathi et al. [18, 19] have reported the generation and characterization of nano-aluminum metal particles by explosion of wires of that material in different inert ambiences such as nitrogen, argon, and helium. Sindhu et al. [22] have reported the effect of pressure of the ambient medium on the size distribution and other material characterization features of na-no-aluminum particles obtained in different ambiences. In a novel approach, Sarathi et al. [18, 19] have explored the use of binary mixtures of inert gases for the ambience in influencing the particle size and other characteristics of nano-aluminum powders. All these works have enabled the control of the peak in the size distribution of the produced nano-particles within the 20–50 nm range, over a spread of 1–100 nm. Sindhu et al. [23] have modeled the process of nucleation and coagulation of vaporized aluminum following the explosion of the wire, and have simulated the population dynamics of particles to obtain the late-time size distribution in different ambiences. The model predictions agree well with the experimental results in terms of the size distributions, and also support the partial formation of irregular-shaped particles in a nitrogen ambience due to nitration on the surface of the particles. A similar modeling effort including a Monte-Carlo simulation accounting for exothermic nature of particle coalescence, not restricted to aluminum, has also been reported earlier by Mukherjee et al. [13].

Ivanov et al. [4] have reported the production of ultra-fine powders of different metals including aluminum by the electrical wire explosion process. They have reported the generation of aluminum particles of mean sizes in the 30–50 nm range. Lee et al. [11] have reported the production of nano-aluminum powder by a similar technique in the size range of 80–120 nm.

Many researchers have investigated the nano-aluminum particle's behavior. Different methods are adopted for production of nano-aluminum, such as the pulsed plasma technique, vapor condensation, and so on, besides, the electro-explosion technique. The sizes of the above particles are varied over a wide range of mean sizes from 24–500 nm, mostly in relation to the method of production. Particles produced by the electrical wire explosion method popularly termed as 'Alex' and widely reported with the typical size of ~100–200 nm, except for what is reported by Ivanov et al. [4] so far.

12.3.2 Nano-Aluminium Characterization

Thermal analyses such as thermo-gravimetric analysis (TGA), differential thermal analysis (DTA), and differential scanning calorimetry (DSC) of nano-aluminum samples have been performed and contrasted against the characteristics of micron-sized aluminum samples by several investigators [8–10, 12, 24]. Some other investigators have examined the role of nano-Al in the thermal decomposition of other propellant or explosive ingredients, such as ammonium perchlorate (AP) [11] and RDX [9, 14]. The important, pertinent results of these investigations are that nano-aluminum melts at a lower temperature than bulk aluminum; oxidation of nano-aluminum commences at quite low temperatures, and the low temperature decomposition of AP is accelerated in the presence of nano-aluminum. The last result supports the observation of extraordinarily high burning rates of pressed pellets of dry mixtures of AP and ultra-fine aluminum reported earlier [15]. Sigman et al. [20] reported that the LPDL of dry-pressed pellets of aluminum including ultra-fine aluminum (Alex) mixed in AP is increased with decrease in the Al particle sizes. They showed that Alex has high oxide content owing to its high specific surface area and the oxide skin is either fully crystalline or amorphous, making it difficult for molten aluminum inside to crack through it, unlike in large Al particles, where interfaces between crystalline and amorphous islands on the oxide skin are prone to crack formation. DeSena and Kuo [1] examined the possibility of stored energy due to rapid solidification in the formation of ultra-fine exploded aluminum particles and found it to be negligible.

Dubois et al. [2] have reported the coating of nano-aluminum and boron particles by grafting polyethylene type of polymers and polyurethane on the metal powders, and characterization of the resulting materials. Jayaraman et al. [5–7] have produced the nano-aluminum particles using the top-down approach method of exploding wire technique.

12.4 PRODUCTION OF NANO ALUMINUM USING EXPLODING WIRE TECHNIQUE

The major factor determining the particle size of the nano-powder obtained in the wire explosion process is the extent of superheating of the evaporated material. The particle size produced by the wire explosion process reduces substantially with increasing the superheating of the metal. The extent of superheating is determined by the factor $k = W/Ws$, where W is the energy injected into the evaporating wire sand Ws is the sublimation energy for the wire, obviously k diminishes with decreasing diameter of the wire. The experimental setup consists of an electrical circuit, wire explosion chamber, and particle collection chamber. The electrical section consists of diodes, capacitors ($3\mu F$), a switch, and the controlling device. The wire explosion chamber (Figure 12.1) consists of provisions for fixing the wire, the inlet and exit of the vacuum and required gas medium, pressure gauges, pressure valves, and the main chamber. The collection chamber consists of the prefilter, main filter, and postfilter.

The filters are mounted at the bottom of the exploding chamber, primarily to collect the nAl particles. Initially, multiple aluminum wires of 0.42mm in diameter and 140 mm in length each are connected between the electrical contacts and the chamber is evacuated in the range of 700 mm of Hg. Then, the chamber is purged and filled with argon or nitrogen gas (>99% purity) at 1 bar gauge pressure to prevent oxidation of the powder. The capacitor is charged to 24kV, the switch is closed, and the capacitor is discharged. After explosion of all the wires is completed, the inert gas is sent out through the top of the chamber and the nAl powder is collected at the filter. Superheating of the evaporated material determines the particle size in the wire explosion process. The sublimation energy diminishes with the reduction in diameter of the wire [18, 19]; for aluminum, it is 33 J/mm³. For the present experimental conditions, the value k is maintained at 1.14.

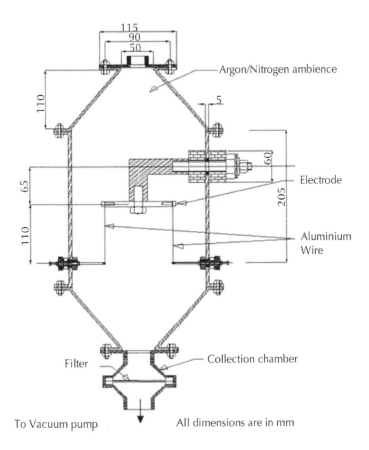

FIGURE 12.1   Schematic of the exploding wire chamber (all dimensions are in mm).

12.5   CHARACTERIZATION OF NANO-ALUMINUM

12.5.1   Wide Angle X-ray Diffraction

The nano-aluminum particles are produced by the wire explosion process in two ambient gas media, namely, nitrogen and argon, at a pressure 0.1 MPa. Figure 12.2 shows the WAXD spectra of the nano-aluminum produced both in argon and nitrogen ambiences. It is observed that the nano-aluminum powder obtained in argon ambience has peaks at 38.5°, 44.7°, 65.1°, and 78.2°, which are the characteristic peaks that occur in the standard XRD pattern of bulk aluminum. The WAXD pattern has not reflected any peaks of the oxide layer of the nano-aluminum particles produced in argon ambience. The nano-aluminum powder generated in the nitrogen ambience, on the other hand, shows additional minor peaks at 33.2°, 36.0°, 37.9°, 49.8°, 59.3°, 66.0°, and 69.7°, which are characteristics of hexagonal aluminum nitride.

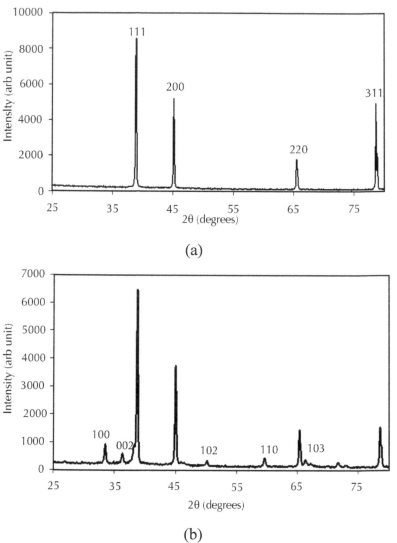

FIGURE 12.2 WAXD pattern of nano-aluminum powder obtained in different ambiences. (a) Argon ambience. (b) Nitrogen ambience.

Note that the wire explosion process is basically an evaporation-condensation technique. During the explosion process in argon ambience, the aluminum vapor does not chemically react with argon, and it alone condenses, solidifies, and forms as particles. But, in case with nitrogen ambience, the local temperature is high enough to initiate chemical reaction between nitrogen and the aluminum vapor. So, some amount of aluminum nitride is formed during particle formation in nitrogen ambience. This implies that the purity of the nano-Al powder produced in argon ambience is more than that of the nano-Al powder produced in nitrogen ambience.

12.5.2 Transmission Electron Microscopy (TEM) Analysis

Figure 12.3 shows the TEM image of nano-aluminum particles produced by exploding wires in argon and nitrogen ambiences. It can be seen that the shape of the particles produced in argon ambience is spherical whereas, some of the particles produced in nitrogen ambience are highly irregular—specifically polygonal—in shape.

The higher cooling rate subsequent to the explosion causes the aluminum vapor particles to nucleate, condense, coagulate, and coalesce into spherical particles. The surface tension of the molten aluminum tends to spheridize the particle during solidification. As observed earlier, a small amount of aluminum nitride is present in the aluminum powder produced in nitrogen ambience. The aluminum vapor reacts with nitrogen and forms aluminum nitride, but the activation energy of this reaction is sufficiently high so that it cannot be sustained with the energy released from aluminum nitride formation. The aluminum nitride covers the molten aluminum particles and solidifies over the aluminum in the core. This arrests the coalescence of the coagulated nuclei and results in nano-aluminum particles of non-spherical shape [23].

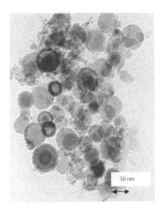

FIGURE 12.3 *(Continued)*

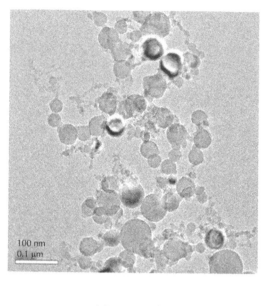

(a)   (b)

(c)

FIGURE 12.3   Typical TEM imaging of nano-aluminum particles produced in different ambiences. a Argon ambience (high magnification). b Argon ambience (low magnification). c Nitrogen ambience.

12.5.3 Particle Size Distribution

The particle size distributions of the nano-powder produced in an argon ambience for two different conditions of wire diameter and pressure are shown in Figure 12.4. The peak size in the distributive distribution as well as the 50% size in the cumulative distribution seen in Figure 12.4a is ~52 nm for the wire diameter of 0.46 mm and applied voltage of 24 kV at 0.1MPa of gauge pressure. By increasing the wire diameter to 0.51 mm and the ambient pressure to 0.3 MPa gauge at the same voltage, the peak size is increased to 65 nm (Figure 12.4b).

Due to increase in the pressure of the ambient medium, the collision rate between molecules of the aluminum vapor is increased. Also, superheating of the evaporated material influences the particle size in the wire explosion process. The increase in wire diameter reduces the volumetric energy applied on the wire, thereby reducing the extent of superheating. The vapor from the thinner wire can reach to higher temperature. These factors contribute to increase in the particle size of the nano-Al powder with the thicker wire.

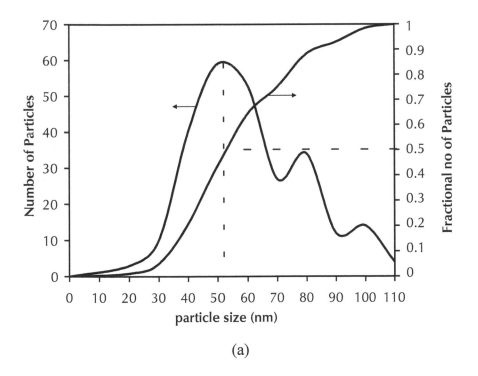

(a)

FIGURE 12.4 *(Continued)*

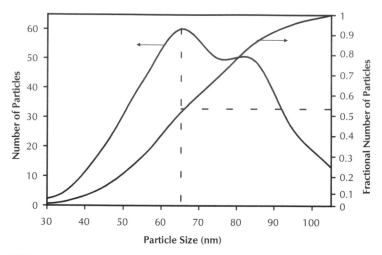

FIGURE 12.4 Distributive (left ordinate) and cumulative (right ordinate) size distributions of nano-aluminum particles obtained in argon ambience. a Wire diameter = 0.46 mm, ambient pressure = 0.1 MPa. b Wire diameter = 0.51 mm, ambient pressure = 0.3 MPa.

12.5.4   Energy Dispersive Analysis by X-Rays (EDAX)

The atomic composition of the aluminum powder produced by wire explosion is determined from the energy dispersive analysis by x-rays. A typical EDAX pattern obtained for the nano-aluminum particles produced in argon ambience is shown in Figure 12.5. It indicates that 93.5% of atomic aluminium and 6.5% of atomic oxygen by weight of the nano-aluminium powder is produced. This reveals that 86.2% of aluminum and 13.8% of Al_2O_3 contained in the sample, assuming only those two compounds to be present in the samples. The high surface area of the nano-aluminum powder causes the relatively high content of aluminum oxide.

12.5.5   Thermo-Gravimetric and Differential Thermal Analyze (TG-DTA)

Figure 12.6 shows the TG-DTA results of the nano-aluminum powder obtained in argon and nitrogen ambiences at a pressure of 0.1MPa. The thermal analyses are, however, performed in nitrogen ambience. The DTA (Figure 12.6a) indicates an endothermic peak at 656°C, which corresponds to the melting of the nano-aluminum powder produced in argon ambience whereas 658°C for the particles produced in nitrogen ambience. A slight reduction in melting point is observed with the nano-sized aluminum powder, relative to that of bulk aluminum, 660°C, as observed with nano-particles. In the TGA (Figure 12.6b), a gradual weight loss of about 2–4% is observed in the 100–400°C range. This is predominantly observed with the nano-powder produced in nitrogen ambience, where certain amount of aluminum nitride is present. Aluminum nitride hydrolyzes in moist air at room temperature and forms aluminum hydroxide. Aluminum hydroxide loses water at 300°C [11], which is also reflected in the DTA as being exothermic. Above 550°C, an increase in weight occurs for both the nano-Al powders. This is due to the reaction of the nano-aluminum powder with the nitrogen ambience in the thermal analysis tests to form aluminum nitride. The weight gain is substantial for the nano-particles produced in argon ambience relative to those from nitrogen ambience. In the latter, aluminum nitride has already been formed in the outer layer of the particles during the explosion process. As such, the extent of nitride formation during testing is limited for these particles relative to the ones produced in argon ambience.

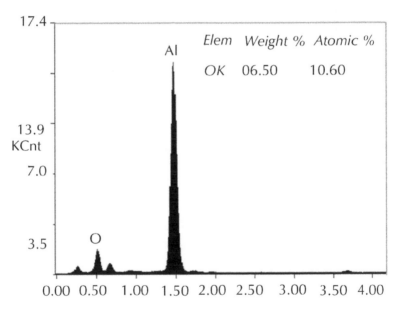

FIGURE 12.5 EDAX results of nano-aluminum produced in argon ambience.

Since the aluminum exploded in argon atmosphere is observed to be purer than that produced in nitrogen atmosphere in all the above tests.

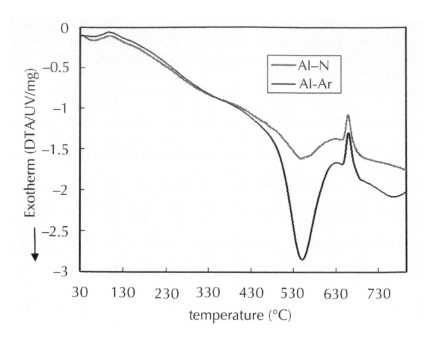

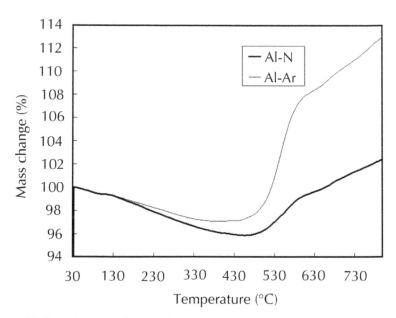

Figure 12.6   TG-DTA of nano-aluminum obtained in argon and nitrogen ambiences. a DTA curve. b TG curve.

12.6 CONCLUSIONS

An experimental study of production of nano-aluminum using the exploding wire technique is performed. The production of nano-aluminum particles through wire explosion process is feasible. The energy applied on the aluminum wire is of the order of the sublimation energy of aluminum. A cylindrical explosion chamber has been developed to meet the requirement on the productivity of nano-aluminum powder. Production of nano-aluminum powder has been attempted in both nitrogen and argon ambiences. The nano-aluminum is characterized using physico-chemical diagnostic methods. The nano-Al powder produced from argon ambience is relatively pure; it contains ~86.2% of Al and ~13.8% of Al_2O_3. Spherical particles are formed in argon ambience versus some polygonal particles representing aluminum nitride in nitrogen ambiences. The particle size distribution peaks around 45–55 nm, with a mean size of 52 nm, for the nano-Al particles. A slight reduction (~4–6°C) in the melting point is observed for the nano-Al relative to bulk aluminum.

KEYWORDS

- **Argon ambience**
- **Exploding wire technique**
- **Monte-Carlo simulation**
- **Nano-aluminium**
- **Transmission electron microscopy**

REFERENCES

1. DeSena, J. T. & Kuo, K. K. (1999). Evaluation of stored energy in ultrafine aluminum powder produced by plasma explosion. *Journal of Propulsion and Power, 15,* 794–800.
2. Dubois, C., Lafleur, P. G., Roy, C., Brousseau, P., & Stowe, R. A. (2007). Polymer-grafted metal nano-particles for fuel applications. *Journal of Propulsion and Power, 23,* 651–658.
3. Ivanov, V., Kotov, Y. A., Samatov, O. H., Bohme, R., Karow, H. V., & Schumacher, G. (1995). Synthesis and dynamic compaction of ceramic nano-powders by techniques based on electric pulsed power. *Nano-Structured Materials, 6,* 287–290.

4. Ivanov, Y. F., Osmonoliev, M. N., Sedoi, V. S., Arkhipov, V. A., Bondarchuk, S. S., Vorozhtsov, A. B., Korotkikh, A. G., & Kuznetsov, V. T. (2003). Productions of ultra-fine powders and their use in high energetic compositions. *Propellants, Explosives and Pyrotechnics, 28,* 319–333.

5. Jayaraman, K., Anand, K. V., Chakravarthy, S. R., & Sarathi, R. (2007). *Production and characterization of nano-aluminum and its effect in solid propellant combustion (AIAA-2007-1430)*, 45th AIAA Aerospace Sciences Meeting and Exhibit, Reno, NV, USA, Jan 2007.

6. Jayaraman, K., Anand, K. V., Chakravarthy, S. R., & Sarathi, R. (2007). *Production of Nano-Aluminum and its Effect on Burning Rates in Solid Propellant Combustion*, 6th Asia-Pacific Combustion Conference, Nagoya, Japan, May 20–23, 2007.

7. Jayaraman, K., Anand, K. V., Bhatt, D. S., Chakravarthy, S. R., & Sarathi, R. (2009). Production, characterization, and combustion of nano-aluminum in composite solid propellants. *Journal of Propulsion and Power, 25,* 471–481.

8. Johnson, C. E., Fallis, S., Chafin, A. P., Groshens, T. J., Higa, K. T., Ismail, I. M. K., & Hawkins, T. W. (2007). Characterization of nanometer- to micron-sized aluminum powders: size distribution from thermogravimetric analysis. *Journal of Propulsion and Power, 23,* 669–682.

9. Kwok, Q. S. M., Fouchard, R. C., Turcotte, A. M., Lightfoot, P. D., Bowes, R., & Jones, D. E. G. (2002). Characterization of aluminum nano-powder compositions. *Propellants, Explosives, Pyrotechnics, 27,* 229–240.

10. Kwon, Y. S., Moon, J. S., Ilyin, A. P., Gromov, A. A., & Popenko, E. M. (2004). Estimation of the reactivity of aluminum superfine powders for energetic applications. *Combustion Science and Technology, 176,* 277–288.

11. Lee, G. H., Park, J. H., Rhee, C. K., and Kim, W. W., (2003) Fabrication of Al Nano-Powders by Pulsed Wire Evaporation (PWE) Method, Journal of Industrial and Engineering Chemistry, 9, 71-75.

12. Mench, M. M., Kuo, K. K., Yeh, C. L., & Lu, Y. C. (1998). Comparison of thermal behavior of regular and ultra-fine aluminum powders (ALEX) made from plasma explosion process. *Combustion Science and Technology, 135,* 269–292.

13. Mukherjee, D., Sonwane, C. G., & Zachariah, M. R. (2003). Kinetic Monte-Carlo simulation of the effect of coalescence energy release on the size and shape evolution of nano particles grown as an aerosol. *Journal of Chemical Physics, 119,* 3391–3404.

14. Pivkina, A., Ulyanova, P., Frolov, Y., Zavyalov, S., & Schoonman, J. (2004). Nano materials for heterogeneous combustion. *Propellants, Explosives, Pyrotechnics, 29,* 39–48.

15. Romonadova, L. D., & Pokhil, P. K. (1970). Action of silica on the burning rates of ammonium perchlorate compositions. *Fizika Goreniya i Vzryva, 6,* 285–290.

16. Sabari Giri, V., Sarathi, R., Chakravarthy, S. R., & Venkataseshaiah, C. (2003). Studies on production and characterisation of nano-al$_2$o$_3$ powder using wire explosion technique. *Materials Letters, 58,* 1047–1050.

17. Sarathi, R., Chakravarthy, S. R., & Venkataseshaiah, C. (2004). Studies on generation and characterization of nano-alumina powder using wire explosion technique. *International Journal of Nanoscience, 3,* 819–827.

18. Sarathi, R., Sindhu, T. K., & Chakravarthy, S. R. (2007). Generation of Nano-aluminum powder through wire explosion process and its characterization. *Materials Characterization, 58,* 148–155.
19. Sarathi, R., Sindhu, T. K., & Chakravarthy, S. R. (2007). Impact of binary gas on nano aluminum particle formation through wire explosion process. *Materials Letters, 61,* 1823–1826.
20. Sigman, R. K., Zachary, E. K., Chakravarthy, S. R., Freeman, J. M., & Price, E. W. (1997). *Preliminary characterization of the combustion behavior of alex in solid propellants,* 34th JANNAF Combustion Meeting, West Palm Beach, Florida, USA.
21. Sindhu, T. K., Chakravarthy, S. R., Jayaganthan, R., & Sarathi, R. (2006). Studies on generation and characterization of nano aluminum nitride through wire explosion process. *Metal-Organic and Nano-Metal Chemistry, 36,* 53–58.
22. Sindhu, T. K., Sarathi, R., & Chakravarthy, S. R. (2007). Generation and characterization of nano aluminum powder obtained through wire explosion process. *Bulletin of Materials Science, 30,* 1–9.
23. Sindhu, T. K., Sarathi, R., & Chakravarthy, S. R. (2009). Understanding the nano particle formation by wire explosion process through experimental and modelling studies. *Nanotechnology, 19*(025703), 1–11.
24. Trunov, M. A., Umbrajkar, S. M., Schoenitz, M., Mang, J. T., & Dreizin, E. L. (2006). Oxidation and melting of aluminum nanopowders. *Journal of Physical Chemistry B, 110,* 13094–13099.

CHAPTER 13

RECENT ADVANCES IN AEROSPACE PROPULSION USING NANO METALS OR NANO MATERIALS IN AEROSPACE PROPULSION

JAYARAMAN KANDASAMY

CONTENTS

13.1   METAL COMBUSTION

The combustion of metals has long been of interest to the combustion community because of its high energy densities. It is well known that energetic metals such as aluminum and boron have higher combustion energies and can be employed as energetic additives in propellants and explosives [1] which are currently being studied for underwater propulsion using seawater as the oxidizer. Silicon metal combustion has gained attention which can be used as an energy carrier [2, 3]. Metals will be the important fuels for the establishment of a lunar mission base and the exploration of Mars. Several researchers have been carried out the different sized-aluminum combustion under diverse oxidizing atmospheres [4–8]. Metalized propellant propulsion systems are considered as replacements for the solid rocket booster and liquid sustainer stages to the current launch vehicles. A complete comparison of energy density of various fuels is shown in Figure 13.1 on a mass and volume basis. The energy available from metals in general is much higher than that of conventional fuels per volume.

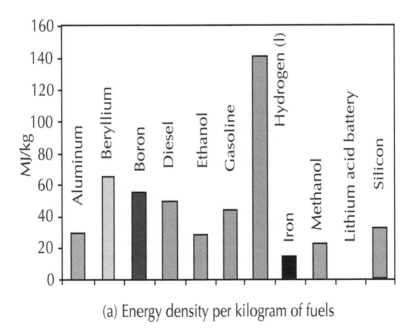

(a) Energy density per kilogram of fuels

FIGURE 13.1   *(Continued)*

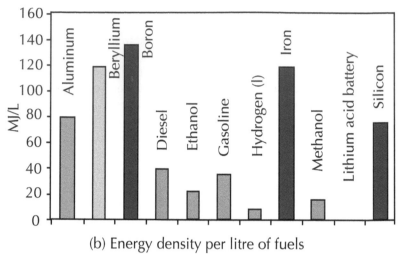

(b) Energy density per litre of fuels

FIGURE 13.1   Energy density comparison per kg and per liter of various metal and liquid fuels.

13.2   NANO METALS AND ITS COMBUSTION PHENOMENA

Application of nanoscale energetic materials to enhance combustion of liquid fuels is gaining interest in recent times. Recent advances in nanoscience and nanotechnology enable production, control and characterization of nanoscale energetic materials, which have shown tremendous advantages over micro-sized materials, which are currently used in the majority of metalized propellants. Nanoscale metal particles have many highly desirable traits mainly because of the high specific surface area, small diffusion scales, which results in dramatically faster reaction rates, and the potential ability to store energy on the surface. At the spatial scale of nanoparticles and nanodomains, thermodynamic properties of materials are altered. This includes changes in the melting point and latent heat of melting, which are of particular significance for reactive nanomaterials. In addition, phase transformations, such as polymorphic transitions between different aluminum oxide phases, are also affected by the reduced dimensions of the oxide films.

In pure metal combustion, high energy release rate and control of reaction rate is not possible. For instance, the oxidation process of bulk metals is a very slow process that is nearly unnoticeable. Pure bulk metals are normally covered by layers of oxides that must be removed before the reaction can proceed rapidly. To achieve a rapid thermal energy release, that is combustion, in most of the metals, one needs an energy source of a temperature of at least 2,000°C to vaporize the oxide layer and expose the bare, reactive metal beneath. It is a well known phenomenon in physics that the properties of most materials are size-dependent. Some relevant properties in metal oxidation and combustion may include the activation energy, oxidation and ignition temperature. Small particles burn much more easily because of their large surface area to volume ratio and short oxidizer diffusion length; the oxidation could generate enough energy to spontaneously ignite and combust particles. There is some evidence that shows the metallic aluminum nanoparticles of 100 nm can be ignited at 250°C, much lower than that in the bulk [9]. Sarou-Kanian et al. [10] and Jayaraman et al. [11] have studied the low temperature reactivity of nano-sized and micro-sized aluminum powders with liquid water respectively. The reduction in reaction and ignition temperature of nano-sized particles opens a big window for potentially wide applications to energetic materials.

Nano-sized energetic material using nano Al particle (nAl) as fuel yield reaction rates several orders of magnitude higher than micron fuelled reactives. Al-based nanoenergetics successfully propagates in much smaller channels than conventional energetic materials allowing them for applications such as micro actuation or micro-thrusters [12]. Nanocomposites have also seen to reduce ignition delay times on the order of two times when compared to the magnitude of micron-scale composite ignition delays [13].

In solid propellant combustion, many researchers have investigated the effect of "ultra-fine" aluminum of particle size ~100 nm or above. The catalytic effects of aluminum on deflagration of ammonium perchlorate (AP) have been reported early in the literature [16]. At present, a lot of research work is being done on the production of nano-/ultra-fine aluminum by various processes and on its performance in solid propellant rockets, guns, explosives, thermites, and so on.

Escot Bocanegra et al. [14] have reported the comparison of the burning times of micron-sized Al uncoated and coated with nickel, with those of uncoated ultra-fine aluminum powders. They find that both the Ni-coated micron-sized aluminum and ultra-fine aluminum exhibit decreased burn times, but the latter burn faster of the two. Currently there are many types of nanoenergetic materials being studied, these include but are not limited to mixed nanopowder composites, multilayered nanofoils, three-dimensional micron-sized nanocomposite powders, liquid, and gels loaded with nanoscaled metal fuels like aluminum, boron, carbon and nanoscale porous silicon composites. Recently, Jayaraman et al. [15] have examined application of aluminum foam to space propulsion applications.

Some work has been done on composite propellants with aluminum particles of ultra-fine size (100–200 nm). In most of the cases, the electric wire explosion method has been adopted for producing ultra-fine Al particles. Only a very few reports are available on solid propellant combustion using actually nano-sized aluminum particles.

Many researchers have reported the effect of ultra-fine aluminum on the burning rates of composite solid propellants, particularly those based on AP. As mentioned earlier, Romonadova and Pokhil [16] reported burning rates of dry-pressed mixtures of AP and ultra-fine aluminum. Dokhan et al. [17] have investigated the burning rates of propellant compositions up to 6.9MPa, with unimodal ultra-fine or micron-sized aluminum particles as well as bimodal combinations of the particles of both sizes. They have reported that the addition of ultra-fine aluminum increases the burning rate of the propellant through ignition of the aluminum particles by the leading edge flames (LEFs) over the course AP/binder interfaces and/ or the flames of the fine AP/binder matrix. Ivanov et al. [18] have also reported increases in burning rates with nano-aluminum by about 1.6–2 times relative to non-aluminized propellant compositions at different pressures and levels of aluminum addition. De Luca et al. [19] and Galfetti et al. [20, 21] have investigated the burning rates of propellants containing ultra-fine aluminum in comparison to those without it over a pressure range up to 7MPa. They found that the burning rates nearly double with addition of ultra-fine aluminum uniformly across the test range of pressures, that is, without any change in the pressure exponent of burning rate averaged over the entire test pressure range. These workers have also tested bimodal blends of aluminum in some propellant formulations, with results that are similar to that of Dokhan et al. [17]. Pivkina et al. [22] have recently reported the production of nano-sized AP as well as nano-sized aluminum containing 10% graphite by the method of mechanical activation, in the size ranges of 35–100 nm for the former and 20–50 nm for the latter. They combusted AP-Al mixtures approximately in the ratio of 76:24 over the 1–6 MPa pressure range, and found that replacement of micron-sized (10μm) AP by nano-sized AP alone did not significantly alter the burning rate or its pressure exponent; replacement of micron-sized (97μm) Al by nano-Al alone increased the burning rate by >5 times, but without altering the pressure exponent; but, simultaneous replacement of both micron-sized AP and Al by nano-sized ones caused over a ten-fold increase in the burning rate relative to their micron-sized counterpart mixtures and also nearly halved the pressure exponent averaged across the entire test pressure range. Popenko et al. [23] have reported increase in burning rates and decrease in pressure exponents of condensed systems of HMX added with small quantities (1.25–5%) of ultra-fine aluminum powders, but notably without any adverse effect on the rheological properties of the model propellants because of such addition. While Dokhan [17] and Galfetti et al. [20, 21] have reported studying the agglomeration of ultra-fine aluminum during combustion of propellant containing such particles, Karasev et al. (2004) have reported further the structure of aggregates of aluminum oxide formed in the combustion of ultra-fine aluminum in air. These studies affirm that agglomerates of ultra-fine aluminum or their combustion products do not exceed a few micrometers in size, an aspect of practical interest in solid rocket propulsion.

Trunov et al. [24] have tried to explain the mechanism of oxidation of nano-aluminum in the framework of the staged oxidation and ignition of micron-sized aluminum particles, and associated crystalline phase changes of the formed oxide, reported in their previous works [25]. Rai et al. [26] examined the oxidation of nano-aluminum particles by mass spectrometry, transmission electron microscopy, and density measurements. The role of diffusion of oxygen across the oxide skin in nano-aluminum particles is highlighted, which is slow prior to melting of the aluminum but rapid thereafter.

Kwon et al. [27] have combusted dusts of ultra-fine aluminum in air to find a prominent role for formation of aluminum nitride from the nitrogen in the air along with aluminum oxide formation. Yuang et al. [28] have combusted particle clouds of bimodal blends of nano- and micron-sized aluminum particles in air, and have reported higher flame speeds of the nano-Al/air flame when compared to micron-sized Al/air flame. For smaller quantities of nano-Al in the blend of both sizes, two distinct flame zones exist, but for larger quantities of it, a merger of the two flame zones is evident.

The effects of surface coatings on nano-aluminum particles and its characterization have been investigated some workers. Jones et al. [29] and Kwok et al. [30] have reported the characterization of nano-aluminum particles of different sizes coated with a thicker oxide layer, a fluoro-polymer, and so on, along with investigation of thermal behavior of such particles in combination with other energetic materials such as RDX, TNT, AP, and so on.

Freeman et al. [32] have reported the various effect of addition of titanium dioxide (TiO_2) with bimodal sizes ($0.02\mu m$ and $0.5\mu m$) in the propellants and fine AP/binder matrixes. These results indicated that the nano-sized TiO_2 retained the plateau region as in baseline propellants with the enhancement of burning rates. Nano-particle additives influenced the burning rates of propellants because of their high surface-to-volume ratios [33].

Jayaraman et al. [34–36] have investigated the burning rate of nano- and micron aluminized propellants. The nano aluminized propellant burning rates are also increased by 100% relative to its micro aluminized counterparts. The results collectively indicate that the propellant burning rate is controlled by the near-surface nAl combustion, which becomes diffusion limited in the elevated pressure range. Jayaraman et al. [37–39] have found that the addition of nano-sized catalysts in the nano- and micro aluminized propellants leads to increase the burning rates significantly for micron-aluminized propellants, but slight increment to nano-aluminized propellants. Also, they have investigated the accumulation of nano-aluminum during combustion of the propellants (2010) and quench collection of nano-aluminum agglomerates from combustion of sandwiches and propellants (2010).

13.3   EXPERIMENTAL TECHNIQUE

13.3.1   Burning Rate Measurements

A 'window bomb', which is a pressure vessel containing two windows—one for illumination of the sample, and another for viewing the combustion event—is used to burn the samples. Nitrogen is used to pressurize up to the test pressure, and the sample is ignited by an electrically heated nichrome wire. A video CCD camera is used to record the combustion event, and the images are replayed frame-by-frame to locate the burning surface of the sample along the overall direction of surface regression in successive frames. A straight line is fitted to the burning surface locations plotted versus the framing time with a correlation coefficient >99%. The slope of the straight line, adjusted for the magnification of the images, gives the burning rate. The uncertainty in the measurement of the burning rate by this technique is estimated as ±3%. Tests are performed in the pressure range of 1–12 MPa.

The propellant contains 87.5% of total solids which includes ammonium perchlorate and aluminum, remaining 12.5% of binder which contains hydroxyl terminated poly butadiene, plasticizer and curing agent. Three types of propellants are incorporated to measure the burning rate of propellants, name non-aluminized, normal (micron-sized) aluminized and nano-aluminized.

13.3.2   Particle Size Distribution Measurements

The nominal diameters, viz., the Sauter mean diameter (SMD or D_{32}) and the arithmetic mean diameter (AMD or D_{10}) of the agglomerates are defined as follows.

$$D_{32} = \sum (n_i d_i^3) / \sum (n_i d_i^2)$$

$$D_{10} = \sum (n_i d_i) / \sum (n_i)$$

where

d_i—nominal diameter of ith particle

n_i—number of particles with diameter d_i.

The repeatability ε in the agglomerate size measurement is expressed as follows:

$$\varepsilon = 100 \times \sqrt{\frac{\sum_{i=1}^{M}[\{|\Delta N_i|^2_{\max} / Avg(N_{i_1}, N_{i_2}, ..)\}d_i^2]}{\sum_{i=1}^{M}[Avg(N_{i_1}, N_{i_2}, ...)d_i^2]}}$$

where,

d_i- nominal diameter of ith agglomerate size band.

$N_{i_1}, N_{i_2}, ..$ and so on—the % cumulative particle densities for repeats at d_i.

$|\Delta N_i|$—largest difference in N_i between any two repeat runs at a given d_i.

M- number of repeated runs.

It was ascertained that the maximum deviation in repeatability in the agglomerate size determination is within 10% of the nominal diameters.

13.4   EFFECT OF ADDITION OF NANO-ALUMINUM PARTICLES IN COMPOSITE PROPELLANT COMBUSTION

13.4.1   Burning Rates of Normal and Nano-Aluminized Propellants

The three baseline propellant formulations afford comparisons based on aluminized versus non-aluminized propellants and the effect of the size of aluminum, as in nano-aluminum versus micro-aluminum. Besides this, other possible comparisons are the effect of the fine AP/binder ratio, the fine AP size, and the type of curing agent. The burning rates of all the nine formulations are shown in Figure 13.2.

The two baseline formulations shown in Figure 13.2, the burning rates of nano-aluminized propellants are increased by 100% when compared to those of the corresponding non-aluminized propellants. The nano-aluminum burns very close to the burning surface of the propellant and transfers additional heat to the propellant by both conduction and radiation. The plateau and mesa burning rate trends exhibited by the corresponding non-aluminized and normal-aluminized propellants are washed out in the nano-aluminized propellants, but the burning rate trends of the latter show markedly low pressure-exponents, regardless of the trends with the former in the same pressure range, and regardless of the other parametric variations in the two baseline formulations.

On the contrary, the addition of normal aluminum to the propellant tends to slightly reduce the burning rate. The plateau and mesa burning rate trends of the non-aluminized formulations are altered, but not washed out. This is due to the removal of coarse AP particles to accommodate the inclusion of normal aluminum at the same total-solids loading. So, with the same coarse AP size, the distance between adjacent coarse AP particles is increased in the aluminized formulations. This alters the interaction between the leading edge portions of the diffusion flame between the AP/binder gaseous decomposition products, and shifts the pressure ranges of plateau burning rates in the normal-aluminized formulations relative to those in the non-aluminized formulations. Effectively, this indicates that the physical mechanism controlling the burning rate remains the same with addition of normal aluminum to the propellant.

On the other hand, the overall increase in the burning rates with the addition of nano-aluminum, the independence of the nano-aluminized formulation trends from those with their other baseline counterparts, and the commonality in those trends across the different baseline formulations, all point to the fact that the burning rate-controlling mechanism is altered by the addition of nano-aluminum to the propellants. In this case, the burning rate is controlled by the heat transfer from the near-surface ignition and near-complete combustion of the nano-aluminum to the propellant burning surface. The nano-aluminum combustion is diffusion-limited at elevated pressures, resulting in low pressure-exponents of the nano-aluminized propellant burning rates in that pressure range.

Comparison of Figures 13.2a and b shows a 30–40% increase in the burning rate with decrease in the fine AP size from 5 to 20μm in the nano-aluminized formulations, as expected. In the formulations without the nano-aluminum, the pressure-exponents are lower with the 5 μm fine AP than the 20 μm fine AP, due to better influence of the binder-melt flow on the former size.

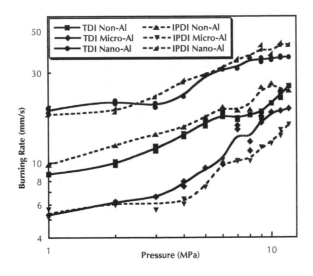

a) Baseline 1

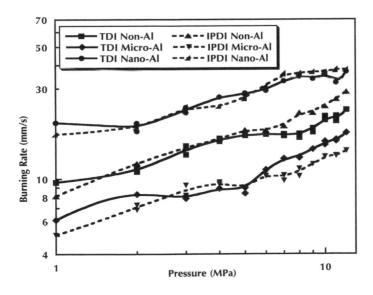

b) Baseline 2

FIGURE 13.2 Burning rates of monomodal aluminized propellant formulations.

The burning rates with the two different curing agents are compared as shown in Figure 13.2. It can be seen that the curing agent does not greatly alter the burning rate trends of the nano-aluminized formulations, unlike with the other formulations. This is because the curing agent affects the binder melt-flow characteristics, and hence the plateau-burning behavior of the non-aluminized and normal-aluminized propellants, but such a mechanism is predominated by the nano-aluminum ignition and combustion in the nano-aluminized propellants, regardless of the type of curing agent and its influence on the binder melt-flow.

13.4.2 Effect of Aluminum Content

With the low content of nano-aluminum (10%) is not enough to gasify the binder melt, it leads to mid-pressure extinction (1–7MPa) in the matrix as seen in Figure 13.3. At higher pressures, the flame is so close to the surface that the heat transfer rates are considerably high; the binder melt has little effect on the pyrolysis rates of the matrix surface, thereby matrix burning is occurred. Micro- aluminum content variation in the matrixes (10–18%) do not vary the mid pressure extinction pressure range, but addition of 15% micro-aluminum in matrix shows wider mid-pressure extinction pressure range than non aluminized ones.

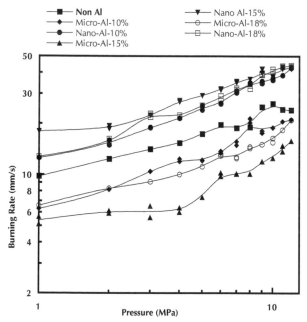

FIGURE 13.3 Burning rates of propellants of various aluminum contents, fine AP (5μm)/binder = 65/35, Coarse AP = 450 μm, IPDI cured.

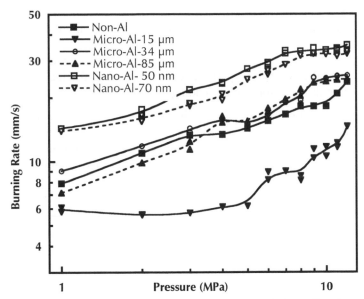

FIGURE 13.4   Burning rates of propellants of various Al sizes, fine AP (20μm)/ binder = 60/40, Coarse AP = 450 μm, IPDI cured.

Addition of nano-aluminum beyond 10% in the propellants does not show any significant increase in burning rates as seen Figure 13.3. The requirement of heat feedback from the flamelet reached the saturation level to pyrolyse the propellant ingredients with the number density of nano-aluminum particles available in 10% addition in the propellants. Addition of 15% micro- aluminum shows lower burning rate compared to 10% and 18% addition of micro- aluminum in the propellants.

13.4.3   Effect of Aluminum Size

Burning rates of propellants are decreased by 10–20%, when the size of the nano-aluminum is increased from 55 to 70nm as seen in Figure 13.4. This is due to the decrease in surface area to volume ratio of the aluminum particles; it decreases the reactivity of the particles. With the micron-sized aluminum, the burning rate increases with the size of the aluminum particles. The increase of aluminum size results decreases the number density, the mobility of particles, and the possibility to accumulate, thereby increases the burning rates of the propellants.

The nano-aluminum burns very close to the burning surface of the propellant and transfers additional heat to the propellant by both conduction and radiation. The plateau and mesa burning rate trends exhibited by the corresponding non-aluminized and micro-aluminized propellants are washed out in the nano-aluminized propellants. On the contrary, the addition of micro-aluminum to the propellant tends to slightly reduce the burning rate. The plateau and mesa burning rate trends of the non-aluminized formulations are altered, but not washed out. This is due to the removal of coarse AP particles to accommodate the inclusion of micro-aluminum at the same total-solids loading.

The burning rate is controlled by the heat transfer from the near-surface ignition and near-complete combustion of the nano-aluminum to the propellant burning surface. The nano-aluminum combustion is diffusion-limited at elevated pressures, resulting in low pressure-exponents of the nano-aluminized propellant burning rates in that pressure range.

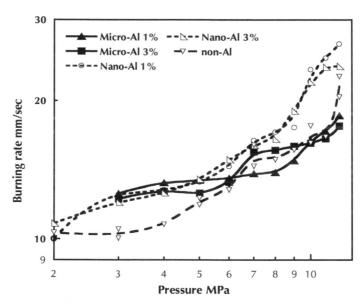

FIGURE 13.5   Effect of micro- and-nano-aluminum on burning rate of propellants at different Al content levels.

13.4.4 Propellants with Smaller Content of Aluminum

Figure 13.5 shows the burning rates aluminized propellants with 1% and 3% of nano and micro-aluminized propellants. At lower pressures the aluminized propellants show lower burning rates than its non aluminized counterparts. When the pressure is increased, the nano-aluminized propellant's is drastically increased. The accumulation and heat sink effect caused by the aluminum particles at low pressures are overcome by the closer stand-off distance of the flame at higher pressures.

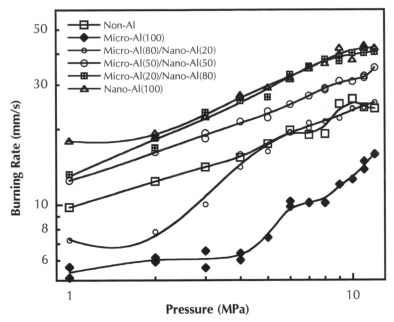

FIGURE 13.6   Burning rates of propellants with bimodal micro-/nano-aluminum.

13.4.5 Bimodal-Aluminized (Micro-/Nano-) Propellants

Figure 13.6 shows the burning rates of propellants with different combinations of micro- and nano-aluminum, ranging from 100% micro- to 100% nano-aluminized formulations. The trends are clearly consistent. When 100% micro- aluminum is used in place of coarse AP in the non-aluminized formulation, the burning rate is decreased, as observed earlier, and a 20% inclusion of nano-aluminum does not alter this trend. A 50-50 share between the two aluminum particles somewhat increases the burning rate over the non-aluminized formulation, but there is hardly any further gain in burning rate between a 80-20 versus 100-0 combination of nano- and micro-aluminum. These results show good control over the burning rate for a range of combination of bimodal aluminum, which could be effectively utilized in combustion instability mitigation.

Figure 13.6 shows a function of the nano- versus micro-aluminum in the bimodal-aluminized propellant formulations. It can clearly be seen that the burning rate monotonically decreases with decreasing content of nano-aluminum *vis-à-vis* micro-aluminum at all pressure levels. It also shows the self-quenched matrix. The Al agglomerate sizes are more than the parent aluminum sizes for both the cases.

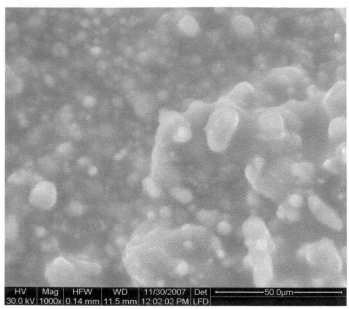

HV	Mag	HFW	WD	11/30/2007	Det	————50.0µm————
30.0 kV	1000x	0.14 mm	11.5 mm	12:02:02 PM	LFD	

FIGURE 13.7   Surface features of self extinguished bi-modal (50/50) aluminized matrix with fine AP (5µm)/HTPB – IPDI= 65/35, at 50 bar.

13.5   QUENCH COLLECTION OF AGGLOMERATES FROM THE PROPELLANTS

13.5.1   Effect of Aluminum Content of Micro-Aluminized Propellants

The influence of aluminum content on the agglomerate sizes can be seen in Figure 13.7. As mentioned earlier, 10, 15 and 18% Al were used in the propellant to study the effect on the agglomerate size trends. It can be seen that the agglomerate sizes for the propellant with 10% mostly increase with pressure, whereas there is a sharp decline in the sizes for the propellants with 15% and 18% Al, followed by a somewhat non-monotonic trend at a low level with increase in pressure. Consequently, over most of the test-pressure range, the propellant with 10% Al produces larger agglomerate sizes when compared to those with higher Al content; there is hardly any variation in the agglomerate sizes with subsequent increase in the Al content from 15 to 18%.

13.5.2   Effect of Aluminum Size of Micro-Aluminized Propellants

Three aluminum sizes of 15μm, 34μm and 85μm were used to study the effect of aluminum size on the propellant's agglomeration behavior. The amount of Al in the propellant is fixed at 15%. The arithmetic and SMDs of agglomerate sizes are shown in Figure 13.8 for these propellants. In general, the dependence of the agglomerate size on pressure is mostly weak; there is a slight decrease for the propellant with the smallest Al size tested, that is, 15μm, whereas the pressure dependence for the other two Al sizes tested, that is, 34μm and 85μm, is mixed and unclear, if not marginal. It can be clearly seen that the propellant containing the 15μm Al shows large agglomerates, whereas, the sizes of those from the other two higher sizes of the Al are nearly the same (around 55μm SMD or 35μm AMD). It is important to note that the propellant containing the 34μm Al still shows mild agglomeration in that its agglomerate sizes are slightly greater than the parent Al size (but not hugely greater, as with the 15μm Al). On the other hand, this signifies a sort of reverse process for the propellant with the 85μm Al, that is, the agglomerate size is actually smaller than the parent Al size. Indeed, most of the quenched agglomerates from the propellant with 85μm Al were alumina particles. This shows that, with 85μm Al, the Al particles mostly emerge from the propellant burning surface as single particles with very little agglomeration, actually resulting in effective sizes that were lower than the parent size. On the whole, these results clearly show a decrease in the tendency for agglomeration with increase in the parent Al size.

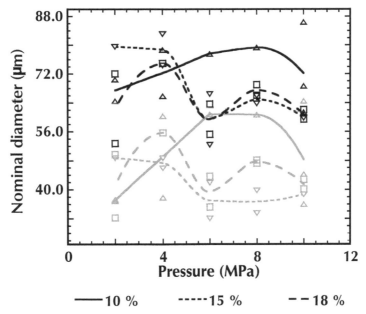

FIGURE 13.8   Effect of aluminum content on the agglomerate sizes of IPDI—cured propellants with a fine AP/binder ratio of 65/35, the coarse and fine AP and Al sizes being 450µm, 5µm, 15µm respectively. (Gray—AMD and Black—SMD).

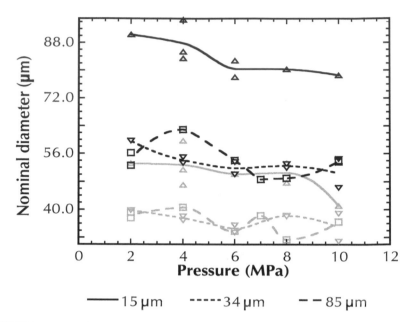

FIGURE 13.9   Effect of aluminum size on the agglomerate sizes of IPDI-cured propellants with a fine AP/binder ratio of 60/40, containing 15% Al, with the coarse and fine AP sizes being 450μm and 20μm, respectively (Gray—AMD and Black—SMD).

13.5.3   Nano-Aluminized Formulations

Figure 13.9 shows the surface features of the quenched aluminized matrix samples. The aluminum agglomeration occurs in both the size of the Al particles. The possibility of aluminum clustering size is more with micron-sized aluminum than nano-sized aluminum as shown in Figure 13.10a, b.

Figure 13.10 shows the SEM micrographs of nano-Al agglomerates collected from the quench collection apparatus at different pressures, namely 40, 60 and 80 bars.

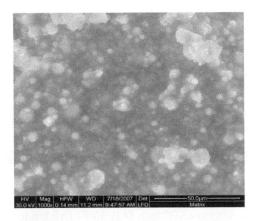

(a) Nano-aluminzed   (b) Micro-aluminized

FIGURE 13.10   Surface features of self extinguished matrixes.

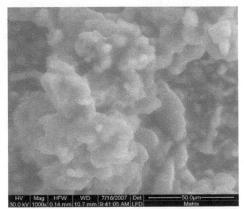

(a) Quenched at 40 bar

FIGURE 13.11   *(Continued)*

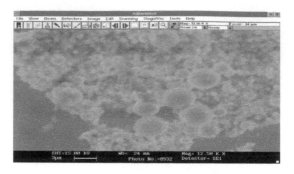

(b) Quenched at 80 bar

FIGURE 13.11   The SEM micrographs of quench collected nano-Al samples from the propellant; 450μm cAP,75μm fAP, 50/50 fAP/binder ratio, 87.5% total solids and 15% of nano-Al by mass.

The maximum size of the nano-Al agglomerate is not exceeding 5μm as observed in previous matrix SEM micrographs. Even if agglomeration occurs in nano-aluminized propellants, nano-aluminum would not cause two-phase flow losses thrust, because of its product size. Figure 13.11 shows SEM micrographs of quench-collected samples of nano-Al from a propellant with 53 μm fine AP, the rest of the formulation parameters remaining the same, at 40 and 80 bar. It can be found that there is significant agglomeration to the extent of 1–5 μm in this case also. There is no clear effect of pressure on the agglomerate size. Figures 13.12 and 13.13 show the SEM micrographs of nano-Al agglomerates emerging from a propellant with 250μm coarse AP and 75μm fine AP, the rest of the formulation parameters being the same as before, at 40 and 80 bar. Figure 13.14 shows the TEM images of same, but for 350μm coarse AP. From these Figure and several others obtained for each case, the size distribution of the nano-aluminum agglomerates has been deduced. From the above Figure, three aspects are apparent: (a) in most cases, the extent of agglomeration is of the order of 0.4μm, that is, 400nm, which itself is about 10 times the original nano-aluminum size; (b) in all those cases, a small fraction of the agglomerates are significantly larger in size, in the 1–5 μm range, which is what prominently shows up in the SEM/TEM images; (c) with the largest coarse AP (450 μm) and fine AP (75 μm), an appreciably larger fraction of agglomerates are of significant size, that is, ~2–3μm. This shows that large parts of the burning surface are available amidst the AP particles of large sizes that cause significant agglomeration to occur, with the ignition sources from the flames attached to the AP particles being farther away, in this case. In all other cases, the high burning rates witnessed earlier have made it possible for relatively fewer nano-Al particles to accumulate before being ignited into relatively smaller agglomerates.

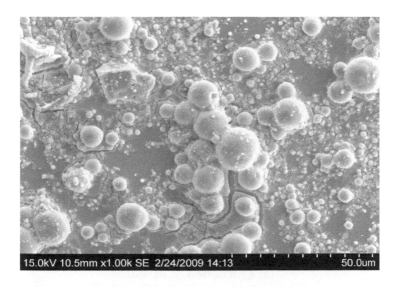

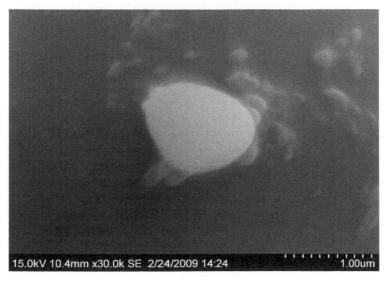

b) Quenched at 80 bar

FIGURE 13.12   The SEM micrographs of quench collected nano-Al samples from the propellant; 450 μm cAP, 53μm fAP, 50/50 fAP/binder ratio, 87.5% total solids and 15% of nano-Al by mass.

(a) **Quenched** at 40 bar

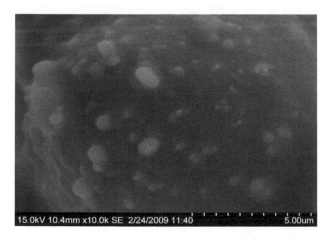

(b) Quenched at 80 bar

FIGURE 13.13 The SEM micrographs of quench collected nano-Al samples from the propellant; 250μm cAP, 75μm fAP, 50/50 fAP/binder ratio, 87.5% total solids and 15% of nano-Al by mass.

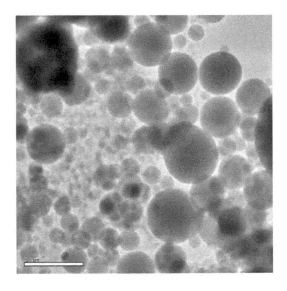

(a) Quenched at 40 bar

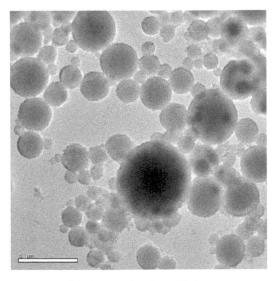

(b) Quenched at 80 bar

FIGURE 13.14 The TEM micrographs of quench collected nano-Al samples from the propellant; 350μm cAP, 75μm fAP, 50/50 fAP/binder ratio, 87.5% total solids and 15% of nano-Al by mass.

13.6 CONCLUSIONS

The burning rates of nano-aluminized propellants are increased by 100% when compared to those of the corresponding micro-aluminized and non-aluminized propellants. The burning rate is controlled by the heat transfer from the near-surface ignition and near-complete combustion of the nano-aluminium to the propellant burning surface. Addition of nano-aluminum beyond 10% in the propellants does not show any significant increase in burning rates. The aluminum agglomerates emerging from the burning surface of propellants have been quench-collected and analyze their size distribution. Plateau burning rate effects significantly affect the agglomerate size distribution variation with pressure. Aluminum content and size directly affect the agglomerate size, as expected, but not monotonically. The coarse AP size, and to a lesser extent, the fine AP size, affect the aluminum agglomerate size, which is of great significance in motor design. Nano-aluminum exhibits significant agglomeration, but only a small fraction of the agglomerates are in the 1–5μm range, except when large-sized coarse and fine AP particles are used in the formulation; even then, fraction of the above-sized agglomerates is larger, but the largest size of agglomerates does not exceed ~5μm. This aspect is expected to be benign for reduced smoke propellant applications from exhaust signature point of view, without sacrificing the energetics of the propellant in the form of using non-aluminized formulations normally adopted when micron-sized aluminum is the only option available.

KEYWORDS

- Aerospace
- Nano-Aluminum
- Nanoscale
- Propellants
- Propulsion

REFERENCES

1. Dreizin, E. L. (2000). Phase changes in metal combustion. *Progress in Energy and Combustion Science, 26,* 57–78.

2. Auner, N. & Holl, S. (2006). Silicon as energy carrier—Facts and perspectives. *Energy, 31,* 1395–1402.

3. Bardsley, W. E. (2008). The sustainable global energy economy: Hydrogen or silicon. *Natural Resources Research, 17,* 197–204.

4. Sarou-Kanian, V., Rifflet, J. C., Millot, F., & Gökalp, I. (2006). Aluminum combustion in wet and dry CO_2: Consequences for surface reactions. *Combustion and Flame, 145,* 220–230.

5. Sarou-Kanian, V., Rifflet, J. C., Millot, F., Veron, E., Sauvage, T., & Gökalp, I. (2005). On the role of carbon dioxide in the combustion of aluminum droplets. *Combustion Science and Technology,* 2299–2326.

6. Sarou-Kanian, V., Rifflet, J. C., Millot, F., Matzen, G., & Gökalp, I. (2005). Influence of nitrogen in aluminum droplet combustion. *Proceedings of the Combustion Institute, 30,* 2063–2070.

7. Shafirovich, E., Escot Bocanegra, P., Chauveaua, C., Gökalp, I., Goldshleger, U., Rosenband, V., & Gany, A. (2005). Ignition of single nickel-coated aluminum particles. *Proceedings of the Combustion Institute, 30,* 2055–2062.

8. Escot Bocanegra, P., Davidenko, D., Sarou-Kanian, V., Chauveau, C., & Gökalp, I. (2010). Experimental and numerical studies on the burning of aluminum micro and nanoparticle clouds in air. *Experimental Thermal and Fluid Science, 34,* 299–307.

9. Yetter, R. A., Risha, G. A., & Son, S. F. (2009). Metal particle combustion and nanotechnology. *Proceedings of the Combustion Institute, 32,* 1819–1838.

10. Sarou-Kanian, V., Ouazar, S., Escot Bocanegra, P., Chauveau, C., & Gökalp, I. (2007). *Low temperature reactivity of aluminum nanopowders with liquid water.* Proceedings of the European Combustion Meeting.

11. Jayaraman, K., Chauveau, C., & Gökalp, I. (2011). *Effects of aluminum particle size, galinstan content and reaction temperature on hydrogen generation rate using activated aluminum and water.* International Conference on Hydrogen Production (ICH2P-2011), June 18–22, Greece.

12. Trunov, M. A., Schoenitz, M., & Dreizin, E. L. (2005). Ignition of aluminum powders under different experimental conditions. *Propellants Explosives Pyrotechnics, 30,* 36–43.

13. Granier, J. J. & Pantoya, M. L. (2004). Laser ignition of nanocomposite thermites. *Combustion and Flame, 138,* 373–383.

14. Escot Bocanegra, P., Chauveau, C., & Gökalp, I. (2007). Experimental studies on the burning of coated and uncoated micro- and nano sized aluminum particles. *Aerospace Science and Technology, 11,* 33–38.

15. Jayaraman, K., Chauveau, C., Gökalp, I., & Calabro, M. (2011). Aluminum foams for space propulsion applications, Eucass. St Petersburg, Russia.

16. Romonadova, L. D. & Pokhil, P. K. (1970). Action of silica on the burning rates of ammonium perchlorate compositions. *Fizika Goreniya i Vzryva, 6,* 285–290.

17. Dokhan, A., Price, E. W., Seitzman, J. M., & Sigman, R. K. (2002). *The effects of bimodal aluminum with ultra-fine aluminum on the burning rates of solid propellants.* Proceedings of the Combustion Institute, *29,* 2939–2945.

18. Ivanov, Y. F., Osmonoliev, M. N., Sedoi, V. S., Arkhipov, V. A., Bondarchuk, S. S., Vorozhtsov, A. B., Korotkikh, A. G., & Kuznetsov, V. T. (2003). Productions of ultra-

fine powders and their use in high energetic compositions. *Propellants Explosives Pyrotechnics, 28,* 319–333.

19. De Luca, L. T., Galfetti, L., Severini, F., Meda, L., Marra, G., Vorozhtsov, A. B., Sedoi, V. S., & Babuk, V. A. (2005). Burning of nano-aluminized composite rocket propellants. *Combustion Explosion and Shock Waves, 41,* 680–692.

20. Galfetti, L., De Luca, L. T., Severini, F., Meda, L., Marra, G., Marchetti, M., Regi, M, & Bellucci, S. (2006). Nano-particles for solid propellant combustion. *Journal of Physics: Condensed Matter, 18,* S1991–S2005.

21. Galfetti, L., De Luca, L. T., Severini, F., Colombo, G., Meda, L., & Marra, G. (2007). Pre and post-burning analysis of nano-aluminized solid rocket propellants. *Aerospace Science and Technology, 11,* 26–32.

22. Pivkina, A. N., Frolov, Yu. V., & Ivanov, D. A. (2007). Nano sized components of energetic systems: Structure, thermal behavior, and combustion. *Combustion Explosion and Shock Waves, 43,* 51–55.

23. Popenko, E. M., Gromov, A. A., Shamina, Yu. Yu., Il'in, A. P., Sergienko, A. V., & Popok, N. I. (2007). Effect of the addition of ultrafine aluminum powders on the rheological properties and burning rate of energetic condensed systems, combustion. *Explosion, and Shock Waves, 43,* 46–50.

24. Trunov, M. A., Umbrajkar, S. M., Schoenitz, M., Mang, J. T., & Dreizin, E. L. (2006). Oxidation and melting of aluminum nanopowders. *Journal of Physical Chemistry B, 110,* 13094–13099.

25. Trunov, M. A., Schoenitz, M., & Dreizin, E. L. (2005). Ignition of aluminum powders under different experimental conditions. *Propellants Explosives Pyrotechnics, 30,* 36–43.

26. Rai, A., Park, K., Zhou, L., & Zachariah, M. R. (2006). Understanding the mechanism of aluminum nano particle oxidation. *Combustion Theory and Modelling, 10,* 843–859.

27. Kwon, Y. S., Gromov, A. A., Ilyin, A. P., Popenko, E. M., & Rim, G. H. (2003). The mechanism of combustion of superfine aluminum powders. *Combustion and Flame, 133,* 385–391.

28. Yuang, Y., Risha, G. A., Yang, V., & Yetter, R. A. (2007). Combustion of bimodal nano/micron-sized aluminum particle dust in air. *Proceedings of the Combustion Institute, 31,* 2001–2009.

29. Jones, D. E. G., Turcotte, R., Fouchard, R. C., Kwok, Q. S. M., Turcotte, A. M., & Abdel-Qader, Z. (2003). Hazard characterization of aluminum nano-powder compositions. *Propellants Explosives Pyrotechnics, 28,* 120–131.

30. Kwok, Q. S. M., Badeen, C., Armstrong, K., Turcotte, R., Jones, D. E. G., & Gertsman, V. Y. (2007). Hazard characterization of uncoated and coated aluminum nano-powder compositions. *Journal of Propulsion and Power, 23,* 659–668.

31. Bocanegra, P. E., Chauveau, C., & Gokalp, I. (2007). Experimental studies on the burning of coated and uncoated micro- and nano sized aluminum particles. *Aerospace Science and Technology, 11,* 33–38.

32. Freeman, J. M., Price, E. W., Chakravarthy, S. R., & Sigman, R. K. (1998). *Contribution of monomodal AP/HC propellants to bimodal plateau burning propellants,* AIAA Paper 98–3388.

33. Small, J. L., Stephens, M. A., Deshpande, S., Petersen, E. L., & Seal, S. (2005). *Burn rate sensitization of solid propellants using a nano-titania additive*, 20th International Colloquium on the Dynamics of Explosions and Reactive Systems, McGill University, Montreal, Canada.
34. Jayaraman, K., Anand, K. V., Chakravarthy, S. R., & Sarathi, R. (2007). *Production and characterization of nano-aluminum and its effect in solid propellant combustion*, AIAA-2007–1430, 45th AIAA Aerospace Sciences Meeting and Exhibit, Reno, NV, USA, Jan 2007.
35. Jayaraman, K., Anand, K. V., Chakravarthy, S. R., & Sarathi, R. (2007). *Production of nano-aluminum and its effect on burning rates in solid propellant combustion.* 6th Asia-Pacific Combustion Conference, Nagoya, Japan.
36. Jayaraman, K., Anand, K. V., Bhatt, D. S., Chakravarthy, S. R., & Sarathi, R. (2009). Production, characterization, and combustion of nano-aluminum in composite solid propellants. *Journal of Propulsion and Power, 25,* 471–481.
37. Jayaraman, K., Anand, K. V., Chakravarthy, S. R., & Sarathi, R. (2009). Effect of nano-aluminum in plateau-burning and catalyzed solid propellant combustion. *Combustion and Flame, 156,* 1662–1673.
38. Jayaraman, K., Chakravarthy, S. R., & Sarathi, R. (2010). Accumulation of nano-aluminum in the combustion of composite solid propellant mixtures. *Combustion, Explosion and Shockwaves, 46,* 21–29.
39. Jayaraman, K., Chakravarthy, S. R., & Sarathi, R. (2011). *Quench-collection of nano-aluminum agglomerates from combustion of sandwiches and propellants.* Proceedings of the Combustion Institute, *33,* 1941–1947.

CHAPTER 14

NANOTECHNOLOGY IN MEDICAL APPLICATIONS

RAMESH CHANDRA PANDEY, VIJAY KUMAR SAXENA, and VISHWAS SHARMA

CONTENTS

14.1 INTRODUCTION

Emerging diseases, mechanized way of living, critically polluted realms of earth, changing climatic regimens are the new challenges that mankind is encountering. To sustain and promote the health of humankind, new tools and technology is being developed in the cartel to blow away the nontraditional threats as well as the diversifying nature of known diseases. The branch of nanotechnology dealing with diagnosis, treatment, vaccination or preventive measures to improve health are referred to as 'Nanomedicine'. Precisely, nanomedicine may be defined as the development of engineered nanoscale (1–100 nm) structures and devices for better diagnostics and medical intervention in curing or replacing damaged tissues [1]. In 2003, NIH roadmap's nanomedicine initiatives envisioned that the cutting edge area of research will begin yielding medical benefits within as early as 10 years. It is very clearly evident now that how quickly this particular field has occupied the worldwide research arena amassing a huge percentage of total research budgets of many developed countries and its total impact on the development of other discipline is burgeoning. It is becoming an industry in itself which can be understood from the fact that there are more than 200 companies active in this field of which 59 are start-ups and small and medium enterprises that focus on the development of nanotechnology-enhanced pharmaceuticals and medical devices. There are more than 40 major pharmaceutical companies and medical devices corporations that employ nanotechnology to develop nanomedicine products and run development projects in this field [2]. Nanotechnology market is expected to be exceeding 30 billion dollars by 2015 according to a research by Global industry analyst, a U.S. based world leader in business intelligence and strategy support. The aim of nanomedicine can be elaborated as to carry out the comprehensive monitoring, repairing and improvement of all human biological systems after understanding the etiology at the molecular level using nanoengineered devices and nanostructures to achieve an overall medical fitness.

Utilization of nanotechnology to biomedical sciences means creation of materials and devices which should interact with the body system at sub-cellular level at high degree of specificity. It is like having a nano-scale messenger to pass on the desired message in a sophisticated and controlled way at the cellular and molecular level with high degree of precision and accuracy. Targeted and non-targeted drug delivery can be seen as a major thrust area of nanomedicine which has slowly evolved over a period of time. Revolution of the concepts and approaches in the nano-based drug delivery systems in past, present and future expectations are presented in the Table 14.1.

TABLE 14.1 Examples of drug delivery technologies in relation to the current nanotechnology revolution. (Adapted from [3]).

	Past nanotechnology	**Transition period (present)**	**Mature nanotechnology (future)**
Technology	Emulsion-based preparation of nano/micro particles	Nano/micro fabrication	Nano or micro manufacturing
Examples	Liposomes Dendrimers Nanocrystals Nanoparticles Microparticles Polymer micelles	Layer-by-layer assembled systems Microneedle transdermal delivery Microdispensed particles Microchip systems	Nano/micro machines for scale-up production

In general, breakthroughs in the field of nanotechnology can be categorized as the achievements in the field of imaging, diagnostics, preventive medicine and therapeutics. The branch of nanomedicine has diversified to the extent that it is now considered as a scientific discipline. Here, in this chapter, some important developments in the usage of nanomolecules in medical and pharmaceutical applications are briefly discussed. It has been tried to highlight the some of the nanotechnology applications with their success stories which have been realized in market and approved by regulatory authorities.

14.2 LIPOSOMES

Liposomes are non-polymeric structures, typically spherical nano-vesicles, formed by phospholipids bilayer. During preparation of liposomes, amphiphilic phospholipid molecules self-assemble to form a bilayer enclosing a part of medium. The medium containing therapeutic agents gets encapsulated between the lipid bilayer which is then used as therapeutic delivery system. The use of liposomes as therapeutic delivery system employs the concept of incorporating the active substances such as drugs, vaccines, genetic materials as well as small molecules in the nano-vesicles. Structural features of liposomes such as its size, composition, lamellarity, and surface properties determine its target specificity. Depending on size, liposomes may be classified as small (<100 nm), intermediate (100–250 nm) or large (>250 nm).

Size and lamellarity of liposomes may be controlled by optimizing the liposome preparation methods. To increase the target specificity, liposomes can be engineered to interact with particular population of cells. Conjugations with polyethylene glycol (PEG) or specific antibody based constructs result in the formation of targeted liposomes such as PEGylated liposome and immuno-liposomes, respectively (reviewed in [4]). While targeted liposomes are specifically internalized by cells through endocytic mechanism, untargeted liposomes may be taken up by cells by adsorption, endocytosis and lipid exchange. In addition, PEGylated liposomes exhibit longer circulation time by escaping clearance by reticulo-endothelial system of immune system and they are called stealth-liposomes.

The extensive work on liposome mediated delivery resulted in realization of the first FDA approved liposome based drug "Doxil" for Kaposi's sarcoma in 1995, that is within 3 decades of discovery of the first liposome in 1965 by Bangham [5]. Currently, there are more than ten liposome-based drugs approved for clinical use and still more are in various stages of clinical trials (Table 14.2). Similarly, there are number of antimicrobial and anticancer drugs which are under clinical trial (reviewed in [5]).

TABLE 14.2   Liposome-based drugs on market. (Modified from [5]).

Product name	Route of injection	Drug	Particle type/size	Drug form/storage time	Approved indication
Ambisome	Intravenous	Amphotericin B	Liposome	Powder/36 months	Sever fungal infections
Abelcet	Intravenous	Amphotericin B	Lipid complex	Suspension/24 months	Sever fungal infections
DaunoXome	Intravenous	Daunorubicin	Liposome	Emulsion/12 months	Blood tumors
Doxil	Intravenous	Doxorubicin	PEGylated liposome	Suspension/20 months	Kaposi's sarcoma, ovarian/breast cancer

TABLE 14.2 *(Continued)*

Myocet	Intravenous	Doxorubi-cin	Liposome	Powder/18 months	Combination therapy with cyclophosphamide in metastatic breast cancer
Visudyne	Intravenous	Verteporfin	Liposome	Powder/48 months	Age-related molecular degeneration, pathologic myopia, ocular histoplasmosis
Depocyt	Spinal	Cytarabine	Liposome	Suspension/18 months	Neoplastic meningitis and lymphomatous meningitis
DepoDur	Epidural	Morphine sulfate	Liposome	Suspension/24 months	Pain management
Epaxal	Intramuscular	Inactivated hepatitis A virus (strain RG-SB)	Liposome	Suspension/36 months	Hepatitis A
Inflexal V	Intramuscular	Inactivated hemaglutinine of Influenza virus (strains A and B)	Liposome	Suspension/12 months	Influenza

14.3 ENGINEERED PLGA NANOPARTICLES

The polymer of lactic and glycolic acid called poly(D,L-lactide-co-glycolide) (PLGA) exhibits characteristics such as biocompatibility, biodegradability and easy to manipulate surface properties. Along with these features, pharmaco-kinetic and pharmaco-dynamic properties of PLGA make it a suitable drug carrier. This nanoparticle has been approved by United States-Food and Drug Administration (US-FDA) as a therapeutic carrier system for human use. Particularly, PLGA has gained attention in controlled-release applications. They are uptaken by target cells by receptor mediated endocytosis which later forms endolysosomes. The change in pH of endo-lysosomal compartment causes release of the drug molecule from the PLGA (reviewed in [6]). The PLGA nanoparticles loaded with therapeutic agents slowly release the drug over extended period of time up to days and weeks at the target site. For example, PLGA containing drugs Triptorelin pamoate (Decapeptyl) and Leuprolide acetate (Lupron Depot) are FDA approved anticancer agents available in market [7]. Recently, multifunctional PLGA nanoparticles carrying multiple functional groups and ligands/drugs, are being designed which can be used for diagnostics/imaging as well as therapeutics in dual purpose applications (Figure 14.1).

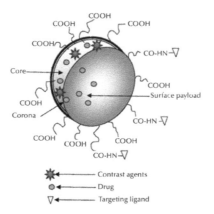

FIGURE 14.1 Schematic representation of a multifunctional PLGA nanoparticle. (Modified from [6]).

However, there are also challenges associated with use of PLGA as therapeutics which limits it applications. For instance, due to acidification of endolysosomal compartment or accumulation of acidic monomers lactic and glycolic acids within the nano-carrier device may be detrimental to proteinaceous agents. Few attempts are presented to improve the efficiency of PLGA as therapeutic carrier system: 1) The conjugation of PLGA with an integrin-binding RGD (Arg-Gly-Asp) peptide to develop micro-capsule-based system as a tumor ultrasound contrast agent to discriminate malignant cells from the benign tumors, 2) The addition of hydrophobic tricaprin additives along with low molecular weight PEG-1000 to improve the release of proteins such as BSA, 3) Encapsulation of peptide (insulin, calcitonin) by chitosan and carbopol-coated mucoadhesive PLGA nanoparticles are the effective carriers to improve oral and pulmonary mucosal delivery due to their prolonged retention and excellent penetration into the mucus layer (reviewed in [7]).

14.4 DENDRIMERS

Dendrimers are branched tree-like nanomolecules which are implicated in drug delivery systems. Drug carrying capacity of dendrimers is achieved by either attaching the active substance to the end of the branches (or termini) or enclosing the active compound in the cavities formed by its branches (reviewed in [8]). Design of the dendrimers such as selection of scaffold building blocks, nature and degree of branching, linker groups, surface charge and functional groups determines pharmaco-kinetic properties, valency (number of drug molecules being carried by each dendrimer), drug release and finally its utility.

After being taken up by target cells the active ingredient conjugated to the dendrimers are cleaved by enzymatic reaction within the cells, releasing the drug molecules. Figure 14.2 represents an example of uptake and release of N-acetyl-cysteine from PAMAM dendrimer which are linked by disulfide bond.

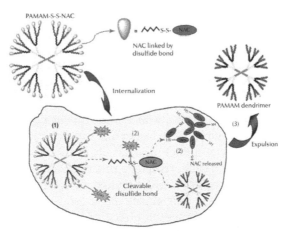

FIGURE 14.2 Schematic representation of the PAMAM dendrimer linked to N-acetylcysteine by disulfide bonds. The dendrimer delivers the drug intracellularly by the cleavage of the disulfide linkage owing to the thiol exchange redox reactions initiated by the intracellular glutathione. The dendrimer carrier is then excreted by the cell. (Modified from.. [9]).

Depending on the extent of branching, they are classified as the first generation (G1), that is with one adapter branch, second generation (G2), that is with two adapter branch or higher generation dendrimers [8]. 2,6-bis(hydroxymethyl)-*p*-cresol and 2,4-bis(hydroxymethyl) phenol are examples of basic building blocks which forms skeleton of dendrimers and connected to drug molecules through appropriate linker molecules. Stimulus/trigger induced disassembly of branched skeleton leading to break-down to building blocks and finally releasing the active substance is called cascade-release or self-immolative or dendritic-amplification. In a recent publication, Wang et al. [8], have reviewed the basic strategies involved in design of dendrimers and improvement of its functions. Poly(amido amine) or PAMAM and poly(propylene-imine) or Astromol (PPI) are the examples of two commercially available dendrimers. A dendrimers based anti-HIV drug SPL7013 gel (VivaGel) has recently completed its clinical trial and marked the success story of dendrimers based approach [10]. Similarly, PAMAM and ethylene-diamine (EDA) based production of dendrimer-chelant-antibody constructs, and boronated dendrimer-antibody conjugates (for neutron capture therapy) are being used as cancer therapeutics [11].

14.5 GOLD NANOPARTICLES

Use of inorganic metal such as gold are documented in Chinese and Indian traditional medicine dating back to 2500–2600 BCE in a form called as "*Swarna Bhasma*" to treat various systematic and infectious diseases (reviewed in [12]). However, use of near-atomic sized gold nanoparticles in nanomedicine relies on its specific opto-electronic properties such as absorption spectra and scattering intensity, which are different from gold as metal. Furthermore, higher affinity of gold nanoparticles to functional groups such as thiols, amines, disulfides permits its conjugation with therapeutic substances or contrast agents and conjugated gold nanoparticles with DNA, peptides, proteins and biological ligands helps to achieve specific activity and target specificity [12].

Fabrication of gold nanoparticle with organic monolayer imparts stability to inner core and allows customizing the interaction with environment. By combination of appropriate characteristics of nanoparticle and the associated monolayers impregnated with desired ligands, receptors, or antibodies, it is possible to modulate biocompatibility, biodistribution, retention time, pharmaco-dynamic and pharmaco-kinetic properties to differentiate the target cells from non-target cells at the site of action. Recently, therapeutic effects of naked gold nanoparticles were shown to inhibit VEGF induced proliferation of endothelial cells *in vivo*. Proliferating endothelial cells facilitates angiogenesis in growing tumors and angiogenesis inhibition are additional approach in cancer therapeutic regimen. A gold nanoparticle bound to cytokine TNF has been tested against cancer was found to be more effective and less toxic than unconjugated TNF (reviewed in [13]).

14.6 QUANTUM DOTS

Quantum dots (QDs) are among another most promising tools in the nanomedicine having several medical applications. They are employed for making nanodiagnostics, imaging, and targeted drug delivery. Quantum dots are the nanocrystals which are composed of a core of a semiconductor material, enclosed within a shell of another semiconductor material such that they together produce a large spectral band gap (as depicted in Figure 14.3), resulting in unique optical properties. These optical properties includes broad range excitation, size tunable narrow emission spectra and high photo stability [14]. When the QD is hit by incident light, it absorbs a photon with a higher energy than that of the band gap of the composing semiconductor, an exciton or electronic whole pair is created. Consequently, the probability of absorption at higher energies will be increased, creating a broadband absorption spectrum. When the exciton returns to a lower energy level, a narrow, symmetric energy band emission occurs. Quantum dots emit quite long fluorescent signal (10–40 ns) creating a strong and stable fluorescent signal. The QDs emission typically ranges from 450 nm to 850 nm and the QDs can be tuned to have desired emission spectra. Application of quantum dots can be understood by their recent use in which antibody-conjugated QDs were used to detect prostate cancer cell marker PSA and the QD conjugates detected the tumor site in mice transplanted with human prostate cancer cells [15]. Water soluble QDs as injectables were used for imaging skin and adipose tissues in mice [16].

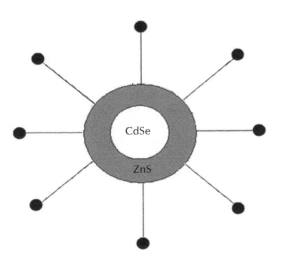

FIGURE 14.3 Schematic representation of CdSe/ZnS quantum dot. (Modified from [14]).

14.7   ANTIMICROBIAL COATING

It refers to the labeling of important surfaces with materials that are able to minimize the persistence and spread of pathogenic microbes. They should have ability to bind microbe and inhibit their growth. Some coating do bring about partial sterilization by minimizing the ability to bind and grow on the coated surfaces specially the instruments, equipments, surgical and so on, which are exposed to the patient's body fluid. Lysozyme is an important natural enzyme forming a part of innate immune system and has antimicrobial effects. In the search of an inert material that can hold upon enzyme in functional state for long, scientists have tried a first successful merger of lysozyme with single walled carbon nanotubes (SWNTs) [17]. Another type of coating involves merging carbon nanotubes with a naturally occurring enzyme 'lysostaphin' which is used by non-pathogenic strains of the bacteria to defend against *Staphylococcus aureus*, including Methicillin-resistant *S. aureus* (MRSA). The resulting nanotube-enzyme "conjugate" can be mixed with any number of surfaces to prevent the spread of MRSA [18]. Some application of antimicrobial coatings is also being employed as part of several water purification and cleansing systems. The EPS formation which occurs in biofilm can be checked by synthesizing protein resistant material by functionalizing surfaces with surfactants [19], protein resistant polymers [20] and biofilm degrading enzyme [21].

14.8   NANOFILTERS

Nanofilteration which employs the application of nanotechnology in this aspect, is a cross-flow filtration technology using pores in nanometer size and have molecular weight cut-off typically less than 1,000 atomic mass unit. Traditional filtration methods employed in the form of bacterial and viral filters uses granular carbon, porous ceramic, polymeric materials, many of which get clogged up and are difficult to clean and so must be replaced frequently. The main advantages of nanofilters as opposed to traditional filters as water purifiers are that less pressure is required to pass water across the filters; they are more efficient and have large surface areas. Scientists have been trying to use carbon nanotubes in addition to nanoscopic material like alumina fibers for nanofilteration as it can remove all contaminants, including oil, bacteria, viruses, turbidity and as well as other organic contaminants. They have ultra smooth surfaces from inside which makes more efficient in spite of having very low nanometer sized pores. Inclusion of active materials, such as silver nanoparticles or titanium dioxide nanoparticles and UV light sources can bring about enhanced killing of trapped viruses, bacteria and fungi. Recently a team of scientists from Rensselaer Polytechnic Institute, Troy, NY and Banaras Hindu University, India has devised a novel way to accommodate million of long carbon molecules on the inner surface of a quartz tube about a centimeter large, resulting in the development of radially oriented nanotubes, packed tightly to act as a cylindrical filter [22]. Water can squeeze out of the nanomolecular sized gaps in the walls but the bacteria like *E.coli* and virus like Polio virus are struck back inside.

14.9   OTHERS

Similarly, there are number of nanomolecules which are in the stage of development and are used in the field of therapeutics and diagnostics. The concept of 'Nanorobots' or simply 'Nanomachines' are in the developmental stage. Nubot is the term used for 'Nucleic acid robot'. Similarly, biocompatible artificial organs are being developed which can overcome the limitations of conventional transplantation like antigen mismatch between the donor and patient. Currently, attempts are being made to develop artificial bone (and joints), blood vessels, lung and heart. Nanofibres are made up of carbon with diameter >1,000nm and has been suggested to be used in medical imaging. It is believed that such devices may bring revolution in the nanomedicine.

14.10   SUCCESS STORIES IN NANOMEDICINE FOR TREATMENT OF DISEASES

14.10.1   Cancer

The treatment of cancer remains one of the most important objectives of pharma industries and therefore immense improvement has been made in this area of nanotechnology. Because of extensive research some of the nano-drugs are already patented and are in the market whereas many others are in the developmental phase. The drug Abraxane is derived from albumin bound taxane particles and marketed for the treatment of lung cancer and ovarian cancer. The drug Doxil, formed from liposomal doxorubicin is promoted for the treatment of ovarian tumor [23]. The Caelye, based on liposomes, functionalized by polyethethylene glycol (PEG) and encapsulated doxorubicin are also used against the treatment of ovarian cancer [24]. The other drug Oncospar (Enzon) is a PEGylated L-asparaginase and is approved by FDA for the treatment of acute lymphoblastic leukemia [25, 26]. Similarly, MRX952 is a nanoparticle encapsulating camptothecin analogues and Aurolase which is a gold nanoshell for treating head and neck cancer are in preclinical stage. Additionally, some other drugs such as INGN 401 and Cyclosert-Camptothecin are still in phase 1 or phase 2 trials.

14.10.2   Cardio-vascular Diseases

The cardio-vascular diseases (CVDs) are a broad term which includes coronary artery, cerebrovascular and peripheral vascular disease [27]. Among success stories of nano-based therapeutic agents for CVD are Prednisolone phosphate encapsulated in PEGylated 3.5-dipentadecyoxybenzamidine hydrochloride liposomes which showed significant reduction of in-stent neointimal growth in rabbits having atherosclerosis [28]. There are also several *in-vitro* experimental reports on liposomes based compounds which demonstrated anti-atherogenic effects such as liposomes base cyclopentenone prostaglandins and liposomes base serum amyloid A peptide fragments [29, 30]. The nano-drug $\alpha_v\beta_3$-intergin can suppress the growth of new blood vessels and inhibit angiogenesis [31]. Wickline and Lonza group used arginine-glycine-aspartic acid (RGD) perfluorocarbon (PFC) for both therapeutic and imaging purpose [32]. Several other studies have highlighted the role of direct conjugation of enzymes to specific immunoglobulins. For instance, angiotensin converting enzyme (ACE), platelet endothelial cell adhesion molecule (PE-CAM) and intracellular adhesion molecule (ICAM) [27]. Additionally, siRNA against NOX2, also called HBOLD7 showed promising results when applied to the arterial wall after angioplasty [33].

Imaging

The wide ranges of nano-particle are used for molecular and cellular imaging purpose. These nano-particles have been differently categorized into fluorescent, paramagnetic, acoustic, light scattering and super paramagnetic groups. Each of these groups consists of nano-particles as a contrasting agent. The first group i.e. fluorescent particles includes quantum dots as a contrasting nano-particle in fluorescence tomography. Further, paramagnetic group comprises polyamidoaminedendrimers gold ions, diaminobutanedendrimers gold ions and iron-oxide microparticles for magnetic resonance imaging (MRI). The other group acoustic covers liposomes used in ultrasound imaging. Finally, light scattering and super paramagnetic groups include gold nanoshells used in optical coherent tomography and ultra small particulars of iron oxide (UPSIO) used in MRI, respectively [27].

14.11 CONCLUSION

In summary, since the concept of nanomedicine was conceived, within short time span, encouraging results from extensive research in this field have led to therapeutic and diagnostic products which are being realized in market and approved by authorities. Accompanying the fast growth of the nano-based drugs, studies to optimize the drug delivery efficiencies, pharmaco-kinetic and pharmaco-dynamic properties, tissue and organelle specificity and finally to assess the toxicological effects of nanomolecules are the current research focus. Some of the applications with high sensitivity but higher toxicity *in vivo,* are being used in diagnostics or *in vitro* imaging which in turn may assist the drug development research. Foreseeing the limitations related to contemporary medicine and success stories of nanomedicine, it could be imagined that nano-based products may dominate the market in future.

KEYWORDS

- **Cascade-release**
- **Dendrimers**
- **Nanomedicine**
- **Nanorobots**
- **Quantum dots**

REFERENCES

1. Jarm, P. & Kramar, A. Zupanic (Eds.) (2007). *Medicon 2007, IFMBE Proceedings* (16, pp. 1135–1136). Berlin: Springer-Verlag.
2. Wagner, V., Hüsing, B., et al. (2008). *Nanomedicine: drivers for development and possible impacts*. European Commission Joint Research Centre Institute for Prospective Technological Studies.
3. Park, K. (2007). Nanotechnology: what it can do for drug delivery. *Journal of Control Release, 120*(1–2), 12–24.
4. Sofou, S. & Sgouros, G. (2008). Antibody-targeted liposomes in cancer therapy and imaging. *Expert Opinion on Drug Deliv, 5*(2), 189–204.
5. Chang, H. I. & Yeh, M. K. (2012). Clinical development of liposome-based drugs: formulation, characterization, and therapeutic efficiency. *International Journal Nanomedicine, 7,* 49–60.
6. Acharya, S. & Sahoo, S. K. (2001). PLGA nanoparticles containing various anticancer agents and tumor delivery by EPR effect. *Advanced Drug Delivery Reviews, 63*(3), 170–183.
7. Mundargi, R. C., Babu, V. R., et al. (2008). Nano/micro technologies for delivering macromolecular therapeutics using poly(D,L-lactide-co-glycolide) and its derivatives. *Journal of Control Release, 125*(3), 193–209.
8. Wang, R. E., Costanza, F., et al. (2012). Development of self-immolative dendrimers for drug delivery and sensing. *Journal of Control Release, 159*(2), 154–163.
9. Menjoge, A. R., Kannan, R. M., et al. (2010). Dendrimer-based drug and imaging conjugates: design considerations for nanomedical applications. *Drug Discovery Today, 15*(5–6), 171–185.

10. McCarthy, T. D., Karellas, P., et al. (2005). Dendrimers as drugs: discovery and pre-clinical and clinical development of dendrimer-based microbicides for HIV and STI prevention. *Molecular Pharmacology, 2*(4), 312–318.
11. Baker, J. R. Jr. (2009). Dendrimer-based nanoparticles for cancer therapy. *American Society of Hematology Education Program, 18,* 708–719.
12. Bhattacharyya, S., Kudgus, R. A., et al. (2011). Inorganic nanoparticles in cancer therapy. *Journal of Pharmacy Research, 28*(2), 237–259.
13. Arvizo, R., Bhattacharya, R., et al. (2010). Gold nanoparticles: opportunities and challenges in nanomedicine. *Expert Opinion on Drug Delivery, 7*(6), 753–763.
14. Azzazy, H. M. E., Mansour, M. M. H., & Kazmierczak, S. C. (2006). Nanodiagnostics: a new frontier for clinical laboratory medicine. *Clinical Chemistry, 52*(7), 1238–1246.
15. Harma, H., Soukka, T., & Lovgren, T. (2001). Europium nanoparticles and time-resolved fluorescence for ultrasensitive detection of prostate-specific antigen. *Clinical Chemistry, 47,* 561–568.
16. Larson, D. R., Zipfel, W. R., Williams, R. M., Clark, S. W., Bruchez, M. P., Wise, F. W., et al. (2003). Water soluble quantum dots for multiphoton fluorescence imaging *in vivo. Science, 300,* 1434–1436.
17. Nepal, D., Balasubramanian, S., et al. (2008). Strong antimicrobial coatings: single-walled carbon nanotubes armored with biopolymers. *Nano Letters, 8*(7), 1896–1901.
18. Pangule, R. C., Brooks, S. J., et al. (2010). Antistaphylococcal nanocomposite films based on enzyme-nanotube conjugates. *ACS Nano, 4*(7), 3993–4000.
19. Hu, C., Yang, C., & Hu, S. (2007). Hydrophobic adsorption of surfactants on water-soluble carbon nanotubes: A simple approach to improve sensitivity and antifouling capacity of carbon nanotubes-based electrochemical sensors. *Electrochemistry Communications, 9,* 124–128.
20. Lin, Y., Allard, L. F., & Sun, Y. P. (2004). Protein-Affinity of Single-Walled Carbon Nanotubes in Water. *Journal of Physical Chemistry B, 108*(12), 3760–3764.
21. Richards, M. & Cloete, T. E. (2010).In T. E. Cloete, M. de Kwaadsteniet, M. Botes, & J. M. L´opez-Romero (Eds.), *Nanotechnology in water applications*. United Kingdom: Caister Academic Press.
22. Rensselaer (2004). Efficient filters produced from carbon nanotubes through Rensselaer Polytechnic Institute-Banaras Hindu University Collaborative Research. http://news.rpi.edu/update.do?artcenterkey=435
23. Surendiran, A., Sandhiya, S., et al. (2009). Novel applications of nanotechnology in medicine. *Indian Journal of Medical Research, 130*(6), 689–701.
24. Kim, P. S., Djazayeri, S., et al. (2011). Novel nanotechnology approaches to diagnosis and therapy of ovarian cancer. *Gynecologic Oncology, 120*(3), 393–403.
25. Zeidan, A., Wang, E. S., et al. (2009). Pegasparaginase: where do we stand? *Expert Opinion on Biological Therapy, 9*(1), 111–119.
26. Heidel, J. D. & Davis, M. E. (2011). Clinical developments in nanotechnology for cancer therapy. *Pharmacy Research, 28*(2), 187–199.
27. Psarros, C., Lee, R., et al. (2012). Nanomedicine for the prevention, treatment and imaging of atherosclerosis. *Nanomedicine, 8*(Suppl 1), S59–S68.
28. Joner, M., Morimoto, K., et al. (2008). Site-specific targeting of nanoparticle prednisolone reduces in-stent restenosis in a rabbit model of established atheroma. *Arteriosclerosis Thrombosis and Vascular Biology, 28*(11), 1960–1966.

29. Homem de Bittencourt, P. I. Jr., Lagranha, D. J., et al. (2007). LipoCardium: endothelium-directed cyclopentenone prostaglandin-based liposome formulation that completely reverses atherosclerotic lesions. *Atherosclerosis, 193*(2), 245–258.
30. Tam, S. P., Ancsin, J. B., et al. (2005). Peptides derived from serum amyloid A prevent, and reverse, aortic lipid lesions in apoE-/- mice. *Journal Lipid Research, 46*(10), 2091–2101.
31. Winter, P. M., Neubauer, A. M., et al. (2006). Endothelial alpha(v)beta3 integrin-targeted fumagillin nanoparticles inhibit angiogenesis in atherosclerosis. *Arteriosclerosis Thrombosis and Vascular Biology, 26*(9), 2103–2109.
32. Lanza, G. M., Winter, P. M., et al. (2006). Nanomedicine opportunities for cardiovascular disease with perfluorocarbon nanoparticles. *Nanomedicine (London), 1*(3), 321–329.
33. Wang, Y., Li, Z., et al. (2010). Nanoparticle-based delivery system for application of siRNA *in vivo*. *Current Drug Metabolism, 11*(2), 182–196.
34. Safinya, C. R. & Ewert, K. K. (2012). Material chemistry: Liposomes derived from molecular vases. *Nature, 489*(7416), 372.

CHAPTER 15

SELF-ASSEMBLING OF NANOSTRUCTURES

SIVA CHIDAMBARAM, KASI NEHRU, and
MUTHUSAMY SIVAKUMAR

CONTENTS

15.1 INTRODUCTION

Nanomaterials have got the spanking attentions after the remarkable speech of the physicist Richard Feynman at 1959, which is entitled as "There's plenty of room at the bottoms". The nanomaterials have got further increased attentions greater than ever, after they have been realized that the morphology of them also desires the noteworthy properties along with particle size. Every one of the nanostructures is unique in their properties. This makes more attention on the structures of nanomaterials, starting from the synthesis phase to precise study on its various characteristics. Series of studies made on various nanostructures revealed them for variety of noteworthy applications. Diverse of morphological sceneries can be obtained from various synthesizing procedures and is also possible even by varying the parameters in a single method [1–4]. Assorted numbers of nanostructures have been revolutionized in the industries, such as solar cell applications, energy storage, nanoelectronics, printing technology, and so on, based on their respective properties [5–8]. Attracting and distinguished properties of nanostructured materials make the researchers and industrialists to concentrate more on procedures toward producing more quality in nanostructures. Self-assembling process has been recognized as an authentic and guaranteed technique to attain outstanding qualities in both organic and inorganic nanostructures. Self-assembling of materials can result in any dimensionalities (0D, 1D, 2D, and 3D) [9–11]. In this perspective, we will discuss about the various parameters involved in the self-assembling process, various types of conjugations for self-assembling, structuring possibilities of both inorganic and organic materials, applications of self-assembled nanostructures and few others too.

Assembling of tiny building blocks to form the structures by themselves is known as self assembling. In brief, with the available neighboring interactions or forces and without any external directions, the atoms, molecules, particles or any compounds may get arrange among them or assembled spontaneously with any molecules or over the surface of any substrate or over any specified structures, which results the self-aligned and assembled nanostructures. This conjugation or positioning based self-assembling occurs sometimes simply by releasing energy to obtain a stable structure. These exothermic reactions have been in practice for several years of nanoscale industries, but in some cases it may follow endothermic and by that self-assembling is known as static, dynamic self-assembling, respectively. Self-assembling has various definitions by various scientists as in the followings,

"Self assembly is the autonomous organization of components into patterns or structures without human intervention" (by G.M. Whitesides & B. Grzybowski).

"Self-assembly is a spontaneous and reversible process that brings together in a defined geometry randomly moving distinct bodies through selective bonding forces" (by Serge Palacin & Renaud Demadrill).

Scale of Self-Assembly

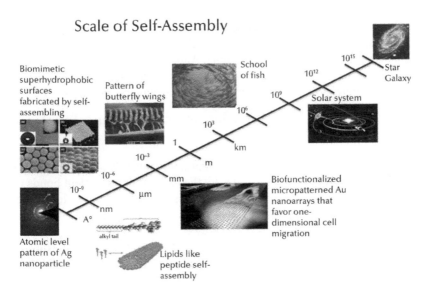

FIGURE 15.1   The length scale relations of various self-assembled bodies.

These spontaneous and automatic arrangements of tiny edificial blocks in an arranged or structural manner are omnipresent in nature itself. Starting from the order of DNA, amino acids, peptides, proteins, entire human body, all are self-assembled structures only. In nanoscale industries self-assembling method has been well suited "to several" applications and this method found to be a solution for several problems. For example, to study the proteins well, protein microarrays are essential. All other methods had met failures in spotting the protein arrays, but self-assembling works better [10–13]. Little or no waste of chemicals during synthesis, opposite arrangement and high quality nature in yield are the few highlighting factors of self-assembling method. These factors are always depicting the inimitabilities of this method among various synthetic routes [14–16]. Like nucleation process in material synthesis process, here, initialization leads the resultant assembling blocks [12]. Figure 15.1 is showing the length scale chart of the various self-assembled structures.

15.2   MATERIALS AND THEIR STRUCTURES

As we know, materials will behave by what they made of and importantly how they composed of. The compositions of a resultant material is depend on the processing ability of synthesizing method, but the architecture is completely depends on nature of the synthetic procedure, which includes interactions between the tiny particles [17]. It is possible to obtain various structures in larger quantities through this self-assembling route, and it has been documented to produce the nanostructures with ever improved monodispersion nature. Nowadays, preparation of various structures like spheres, platonic polyhedral, fine rods, pillars, ellipsoids, discs, tripods or tetrapods, core or shell, nanocages, and dumbbells are the familiar nanostructures in self-assembling. Another advantage of self-assembling is it can process a wide range of materials such as metals, semiconductors, oxides, polymers, inorganic salts, various organic semiconductors and so on [18].

15.3 INTERACTIONS BETWEEN PARTICLES UNDER BROWNIAN MOTIONS

"Chemistry Beyond the Molecule" or "Supramolecular Chemistry" is a subject and study on assembling of molecules by the basis of non covalent bonds [19]. Normally, atoms and molecules arranged among them as particle by means of the chemical forces or chemical bonds. By that they possess strong chemical interactions among them, mostly they are covalent bonds. In contrast to this, the self-assembled structures arranged and hold together by means of hydrogen bond or metal-ligand bond or hydrophobic-hydrophilic interactions or Vander Waals forces and electric or magnetic dipole interactions. The Table 15.1 shows various bonds and interactions exhibit varied strengths [20]. The strengths of hydrogen bond, Vander Waals force, and metal-ligand bond and interactions are several times lesser than that of covalent bond's energy. Only these weak interactions or weak bonds lead the assembling among the molecules here. The assembled nanoparticles are known to be nanoparticle supper lattices. This conversion of all the nanoparticles into super lattices can be occurred as simultaneously or it can follow one by one, which desires the size [15]. Assembling transpires by means of molecular interactions after balancing the attractive and repulsive forces among them. This assembling process normally held out at solution phase in order to provide a free Brownian motion among building blocks and to avoid distracting forces occurred by means of thermal motion. Consequently, interactions among the components and the environment of balancing highly influence the itinerary of the process [16].

TABLE 15.1 Various bonds and their corresponding strengths [20]

Strength of Bonds	
Interaction or Bond Type	**Strength (kJ/mol)**
Covalent bond	100–400
Coulomb	250
Dipole-Dipole	5–50

TABLE 15.1 *(Continued)*

Dipole-Ion	50–200
Cation-π	0–5
π-π	0.50
Hydrogen bond	10–5
Van der waals forces	<5
Metal-ligand	0–400

15.3.1 Based on Hydrogen Bond

It is a weakest bond than all other bonds. Hydrogen bond induced between molecules makes the assembling of building blocks. Basic schemes of living things like double-helix structures of DNA, alpha helix structure of proteins are hydrogen bonding based assembled structures only. The same mechanics is mimicked in laboratories to obtain the assembled structures. For example, hydrogen bonding occurs between an amphiphilic block co-polymer and hydrogen bonding agent (HA) which facilitates to create a hydrogen bond among the molecules. This increases the entire molecular weight and obtains a solid state material. A wide range of structures can be prepared in this hydrogen bond based assembling. Especially reversible cross-linked complex materials can be assembled by means of hydrogen bonds [21]. The hydrogen bonding cannot form simply among the molecules. It needs some essential factors to occur. Those factors are (1) Electrostatic or coulomb energy (ΔE_{cou}), (2) Exchange repulsion (ΔE_{ex}), (3) Dispersion forces (ΔE_{dis}), (4) Charge-transfer energy or covalent bonding (ΔE_{cht}), (5) Polarization energy (ΔE_{pol}) [14, 22].

Hydrogen bonding auspiciously escorts the assembling process as highly directional oriented. This reason allows a user to simplify the parameters for recognition and binding during the assembling process. In 1990s, so many researches have been limited with this hydrogen bond based molecular self-assembling. Hydrogen bonding based assembling and their studies revealed the ways of understanding the molecular recognition. These studies reveal the preferable natures of the hydrogen bonding between the hydrogen bonding agent (HA) to an amphiphilic block copolymer [23–28]. Hydrogen bonding enables the simplified formation of monodispersed spherical solid particles, which are important kind of nanostructures in various applications [29]. The HA simply placed inside the particle, where solid polymer surrounds the core HA in this monodispersed solid particles. Further removal of HA with suitable solvent will result the monodispersed hallow spheres. The controlling parameters involved here allows to attain the possibility to tune the hallow cavity size and surface reactivity. Figure 15.2 shows the schematic representation of the formation of hydrogen bonding in the first stage of the hallow sphere formation. Also, it shows the removal of HAs, which occupied the center part during the assembling process, to form the hallow spheres.

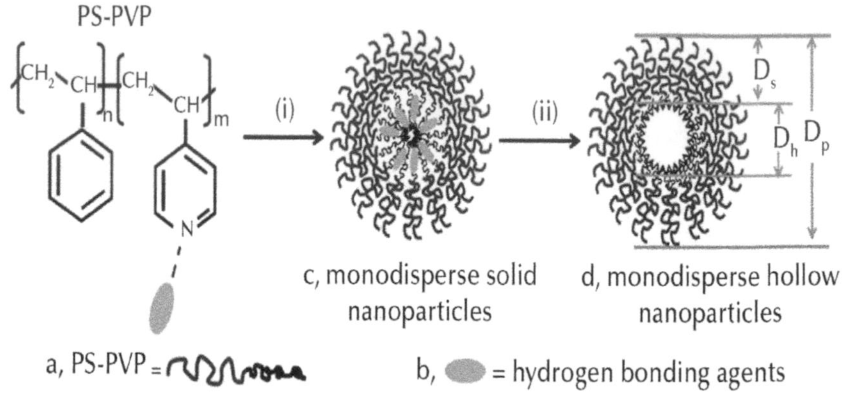

PS-PVP

(i)

(ii)

D_s

D_h D_p

c, monodisperse solid nanoparticles

d, monodisperse hollow nanoparticles

a, PS-PVP =

b, ● = hydrogen bonding agents

FIGURE 15.2 Schematic representation of hollow spheres formation by means of hydrogen bonding formation and deformation (i) Hydrogen-bonding-assisted self-assembly and (ii) Formation of hollow nanoparticles. (Image reproduced from [30] Copyright 2009 American Chemical Society).

In this case, polystyrene-b-polyvinylpyridine (PS-PVP) as amphiphilic molecule and 2-(4′-Hydroxybenzeneazo) benzoic acid (HBBA) as a HA, has been taken at first. PVP is highly soluble in simple alcohols, such as methanol, ethanol. PS chains will not dissolve in alcohols normally. With these considerations, in this case, simple alcoholic wash has been done to remove the HBBA completely. This washing actually effortlessly leads the breaking the hydrogen bonds which was assembled between PVP molecules to HBBAs.

Scanning electron microscopy (SEM) study and transmission electron microscopy (TEM) results reveals the formation of monodispersed particles of PS-PVP. Figure 15.3 shows the SEM image of hydrogen bonding based self-assembled spherical particles before removal of core. Inset in that shows the TEM image of the same, which confirms the solid particle nature and tells the monodispersion of with an average diameter of 35 nm sized particles (Dp). Figure 15.3b is the TEM image of the particles after removal of HBBA. The contrast in the image confirms the cavity in inner side and reveals its parameters like shell thickness (Ds) and diameter of the void space. These hollow spheres allow ~98% of near-IR spectrum to penetrate through them and this value is healthier result for transmitting the light than that of the inorganic coatings [30].

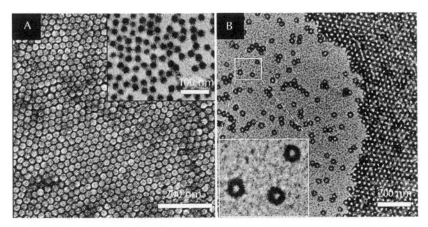

FIGURE 15.3 (a) shows the SEM image of the self-assembled particles by hydrogen bonding and inset is the TEM image of the same. (b) is the TEM image of the hallow spheres.
(Image reproduced from [30]) Copyright 2009 American Chemical Society

Controlling this hydrogen bonding based assembly is almost easy by simple adjustment of hydrogen ion concentration by varying pH level of reaction medium [acidic and basic conditions]. Thus, one can straightforwardly control this assembling progression [31–33]. Recently, this assembling process have been suggested as promising method of obtaining uniform hallow structure assemblies as it is discussed here [30]. Marcus Weck et al. [34] have assembled a polymer network by this method and altered the mechanical properties of assembled molecules. During the rheological changes from viscous to highly solute state of assemblies, the structure or hydrogen bond does not alter in its nature. Zhan-Ting Li et al. [35] obtained highly stable and higher molecular weight copolymers by hydrogen bonding based self-assembling. Polymers can be stacked in a sequence to form a layer by layer structure. Michael F. Rubner et al. [36] had assembled poly (vinyl alcohol)(PVA) layers along with poly-acids, such as poly (acrylic acid)(PAA), Poly (methacrylic acid) to obtain a layer by layer hydrogen assembly. These are few remarkable results to account in the researches of hydrogen bonding based assembling.

15.3.2 Based on Pi-Pi Interaction

One of the significant approaches of self-assembling is interaction based assembling, which is boon for structuring the organic molecules. Organic molecules are getting more attentions in the field of semiconductor electronics from last decade. Organic molecules grown or assembled or deposited over a silica substrate have been replaced most of inorganic semiconductors, due to their efficient charge transportation processes, better flexibility in solid state, unique fragility of materials and simple processing techniques to attain the structures. In the last decade, organic field effect transistors (OFET) have been made by Pentacene molecules by a series of researches. In various aspects, OFET had shown higher mobility with pentacene molecules than any other kind of molecules. Organic single crystals have been reported for their higher charge mobility, which is $20cm^{-1} V^{-1} S^{-1}$, which is 6–7 times higher than the inorganic thin films. Such single crystal formations have to be materialized to be the thin film structures for various technological aspects. The higher mobility of the organic materials needs to be improved further, when they are structured over a substrate for device applications in simple and cost-effective routes. The Figure 15.4 shows a typical block diagram of the electronic device made by organic molecules [37–40].

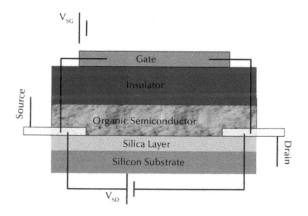

FIGURE 15.4   Block diagram of a typical OLED [38].

Inorganic molecule structuring, especially one dimensionality, has been developed as much as possible via various synthetic routes in the nanoscale industries. But nano level structuring of the organic species has not been explored well in all methodologies like sol-gel processing. The interaction based assembling allows constructing the quantum confined nanostructures. Although all the dimensionalities have been developed by interaction, structuring the one dimensional organic molecules is one step in front of all [41]. Self-assembled organic molecules toward one dimensionally by means of planer -conjugation have been excessively utilized for various nanodevices including electrical sensors, optical sensors, and optoelectronic devices (e.g., Flat panel displays, LED). Almost, only two types of molecules (p-type and n-type) have been employed in this conjugation. The p-type is aryl ethynylene macrocycle (AEM) like molecules and n-type is like perylene tetracarboxylic diimide (PTCDI) molecules. The assembled structures always exhibit as a multifunctional material, due to the contributions of multiple types of chemical species. And thus the structures are composed of lengthier molecules, so the resultant conjugated materials would show a vast amount of void space, that is porosity. These huge void spaces may not be useful. At the same, these porous materials are exhibiting efficient charge transportation, better exciton migration and a strong interfacial interaction with the targets. Behind these superior properties of one dimensionalized organics, the strong interaction occurred between the planer skeletons and also in their non collapsible structures, act as backbone. Several researches have been going on interaction based self assembling, which are not limited only on one dimensionality [42–49].

Organic molecules in a solution can assemble themselves as a nanostructure for better functioning light emission, light gathering, light transmission, charge transmission, and so on. Most of the semiconducting organic molecules and semiconducting moieties are accepted as next generation low-cost solar cell materials. Molecules such as Poly (3-hexylthiophene) (P3HT), an electron donor, and phenyl-C61-butyric acid methyl ester (PCBM), an acceptor have been showing their superior power conversion efficiency values than any other organic species, because of their convenient HUMO and LUMO positions [50–55]. Due to their controlled stacking and directed structuring, assembled nanostructures possess better crystalline nature, which ultimately increases the charge mobility more and more. Crystalline oligomers, especially acenes, shows the hole mobility of 5.0cm2 V-1 s-1, where inorganic element, amorphous silicon, shows only of about 1.0cm2 V-1 s-1. Among the other quantum confinements, one dimensional has reported as better efficiency materials in light energy processes. By executing both tight packing and one dimensional structuring, self-assembling of nanomaterials based on interaction is the key in attaining the stated fabulous properties in structured materials [56–58].

Assembling based on interactions eventually leads the well crystalline nature in the confined structures. Unlike inorganic, organic semiconductors and their charge transport properties are distinctive, which follows a hopping mechanism of polarons among neighboring localized states aided but vibration occurred in lattices. As discussed in previous, charge transfer rate of crystalline organic semiconducting solid parts is huge than the chaos of semiconducting organic species. The improved charge transfer rate is due to the minimization of grain boundaries and reduced lattice defects. Lattice defects are primary reason for trapping of charges (both electron and hole trapping). Very low level of grain boundaries and lattice defects in crystalline domains due to face to face stacking boosts the mobility. At the same time, soluble conjugated molecule species leads the formation of anarchies of film structures. Recently, Wei-Wen Tsai et al. [57] assembled a novel hairpin shaped self-assembling organic species in a series of processes based on π–π connections and clearly studied their stacking process. Figure 15.5

representing the chemical structure of a single hairpin shaped molecule and the colors denotes their various groups of parts.

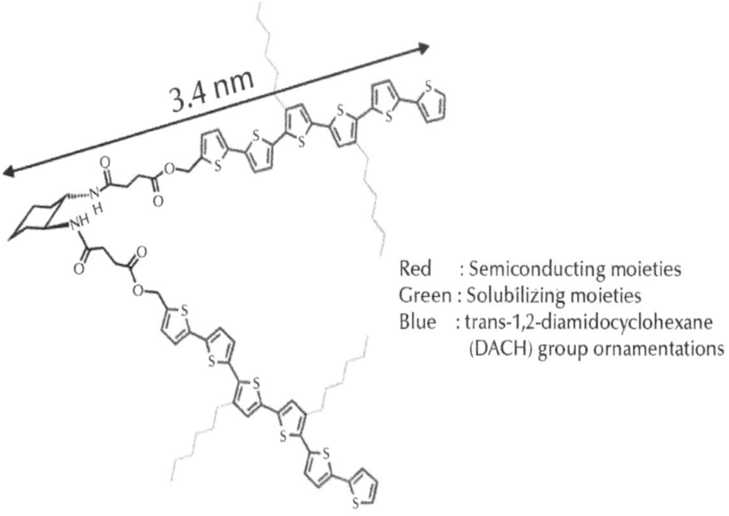

Red : Semiconducting moieties
Green : Solubilizing moieties
Blue : trans-1,2-diamidocyclohexane
(DACH) group ornamentations

FIGURE 15.5 Molecular model of the hairpin-shaped sexithiophene molecule and their functional parts.
(Image reproduced from [57]) Copyright 2010 American Chemical Society

This hairpin assembly contains electronically active sexithiophene moieties. In this study, it is found that hairpin structure is formed in non polar solvents like chlorocyclohexane with π–π interaction between the sexithiophene moieties. It is also found that, this formation is not only with π–π interactions but also due to the hydrogen bonding. The head of the hairpin is formed due to the hydrogen bonding only, which is between amide groups. Their structures show a fibrous shape with uniform diameters of ~3nm. The schemes given in the Figure 15.6 shows the step by step progress in the self-assembling progress. This inset Figure15.6a clearly shows the hydrogen bonding between one molecule to another, which has been occurred among their head groups (blue) and at the same green colored parts, which are H-aggregates of the sexithiophenes, get aligned to form H-aggregates parallel to each other. The inset Figure15.6c shows the formation of bundling of nanofibers facilitated by J-aggregates.

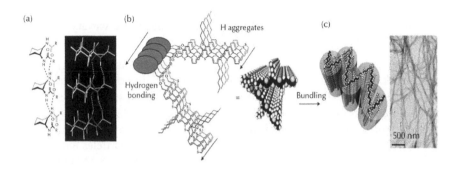

FIGURE 15.6 Design of the self-assembling between the hair-pin shaped sexithiophene shaped molecule. (a) Formation of Hydrogen bonds between the trans-1,2-diamidocyclohexyl groups. (b) Head group conjugation (blue color) and eventual leading of H-aggregates of the sexithiophenes (in green) to form nanofibers. (c) Nanofibers bundling facilitates the formation of J aggregates in the hierarchical manner and their TEM image [58].
(Image reproduced from [57] Copyright 2010 American Chemical Society

Thus discussed in this case, the H-aggregates are leading the formation of nano-fibers formations by bundling the sexithiophene molecules, which are hydrogen bonding based conjugated hairpin shaped molecules. These nanofibers are about an arm of length. Thus by this process of self-assembling based production of semiconducting organic molecules nowadays can greatly destroy the barriers in producing the bulk amount of nanomaterials to implement at industrial scales [57, 58].

15.3.3 Based on Metal-Ligand Conjugation

Metal-ligand bonding based abiological structures had been materialized for the past three decades as a cutting edge technology and also as advancements in various industries. The hydrogen bond's strength is very low (less than 10kJ/mol), which is very lesser than covalent bonds (100–400kJ/mol). And by that, lack of directionality, misalignment in arranging the several tens of interactions as a series are became negative factors that have been found in hydrogen bonding based structuring. Dative metal-ligand bonds are found as a solution to this, because of d-orbital attachment. Third row metal-ligand bond are exhibiting the average energies around 15–25kcal/mol. This can be amended to self repairing and self healing in the formation of assemblies and this metal-ligand bond can be the right substitution for several hydrogen bonds. Studies have been evolved to obtain defined structure and shaped assemblies, such as ladders, grids. The pre-defined self-assembling of metal-lacycles and metallacages and their property analysis had been in progress of various laboratories [59–68]. The accurate rational design predictability and assembling of the same have been achieved with simple alteration of the size and shape of the building blocks. It gives an opening to use variety of transition metals and metal complexes [69–73].

In 1980s, researchers had discovered that n-alkanethiols ($HS(CH_2)_{(n-1)}CH_3$) were get attached impulsively to noble metals. Especially, over the surfaces of gold particle or gold thin films, anchor sulfur organic groups, like thiols (RSH), and disulfides (RSSR) (R denotes alkyl chain) are getting attached. This array of complete protection over the metal surfaces by means of alkanethiol goups forms the monolayer protected clusters (MPCs) against the metal particles and self-assembled monolayer thin films (SAMs) over the metal thin films. SAMs formation of alkanethiol groups allows the further attachment of other compounds over the alkyl chain. This spontaneous assembling of organic ligands to the metal surfaces leads the formation of densely packed SAMs [74]. Further, these structures have been studied well with various characterization techniques, such as scanning tunneling microscopy, infrared spectroscopy, wetting measurements, and electron diffraction techniques. Results of these studies state the followings:

- The interaction forces between alkanethiol to noble metals play a major role in attraction. Among all interaction forces, electron affinity of the sulfur to the gold is the primary reason behind this assembly. S-Au interaction is on the order of 45 Kcal/mol, which might form a semi covalent bond.

- Structure of the monolayer depends on the chemistry of the chain. In a typical alkanethiol group get attach to the surface with an angle of 30° (Au with thiol group). Thus, it forms the monolayer of R30° ($\sqrt{3}$ x $\sqrt{3}$) stable structure over a metal surface.

- Attachment of the alkanethiols follows well ordered assembling nature and they are densely ordered. The standing alkyl groups from the outer surfaces are free to process further (Figure 15.7) [75–80]).

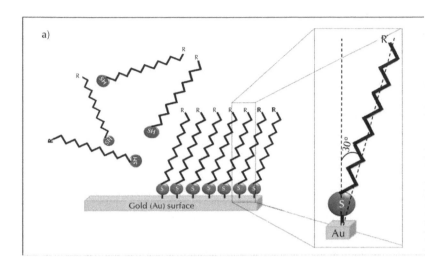

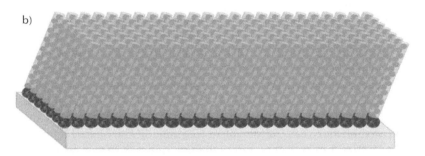

FIGURE 15.7 (a) Pictorial representation of alkanethiol molecule to metal substrate attachment (Inset shows the single molecule attachment) and (b) Computer model of the series of thiol group densely arranged over a gold substrate to form SAMs.

Later, so many organic molecules have been studied for their attachment to various metals. Excellent structures have been obtained with this metal-ligand bond based self-assembling process [81–89]. Recently, eight different compounds have been used and self sorting base assembling has been done by Schmittel et al. [90], and followed by this, three components based self-assembled structures (2D and 3D assembles) have been studied [91]. Considerably, Liu et al. [92] have been reported the way to convert metal-organic squares to porous-like supramolecular assemblies. Lin et al. [93] studied the bonding nature between Cu or Fe metal islands on Au (111) substrate to tripyridyl ligands. They studied the thermodynamics involved in metallasupramolecular assembly (2D) using Scanning Tunneling

Microscope. Gao et al. [94] studied the Ag-acetylide ligands interactions to obtain the self-assembling of rotaxane based structures. Argent et al. [95] investigated and resulted that cadmium ball (Cd_{66}) can also be effectively used as a metal part for self-assembling processes with the ligand H_3L. Thus metal-ligand complexes always look like legends of researches and it seems to be continuing forever.

15.3.4   Assembling of Inorganic Structures

Apart from the various organic nanostructures, structural basis self assembly of nanostructures are also needed to demonstrate for inorganic blocks. Countless works have been chronicled in the accounts of research for self-assembling of inorganic materials. Especially, magnetic materials have been well remarked as superior magnetic property exhibiting blocks, when they are self-assembled together [96, 97]. Uniform spongy-like one dimensional structure is much favored in applications, such as chromatography, drug delivery, and nanoelectronic devices. Chemical self-assembling favors to satisfy these needs ultimately. Highly uniform shaped silica nanofibers can be produced at industrial scales with chemical self-assembling process [98]. Catalytic performance of a catalyst always depends on the surface area of the particles. Large surface area always increases the catalytic performance. Self-assembling process of inorganic nanostructures also prefers the preparation of large surface area possessing architectures. Mesoporous nanocrystal clusters of anatase TiO_2 have been produced by Yadong Yin et al. [99], by means of self-assembling process. These TiO_2 particles were reported for their higher surface area (BET surface area as high as $277m^2/g$) and showed a very good photocatalytic activity. As reported recently, self-assembled silver nanoparticles over the graphene sheets are realized for enhanced output in GaN-based light emitting diodes [100]. Thus inorganic nanostructures delivered from self-assembling process are always showing remarkable improvements in various industries by possessing significant chemical, electrical, electronic, optical, and mechanical properties.

Crystallization of tiny quantum dots, finest nanoparticles can be formed by means of solvent evaporation, which is also a technique of self-assembling process. This is an irreversible process and having the ability to produce almost of 2D and 3D nanostructures. Evaporation of solvents from the tiny solids not only leaves the pot, but also influences in the arrangement of solid crystals, when they are drying. This process simply allows the inorganic structured tiny crystals or quantum dots to assemble together to form the definite bulk structure. Various reports of solvent evaporation based self-assembling have been suggesting that the simplest form of structuring at nanoscale is controlled evaporation of solvent from the solids. Here various parameters are playing the role of structuring, like rate of evaporation of solvent, substrate nature, density of the solids, adhesive nature between the solids, and morphology of the solids [101]. Controlled evaporation of solvent based self assembly has been utilized to get cubic and polyhedron shapes [102]. This method holding the advantages of procedure simplicity, cost-free technique, ability to cultivate a millimeter sized arrays, and nanostructures, possibility to control various properties of thin films (thickness, composition, etc.).

Self-assembling of inorganic one dimensional nanostructures are possible by means of templated self assembling, assembling on nanowires, arraying of nanocrystals using physical confinement and capillary forces. 2D and 3D architectures can be formed by means of arraying over 2D substrates, assembling using gravity sedimentation, assembling during solvent evaporation, assembling inside the micro channels, assembling of colloidal nanoparticles between two plates, assembling of monolayers over vertical substrates, and assembling during shear flow [103].

Various parameters have been involving in these inorganic self-assembling processes. One of the common parameters, temperature, has been tested for assembling of silica nanofibers, which resulted in a better way to optimize the structure. In chemical self-assembling of silica fibers higher temperature or room temperatures does not give uniform structures. But at lower temperatures, it assembles as uniform sized, hexagonal shaped fibers [98]. Like that hydrophobic and hydrophilic SAM natures were utilized to produce changes in the facets of Ag nanocubes. This has been achieved by changing the functionalization of six faces a nano cube with SAMs of different schemes [103]. Various materials can be assembled as various nanostructures for various applications in various assembling routes. Interestingly, nanodots of magnetic materials [104], cobalt nanowire like self aligned islands [105] are assembled over the metal substrate. Tiny silicon nanowires have self-assembled as silicon nanotubes via embedding in polymers and the tubes showed improved electronic properties [106]. Hallow micro capsules preparation are also possible by utilizing self-assembling process [107].

15.4 ADVANTAGES OF SELF-ASSEMBLY

A Range of Structures

Spontaneous arrangement of tiny blocks finally forms various architectures. As per the above discussions, almost all kinds of quantum confined nanostructures are promised to attain in this simple Brownian motion based assembling [11, 15, 16].

Controllability of Architectures

Self-assembling of nanostructures is basically occurring by means of Brownian motion of the particles, electron affinity, weak forces, and weak bonds. But these parameters can be simply tuned by means of temperature, pH of the solution, substrate adhesion nature, solution viscosity, and so on, by which one can easily control the architecture. These wide ranges of parameters allow better controlling on the building assemblies [98, 103].

Quality in Yields

Nowadays, self-assembling is always in front of all other techniques in producing the quality products. Better controllability in the complete nature assembling keeps the uniformity in stacking the smallest assemblies everywhere in the container or in the substrate [98].

Low Cost Designing

Self-assembling of nanostructures are always far better than all other structural mechanisms, especially than the mechanically structuring of the materials. Self structuring does not require any machinery for structuring the materials at nanoscale. Mainly, it does not need any pressure or higher temperatures, which are actually necessary things in hydrothermal or solvothermal techniques. As another example, in electrospinning of fibers, which allows confirmed fiber structuring, higher electric potential (tens of kilo volts) is essential. But self-assembling is always free of these costlier things, thus, this method accepted as low cost structuring. SAMs gives a very fantastic uniformity in monolayer formation, which is expensive to prepare by any other alternative techniques [73, 108].

Bulk Production

Self-assembling is not limited to experimental level. This can be processed for a vast amount. A tonnes of architectures assembling in a single pot is also possible [103].

Substitution to Nature

Bio molecule structures are very equal to natural biological structures in performance basis. These tiny nanostructures are biocompatible than chemically or mechanically driven nanostructures. Nanobiostructures and their molecular dynamics have opened a new gateway to heal the existing problems in medicinal fields to various biological streams. Artificial molecular machines are the next level of pioneering materials in biological industries and they have found simplest preparations via self-assembling processes [13].

15.5 NUISANCE OF ASSEMBLING

Reproducibility

Reproducibility is the primary parameter in desiring a procedure as possible industrial level technique. Self-assembling mainly fails in reproducing the structure. This poor repeatability is very difficult to control in the self-assembling process [11, 12].

Complexes in Structures

Almost all materials produced in this assembling are very complex in their structure. This ultimately results some difficulties in studying the physical and chemical properties of the entire assembly [11, 12].

Unused Parts in Assemblies

In the entire assembly, entire parts may not be useful in the applications. Some of the parts in the structures cannot be used and also they cannot be avoided. In some architecture, most of the volume of the entire construction is void space. These void spaces increase the size of the entire structure enormously [41].

15.6 CONCLUSION

Self assembly is holding more beauties than mechanically derived assemblies. Simply self assembly directs the molecules by means of Brownian motion and arranges them in an order with lesser time consumption. This allows to synthesize the structures precisely even for several grams of molecules. It allows the researcher to obtain a bulk amount of assembled structures. But in some way, it is limited to make them into industrial applications, like in an assembled complex structure apart from useful part it also hold some invaluable parts in larger amount.

KEYWORDS

- **Hydrogen bond**
- **Magnetic dipole interactions**
- **Nanostructures**
- **Scanning electron microscopy**
- **Silver nanoparticles**

REFERENCES

1. Nalwa, H. S. (2011). *Nanostructured materials and Nanotechnology*. California: Academic press.
2. Logothetidis, S. (2012). *Nanostructured materials and their applications Nanoscience and Technology*. Berlin: Springer.
3. Chuang, T. J., Anderson, P. M., Wu, M.-K., & Hsieh, S., (2006). *Nanomechanics of materials and structures*. The Netherlands: Springer.
4. Whitesides, G. M., Kriebel, J. K., & Mayers, B. T. (2005). Self-Assembly and Nanostructured Materials. In W. T. S. Huck (Ed.), *Nanoscale assembly: Chemical techniques*. Berlin: Springer.
5. Wu, H. & Cui, Y. (2012). Designing nanostructured Si anodes for high energy lithium ion batteries. *Nano Today, 7,* 414.
6. Bai, C. & Liu, M. (2012). Implantation of nanomaterials and nanostructures on surface and their applications. *Nano Today, 7,* 258.
7. Gorte, R. J. & Vohs, J. M. (2009). Nanostructured anodes for solid oxide fuel cells. *Current Opinion in Colloid & Interface Science, 14,* 236.
8. Wu, H. B., Chen, J. S., Hng, H. H., & Lou, X. W. (2012). Nanostructured metal oxide-based materials as advanced anodes for lithium-ion batteries. *Nanoscale, 4,* 2526.
9. Gazit, E. (2007). Self-assembled peptide nanostructures: the design of molecular building blocks and their technological utilization. *Chemical Society Review, 36,* 1263.
10. Boncheva, M., Bruzewicz, D. A., & Whitesides, G. M. (2003). Millimeter-scale self-assembly and its applications. *Pure Appllied Chemistry, 75,* 621.
11. Pelesko, J. A. (2007). *Self assembly: The science of things that put themselves together*. Boca Raton: Chapman and Hall/CRC Press.
12. Tompa, P. (2009). *Structure and function of intrinsically disordered proteins*. Boca Raton: Chapman and Hall/CRC Press.
13. Ramachandran, N., Hainsworth, E., Bhullar, B., Eisenstein, S., Rosen, B., Lau, A. Y., Walter, J. C., & LaBaer, J. (2004). Self-Assembling Protein Microarrays. *Science, 305,* 86.
14. Setoa, C. T. & Whitesides, G. M. (1993). Molecular self-assembly through hydrogen bonding: supramolecular aggregates based on the cyanuric acid. Melamine lattice. *Journal of american chemical society, 115,* 905.
15. Sun, S. (2006). Recent advances in chemical synthesis, self-assembly, and applications of fept nanoparticles. *Advanced materials, 18,* 393.
16. Sun, S. (2006). Self-assembled nanomagnets. In D. J. Sellmyer, R. Skomski, & Klumer (Eds.), *Advanced magnetic nanostructures*(p. 239). USA: Springer.
17. Whitesides, G. M. Kriebel, J. K., & Mayers, B. T. (2005). Self-assembly and nanostructured materials. In W. T. S. Huck (Ed.), *Nanoscale assembly chemical techniques* (p. 217). USA: Springer.
18. Bishop, K. J. M., Wilmer, C. E., Soh, S., & Grzybowski, B. A. (2005). Nanoscale forces and their uses in self-assembly. *Small, 14,* 1600.
19. Lehn, J. M. (2002). Toward self-organization and complex matter. *Science, 295,* 2400.

20. Hoeben, F. J. M, Jonkheijm, P., Meijer, E. W., & Schenning, A. P. H. J. (2005). About Supramolecular assemblies of π-conjugated systems. *Chemical Reviews, 105,* 1491.
21. Rieth, R., Eaton, R. F., & Coates, G. W. (2001). Polymerization of ureidopyrimid-inone-functionalized olefins by using late-transition metal ziegler–natta catalysts: Synthesis of thermoplastic elastomeric polyolefins lee. *Angewadte Chemie, International Edition, 40,* 2153.
22. Prins, L. J., Reinhoudt, D. N., & Timmerman, P. (2001). Non covalent synthesis using hydrogen bonding. *Angewadte Chemie International Edition, 40,* 2382.
23. Lehn, J. M. (1988). Supramolecular chemistry: scope and perspectives molecules, supermolecules, and molecular devices. *Angewadte Chemie International Edition, 27,* 89.
24. Chang, S. K. & Hamilton, A. D. (1988). Molecular recognition of biologically inter-esting substrates: Synthesis of an artificial receptor for barbiturates employing six hydrogen bonds. *Journal of American Chemical Society, 110,* 1318.
25. Zimmerman, S. C. VanZyl, C. M., & Hamilton, G. S. (1989). Rigid molecular twee-zers: Preorganized hosts for electron donor-acceptor complexation in organic sol-vents. *Journal of American Chemical Society, 111,* 1373.
26. Rebek, J. (1990). Molecular recognition with model systems. *Angewadte Chemie In-ternational Edition English, 29,* 245.
27. Doig, A. J. & Williams, D. H. (1992). Binding energy of an amide-amide hydrogen bond in aqueous and nonpolar solvents. *Journal of American Chemical Society, 114,* 338.
28. Christopher, T. S. & Whitesides, G. M. (1993). Molecular self-assembly through hy-drogen bonding: supramolecular aggregates based on the cyanuric acid-melamine lat-tice. *Journal of American Chemical Society, 115,* 905.
29. Park, J., Joo, J., Kwon, S. G., Jang, Y., & Hyeon, T. (2007). Synthesis of monodis-perse spherical nanocrystals. *Angewadte Chemie International Edition, 46,* 4630.
30. Sun Z., Bai F., Wu H., Schmitt S. K., Boye D. M., & Fan H. (2009). Hydrogen-bond-ing-assisted self-assembly: monodisperse hollow nanoparticles made easy. *Journal of American Chemical Society, 131,* 13594.
31. Beijer, F. H., Sijbesma, R. P., Vekemans, J. A. J. M., Meijer, E. W., Kooijman, H., & Spek, A. L. (1996). Hydrogen bonded complexes of diaminopyridines and diami-notriazines: Opposite effect of acylation on complex stabilities. *Journal of Organic Chemistry, 61,* 6371.
32. Zimmerman, S. C. & Corbin, P. S. (2000). Heteroaromatic modules for self-assembly using multiple hydrogen bonds. *Structure and Bonding, 96,* 63.
33. Cooke, G. & Rotello, V. M. (2002). Methods of modulating hydrogen bonded interac-tions in synthetic host–guest systems. *Chemical Society Review, 31,* 275.
34. Nair, K. P., Breedveld, V., & Weck, M. (2011). Modulating mechanical properties of self-assembled polymer networks by multi-functional complementary hydrogen bonding. *Soft Matter, 7,* 553.
35. Chen, S. G., Yu, Y., Zhao, X., Ma, Y., Jiang, X. K., & Li, Z. T. (2011). Highly stable chiral $(A)_6$-B supramolecular copolymers: A multivalency-based self-assembly pro-cess. *Journal of American Chemical Society, 133,* 11124.

36. Lee, H., Mensire, R. Cohen, R. E., & Rubner, M. F. (2012). Strategies for hydrogen bonding based layer-by-layer assembly of Poly (vinyl alcohol) with weak polyacids. *Macromolecules, 45,* 347.

37. Miao, Q., Nguyen, T.-Q., Someya, T., Blanchet, G. B., & Nuckolls, C. (2003). Synthesis, assembly, and thin film transistors of dihydrodiazapentacene: An isostructural motif for pentacene. *Journal of American Chemical Society, 125,* 10284.

38. Witte, G. & Woll, C. (2004). Growth of aromatic molecules on solid substrates for applications in organic electronics. *Journal of Material Research, 19,* 1889.

39. Loi, M. A., Como, E. D., Dinelli, F., Murgia, M., Zamboni, R., Biscarini, F., & Muccini, M. (2005). Supramolecular organization in ultra-thin films of α-sexithiophene on silicon dioxide. *Nature Materials, 4,* 81.

40. Muccini, M. (2006). A bright future for organic field-effect transistors. *Nature Materials, 5,* 605.

41. Zang, L., Che, Y., & Moore, J. S. (2008). One-dimensional self-assembly of planar π-conjugated molecules: adaptable building blocks for organic nanodevices. *Accounts of Chemical Research, 41,* 1596.

42. Apperloo, J. J., Janssen, R. A. J., Malenfant, P. R. L., & Fréchet, J. M. J. (2000). Concentration-dependent thermochromism and supramolecular aggregation in solution of triblock copolymers based on lengthy oligothiophene cores and poly (benzyl ether) dendrons. *Macromolecules, 33,* 7038.

43. Apperloo, J. J., Janssen, R. A. J., Malenfant, P. R. L., & Fréchet, J. M. J. (2001). Interchain delocalization of photoinduced neutral and charged states in nanoaggregates of lengthy oligothiophenes. *Journal of American Chemical Society, 123,* 6916.

44. Schenning, A. P. H. J., Kilbinger, A. F. M., Biscarini, F., Cavallini, M., Cooper, H. J., Derrick, P. J., Feast, W. J., Lazzaroni, R., Leclère, P., McDonell, L. A., Meijer, E. W., & Meskers S. C. J. (2002). Supramolecular organization of α,α'-disubstituted sexithiophenes. *Journal of American Chemical Society, 124,* 1269.

45. Jonkheijm, P., Hoeben, F. J. M., Kleppinger, R., Herrikhuyzen, J., Schenning, A. P. H. J., & Meijer, E. W. (2003). Transfer of π-conjugated columnar stacks from solution to surfaces. *Journal of American Chemical Society, 125,* 15941.

46. Leclère. P., Surin, M., Jonkheijm, P., Henze, O., Schenning, A. P. H. J., Biscarini, F., Grimsdale, A. C., Feast, W. J., Meijer, E. W., Müllen, K., Brédas, J. L., & Lazzaroni, R. (2004). Organic semi-conducting architectures for supramolecular electronics. *European Polymer Journal, 40,* 885.

47. Sinnokrot, M. O. & Sherrill, C. D. (2004) Substituent effects in π–π Interactions: Sandwich and T-shaped configurations. *Journal of American Chemical Society, 126,* 7690.

48. Gesquière, A., Jonkheijm, P., Hoeben, F. J. M., Schenning, A. P. H. J., Feyter, S. D., Schryver, F. C. D., & Meijer E. W. (2004). 2D-structures of quadruple hydrogen bonded oligo(p-phenylenevinylene)s on graphite: self-assembly behavior and expression of chirality. *Nano Letters, 4,* 1175.

49. Jonkheijm, P., vander Schoot, P., Schenning, A. P. H. J., & Meijer, E. W. (2006). Probing the solvent-assisted nucleation pathway in chemical self-assembly. *Science, 313,* 80.

50. Peet, J., Heeger, A. J., & Bazan, G. C. (2009). "Plastic" solar cells: Self- assembly of bulk heterojunction nanomaterials by spontaneous phase separation. *Accounts of Chemical Research, 42,* 1700.

51. Li, G., Shrotriya, V., Huang, J., Yao, Y., Moriarty, T., Emery, K., & Yang, Y. (2005). High-efficiency solution processable polymer photovoltaic cells by self-organization of polymer blends. *Nature Materials, 4,* 864.

52. Chen, H. Y., Hou J. H., Zhang S. Q., Liang Y. Y., Yang G. W., Yang Y., Yu, L. P., Wu, Y., & Li, G. (2009) Polymer solar cells with enhanced open-circuit voltage and efficiency. *Nature Photonics, 3,* 649.

53. Park, S. H., Roy, A., Beaupre, S., Cho, S., Coates, N., Moon, J. S., Moses, D., Leclerc, M., Lee, K., & Heeger, A. J. (2009). Bulk heterojunction solar cells with internal quantum efficiency approaching 100%. *Nature Photonics, 3,* 297.

54. Janssen, R. A. J., Hummelen, J. C., & Saricifti, N. S. (2005). Polymer-fullerene bulk heterojunction solar cells. *MRS Bulletin, 30,* 33.

55. Gunes, S., Neugebauer, H., & Sariciftci, N. S. (2007). Conjugated polymer-based organic solar cells. *Chemical Reviews, 107,* 1324.

56. Taddei, M., Costantino, F., Vivani, R., Sangregorio, C., Sorace, L., & Castelli, L. (2012). Influence of $\pi-\pi$ stacking interactions on the assembly of layered copper phosphonate coordination polymers: Combined powder diffraction and electron paramagnetic resonance study. *Crystal Growth and Design, 12,* 2327.

57. Tsai, W. W., Tevis, I. D., Tayi, A. S., Cui, H., & Stupp, S. I. (2010). Semiconducting nanowires from hairpin-shaped self-assembling sexithiophenes. *Journal of Physical Chemistry. B, 114,* 14778.

58. Tevis, I. D., Tsai, W. W., Palmer, L. C., Aytun, T., & Stupp, S. I. (2012). Grooved nanowires from self-assembling hairpin molecules for solar cells. *ACS Nano, 6,* 2032.

59. Stang, P. J. & Olenyuk, B. Self-assembly, symmetry and molecular architecture: coordination as the motif in the rational design of discrete supramolecular metallacyclic polygons and polyhedra. *Accounts of Chemical Research 30,* 502.

60. Caulder, D. L., & Raymond, K. N. (1999). Supermolecules by Design, K. N. *Accounts of Chemical Research, 32,* 975.

61. Leininger, S., Olenyuk, B., & Stang, P. J. (2000). Self-assembly of discrete cyclic nanostructures mediated by transition metals. *Chemical Reviews, 100,* 853.

62. Fujita, M., Umemoto, K., Yoshizawa, M., Fujita, N., Kusukawa, T., & Biradha, K. (2001). Molecular paneling via coordination. *Chemical Communications,* 509.

63. Holliday, B. J., & Mirkin, C. A. (2001). Strategies for the construction of supramolecular compounds through coordination chemistry. *Angewandte Chemie International Edition, 40,* 2022.

64. Seidel, S. R., & Stang, P. J. (2002). High-symmetry coordination cages via self-assembly. *Account of Chemical Research, 35,* 972.

65. Fujita, M., Tominaga, M., Hori, A., & Therrien, B. (2005). Coordination assemblies from a Pd (II)-cornered square complex. *Accounts of Chemical Research, 38,* 369.

66. Pitt, M. A., & Johnson, D. W. (2007). Main group supramolecular chemistry. *Chemical Society Reviews, 36,* 1441.

67. Nitschke, J. R. (2007). Construction, substitution, and sorting of metallo-organic structures via subcomponent self-assembly. *Accounts of Chemical Research, 40,* 103.

68. Lee, S. J., & Lin, W. (2008). Chiral metallocycles: rational synthesis and novel applications. *Account of Chemical Research, 41,* 521.
69. Oliveri, C. G., Ulmann, P. A., Wiester, M. J., & Mirkin, C. A. (2008). Heteroligated supramolecular coordination complexes formed via the halide-induced ligand rearrangement reaction. *Accounts of Chemical Research, 41,* 1618.
70. Northrop, B. H., Zheng, Y. R., Chi, K. W., & Stang, P. J. (2009). Self-organization in coordination-driven self-assembly. *Accounts of Chemical Research, 42,* 1554.
71. Jin, P., Dalgarno, S. J., & Atwood, J. L. (2010). Mixed metal-organic nanocapsules, *Coordination Chemistry Reviews, 254,* 1760.
72. De, S., Mahata, K., & Schmittel, M. (2010). Metal-coordination-driven dynamic heteroleptic architectures. *Chemical Society Reviews, 39,* 1555.
73. Chakrabarty, R., Mukherjee, P. S., & Stang, P. J. (2011). Supramolecular coordination: Self-assembly of finite two- and three-dimensional ensembles. *Chemistry Reviews, 111,* 6810.
74. Nuzzo, R. G. & Allara, D. L. (1983). Adsorption of bifunctional organic disulfides on gold surfaces. *Journal of the American Chemical Society, 105,* 4481.
75. Strong, L. & Whitesides, G. M. (1988). Structures of self-assembled monolayer films of organosulfur compounds adsorbed on gold single crystals: Electron diffraction studies. *Langmuir, 4,* 546.
76. Dubois, L. H., & Nuzzo, R. G. (1992). Synthesis, structure, and properties of model organic surfaces. *Annual Review of Physical Chemistry, 43,* 437.
77. Bain, C. D. & Whitesides, G. M. (1989). Formation of monolayers by the coadsorption of thiols on gold: Variation in the length of the alkyl chain. *Journal of the American Chemical Society, 111,* 7164.
78. Bain, C. D. & Whitesides, G. M. (1989). Modeling organic surfaces with self-assembled monolayers. *Angewandte Chemie International Edition, 28,* 506.
79. Porter, M. D., Bright, T. B., Allara, D. L., & Chidsey, C. E. D. (1987). Spontaneously organized molecular assemblies. 4. Structural characterization of n-alkyl thiol monolayers on gold by optical ellipsometry, infrared spectroscopy, and electrochemistry. *Journal of the American Chemical Society, 109,* 3559.
80. Sondag-Huethorst, J. A. M., Schonenberger, C., & Fokkink, L. G. J. (1994). Formation of holes in alkanethiol monolayers on gold. *The Journal of Physical Chemistry, 98,* 6826.
81. Stang, P. J., Cao, D. H., Saito, S., & Arif, A. M. (1995). Self-assembly of cationic, tetranuclear, Pt(II) and Pd(II) macrocyclic squares. X-ray crystal structure of [Pt2+(dppp)(4,4'-bipyridyl).cntdot.2-OSO2CF3]4. *Journal of the American Chemical Society, 117,* 6273.
82. Stang, P.J., & Whiteford, J. A. (1996). Supramolecular chemistry: Self-assembly of titanium based molecular squares. *Research on Chemical Intermediates, 22,* 659.
83. Manna, J., & Stang, P. J. (1996). Design and self-assembly of nanoscale organoplatinum macrocycles. *Journal of the American Chemical Society, 118,* 8731.
84. Whiteford, J. A., Lu, C. V., & Stang, P. J. (1997). Molecular architecture via coordination: Self-assembly, characterization, and host–guest chemistry of mixed, neutral-charged, Pt–Pt and Pt–Pd macrocyclic tetranuclear complexes. X-ray crystal structure of cyclobis[[cis- Pt(dppp)(4-ethynylpyridine)2][cis-Pd2+(PEt3)22-OSO2CF3]]. *Journal of the American Chemical Society, 119,* 2524.

85. Stang, P. J., Cao, D. H., Chen, K., Gray, G. M., Muddiman, D. C., & Smith, R. D. (1997). Molecular architecture via coordination: Marriage of crown ethers and calixarenes with molecular squares, unique tetranuclear metallamacrocycles from metallacrown ether and metallacalixarene complexes via self-assembly. *Journal of the American Chemical Society, 119,* 5163.

86. Stang, P. J., Persky, N. E., & Manna, J. (1997). Molecular architecture via coordination: self-assembly of nanoscale platinum containing molecular hexagons. *Journal of the American Chemical Society, 119,* 4777.

87. Stang, P. J., Olenyuk, B., Muddiman, D. C., & Smith, R. D. (1997). Transition-metal-mediated rational design and self-assembly of chiral, nanoscale supramolecular polyhedra with unique T symmetry. *Organometallics, 16,* 3094.

88. Stang, P. J. (1998). Molecular architecture: Coordination as the motif in the rational design and assembly of discrete supramolecular species—self-assembly of metallacyclic polygons and polyhedra. *Chemistry—A European Journal, 4,* 19.

89. Newkome, G. R., Cho, T. J., Moorefield, C. N., Cush, R., Russo, P. S., Godínez, L. A., Saunders, M. J., & Mohapatra, P. (2002). Hexagonal terpyridine–ruthenium and –iron macrocyclic complexes by stepwise and self-assembly procedures. *Chemistry–A European Journal, 8,* 2946.

90. Mahata, K., Saha, M. L., & Schmittel, M. (2010). From an eight-component self-sorting algorithm to a trisheterometallic scalene triangle. *Journal of the American Chemical Society, 132,* 15933.

91. Zheng, Y. R., Zhao, Z., Wang, M., Ghosh, K., Pollock, J. B., Cook, T. R., & Stang, P. J. (2010). A facile approach toward multicomponent supramolecular structures: Selective self-assembly via charge separation. *Journal of the American Chemical Society, 132,* 16873.

92. Wang, S., Zhao, T., Li, G., Wojtas, L., Huo, Q., Eddaoudi, M., & Liu, Y. (2010). From metal–organic squares to porous zeolite-like supramolecular assemblies. *Journal of the American Chemical Society, 132,* 18038.

93. Shi, Z., Liu, J., Lin, T., Xia, F., Liu, P. N., & Lin, N. (2011). Thermodynamics and selectivity of two-dimensional metallo-supramolecular self-assembly resolved at molecular scale. *Journal of the American Chemical Society, 133,* 6150.

94. Gao, C. Y., Zhao, L., & Wang, M. X. (2011). Designed synthesis of metal cluster-centered pseudo-rotaxane supramolecular architectures. *Journal of the American Chemical Society, 133,* 8448.

95. Argent, S. P., Greenaway, A., Gimenez-Lopez, M. C., Lewis, W., Nowell, H., Khlobystov, A. N., Blake, A. J., Champness, N. R., & Schröder, M. (2012). High-nuclearity metal–organic nanospheres: A Cd_{66} ball. *Journal of the American Chemical Society, 134,* 55.

96. Park, J., Lee, E., Hwang, N. M., Kang, M., Kim, S. C., Hwang, Y., Park, J. G., Noh, H. J., Kim, J. Y., Park, J. H., & Hyeon, T., (2005). One-nanometer-scale size-controlled synthesis of monodisperse magnetic iron oxide nanoparticles. *Angewandte Chemie International Edition, 44,* 2872.

97. Green, M. (2005). Organometallic based strategies for metal nanocrystal synthesis. *Chemical Communications, 18,* 3002.

98. Kievsky, Y., & Sokolov, I. (2005). Self-Assembly of Uniform Nanoporous Silica Fibers. *IEEE Transactions on Nanotechnology, 4,* 490.

99. Zhang, Q., Joo, J. B., Lu, Z., Dahl, M., Oliveira, D. Q. L., Ye, M., & Yin, Y. (2011). Self-assembly and photocatalysis of mesoporous tio$_2$ nanocrystal clusters. *Nano Research, 4,* 103.

100. Shim, J. P., Kim, D., Choe, M., Lee, T., Park, S. J., & Lee, D. S. (2012). A self-assembled Ag nanoparticle agglomeration process on graphene for enhanced light output in GaN-based LEDs. *Nanotechnology, 23,* 255201.

101. Balandin, A. A., & Wang, K. L. (2006). Methods of Self-Assembling in Fabrication of Nanodevices. *Handbook of Semiconductor Nanostructures and Nanodevices* (Vol 2, pp. 181–213). Valencia: American Scientific Publishers.

102. Zeng, H., Rice, P. M., Wang, S. X., & Sun, S. (2005). Shape-controlled synthesis and shape-induced texture of MnFe$_2$O$_4$ nanoparticles. *Journal of the American Chemical Society, 126,* 11458.

103. Rycenga, M., McLellan, J. M., & Xia, Y. (2008). Controlling the assembly of silver nanocubes through selective functionalization of their faces. *Advanced Materials, 20,* 2416.

104. De, P. P., Olivieri, B., Mariot, J. M., Favre, L., Berbezier, I., Quaresima, C., Paci, B., Generosi, A., Rossi, A. V., Cricenti, A., Ottaviani, C., Luce, M., Testa, A. M., Peddis, D., Fiorani, D., Scarselli, M., De, G. M., Heckmann, O., Richter, M. C., Hricovini, K., & d'Acapito, F. (2012). Ferromagnetic Mn-doped Si0:3Ge0:7 nanodots self-assembled on Si(100). *Journal of Physics: Condensed Matter, 24,* 142203.

105. Schouteden, K., & Haesendonck, C. V. (2010). Narrow Au(111) terraces decorated by self-organized Co nanowires: A low-temperature STM/STS investigation. *Journal of Physics: Condensed Matter, 22,* 255504.

106. Convertino, A., Cuscun`a, M., & Martelli, F. (2012). Silicon nanotubes from sacrificial silicon nanowires: fabrication and manipulation via embedding in flexible polymers. *Nanotechnology, 23,* 305602.

107. Sen, D., Bahadur, J., & Mazumder, S. (2012). Evaporation induced self-assembly of nanoparticles in realizing hollow microcapsules. *AIP Conference Proceedings, 1447,* 239.

108. Inagaki, M., Yang, Y., & Kang, F. (2012). Carbon nanofibers prepared via electrospinning. *Advanced Materials, 24,* 2547.

CHAPTER 16

HYBRID NANOSTRUCTURES FOR PHOTOVOLTAICS

GANESAN MOHAN KUMAR, SIVA CHIDAMBARAM,
JIN KAWAKITA, PARK JINSUB, and RAMASAMY JAYAVEL

CONTENTS

16.1 INTRODUCTION

Harvesting energy directly from sunlight using photovoltaic (PV) technology is well recognized as an essential component for the future global energy requirements. PV based devices are nowadays considered to be economically competitive on par with the other fossil fuels or renewable energy technologies. In addition, the large-scale production of these devices (as a sustainable energy source) could help us to meet a significant portion of our daily energy requirements. These devices were extensively studied since 1950s, when the first crystalline silicon solar cell (efficiency 6%) was developed at Bell Laboratories [1]. Over these years, solar cells have been made from many other semiconducting materials with various device configurations such as single-crystal, polycrystalline, and amorphous thin-film structures [2].

The widespread expansion in the use of inorganic solar cells remains limited due to the high costs imposed by fabrication procedures involving elevated temperature (up to 1,400°C), high vacuum, and numerous lithographic steps. So, in order to seek a cost effective alternate, conducting polymers are preferred. Such polymer-based organic solar cells can be readily processed from solutions and are reported to have solar power efficiencies up to 10.7% [3].

The main reason for the superior efficiency of inorganic devices over that of organic is due to the possession of high intrinsic carrier mobilities in them. Higher carrier mobilities mean that charges are transported to the electrodes more quickly, which reduces current losses via recombination. For many conjugated polymers, electron mobilities are extremely low, typically below $10^{-4}cm^2V^{-1}s^{-1}$, due to the presence of ubiquitous electron traps such as oxygen [4]. Therefore, polymer photovoltaic (PV) devices rely on the introduction of another material for electron transport. The presence of a second material also provides an interface for charge transfer. Compounds such as small conjugated molecules have been blended with polymers at a concentration that enables the formation of percolation pathways for electron transport [4–8]. The efficiency of these devices are limited by inefficient hopping charge transport, and the electron transport by the presence of structural traps in the form of incomplete pathways in the percolation network.

One way to overcome these charge transport limitations is to combine polymers with inorganic semiconductors. Here, charge transfer is favored between high electron affinity inorganic semiconductors and relatively low ionization potential organic molecules and polymers [4–12]. Charge transfer rates can be remarkably fast in the case of organics that are chemically bound to nanocrystalline and bulk inorganic semiconductors, which have a high density of electronic states [13–16]. Because of the nanoscale nature of light absorption and photocurrent generation in solar energy conversion, the advent of methods for controlling inorganic materials on the nanometer scale opens up new opportunities for the development of future generation solar cells.

16.2   WHAT ARE HYBRID MATERIALS?

Conjugated polymers while combined with n-type inorganic semiconductors, they result in a new class of materials called as hybrid materials. These hybrid polymer-inorganic nanocomposites usually combine the properties of both the materials, which involves the solution processing of polymer semiconductors and the high electron mobility of inorganic semiconductors. Several hybrid bulk heterojunction polymer solar cells have been reported including nanodots, nanorods, tetrapods and nanoparticles [17–23].

The discovery of conducting polymers is believed to be the backdrop behind the thrust on the extensive research in organic electronics [24, 25]. The development of a variety of organic-based optoelectronic devices, such as simple diodes, light-emitting diodes (LED's), photodiodes or solar cells, field-effect transistors (FET's), and memory devices [26–40] have been found to provide an appealing alternate to the inorganic-based electronic devices. The main reason for the extensive interest in organic semiconducting materials is their potential for the realization of a low cost, easily processable and flexible renewable energy source.

The main objective discussed in the present chapter is in the control of an inorganic structure by an organic component, which actually reveals an interactive structural hierarchy in the materials. In such materials, the inorganic oxides contribute to the increased complexity and hence, functionality through incorporation as one component in a multilevel-structured material, where there is a synergistic interaction between the organic and inorganic components. Since these interactions between the organic–inorganic hybrid materials arises from the nature of interface between the organic and inorganic components, synthetic and structural studies of materials that exhibit such an interface will contribute to the development of structure-function relationships for these hybrid materials. Controlling the surface passivation of nanocrystals, while using organic ligands allows the particles to be dissolved in a variety of solvents and dispersed in numerous polymers. The ease of processing nanocrystals through solution phase, spin casting, inkjet, and screen-printing offers the possibility to lower the device fabrication costs. The prime purpose of using the semiconducting nanostructures in the hybrid component is due to their physical properties, which can be controlled through modification of the diameter and shapes of these particles.

16.3   WHAT ARE HYBRID SOLAR CELLS?

Optoelectronic devices such as PV cells and LED's capitalize on the variation in band gap of nanocrystals with particle radius to absorb and emit light, respectively, at tunable wavelengths. Further, the concentration of density of states into quantized energy levels may result in high oscillator strengths for high absorptivity or emissivity as compared with bulk values. Conjugated polymers seem to be an attractive alternate to the traditional silicon PV technologies, as they are strong absorbers of visible light and can be deposited onto flexible substrates over large areas using wet-processing techniques like roll-to-roll coating or printing. Additionally, the large exciton-binding energy in a polymeric matrix results in strongly localized electron-hole pairs upon light absorption, giving rise to the small exciton-diffusion length and inefficient exciton dissociation.

Polymers have been previously used in photovoltaic cells, but the low electron mobilities in most of the conjugated polymers restricts them to be used as an hole transporting component in a blend along with materials, such as fullerenes, organic dyes, or nanocrystals. Conjugated polymers share the same processing advantages with nanocrystals while further allowing for mechanical flexibility of the device film. In these composite materials, the photocurrent could be increased by several orders of magnitude with regard to polymer only devices [41–43].

Hybrid solar cells employing conjugated polymers have revolutionized the photovoltaic industry by offering the prospect for large-scale energy conversion applications through cost-effective fabrication techniques. The large-scale use of PV devices for commercial and residential applications tends to drive the present concern for such hybrid systems in solar modules for the most promising and inexpensive large-scale energy conversions [44, 45]. Generally, hybrid polymer–metal oxide systems are selected by considering their ability to provide the direct and ordered path for the photogenerated electrons, which in turn could make the charge conversion process more effective [46]. The recent availability of wide band gap oxide materials with high surface area, and their capability to be synthesized over a wide range of morphological structures seem to offer tremendous opportunities for energy harvesting requirements along with their organic counterparts, due to their high electron mobility and porous nature, which could additionally increase the optical path length in a typical solar cell structure. Increasing global competition in photovoltaics has placed new demands on the fascinating hybrid nanostructures to develop a low-cost technology in reality.

16.4 COMPOSITION OF A NANOHYBRID SYSTEM

16.4.1 N-Type Oxides

The application of semiconductor nanostructures within devices is one of the major focuses of contemporary nanotechnology. In particular, the fabrication of well-aligned arrays of elongated nano crystallites, such as nanorods or nanowires is a subject of increasing interest (ZnO in this case). It is well known that nanostructures could play crucial role in many applications, as the properties of the materials depend closely on their crystal size, morphology, aspect ratio, and orientation. The unique electrical and optoelectronic properties of ZnO could be an added advantage for its use as an electrode in solar cell systems [47, 48]. Recently, the doping of transition-metal ions in ZnO systems has been investigated toward extending their photo response (photon absorption) over the visible range, which could possibly enhance the photochemical conversion rate.

Ordered oxide nanostructures are expected to enhance the performance of various technologically important devices, such as short-wavelength lasers, field-emission devices, Schottky diodes, electroluminescent devices, and sensors [49–63]. However, most of the work on ZnO nanorod arrays to date has focused on the synthetic methodologies, rather than on their practical applications. Hybrid organic or inorganic solar cells are made from composites of conjugated polymers with nanostructured metal oxides, in which the polymer component serves the function of both light absorber and hole conductor, and the metal oxide acts as the electron transporter.

16.4.2   P-Type Polymer

Despite years of intensive research on the basic aspects of polymerization, the field of electrically conducting polymers is still full of controversies [64–66]. Apart from the very first step, the formation of radical cations, there has not been any clear experimental evidence on the main reaction path so far [67, 68]. Such a situation is, of course, due to the complexity involved in the film formation process and the lack of effective approaches in studying the reaction kinetics and characterizing the insoluble polymers. On the other hand, the details of experimental procedures have been, more or less, poorly respected when phenomena were explained and conclusions were drawn. This is very particular in the case of pyrrole, owing to the fact that polypyrrole (PPy) doped with a wide variety of anions [69] can be readily electrosynthesized on electrodes of different materials in both aqueous [70, 71] and organic solutions. Moreover, because of the generally admitted (but not incontrovertible) view that the electropolymerization of aromatics follows the same mechanism and that some similarities exist among pyrrole and thiophene [72–74].

Regarding the p-type material, polypyrrole has been one of the most extensively investigated conducting polymers with rich features, such as ease of oxidizability, high conductivity, and chemical and environment stability [75]. In case of PPy, it has been well established that higher potentials favor the formation of polymers with longer chains and lower redox potentials. However, this does not always hold well, which has been reported through the structural diversity of PPy using voltammetric studies [76]. The PPy prepared at different switching potentials (potentiodynamic) or currents (galvanostatic) displayed different shapes of cyclic voltammograms (CVs). With decreasing switching potential or current, CVs of PPy varied from nearly symmetrical anodic and cathodic waves to those with a strong hysteresis between the anodic and the cathodic scans, and at extremely low potentials or currents, which barely sustain film formation, an additional peak like wave appeared at more negative potential.

16.5 ELECTROPOLYMERIZATION REACTIONS

Electrodeposition of conducting polymer films at the surface of an electrode has opened up several new possibilities in energy conversion and storage applications, along with electro triggered drug delivery, soft actuators, chemical, bio, and gas sensors, biocompatible films, and artificial muscles. Initially, electropolymerization could be described in the context of four generations, which have been distinguished as the era of physicists, the era of electrochemists, the era of polymerists and the era of molecular electronics [77–80]. As the progress in each of these eras resulted from mutual enrichment between these scientific communities, a systematic approach has been made to address these aspects. The era of electrochemists actually starts with the early use of electropolymerization in the 1980s. The second part presents the major milestones reached by the electropolymerization process in the presence of light as the functionalization of surfaces for the electrodeposition of increasingly sophisticated conjugated architectures, endowed with specific functionalities from sensors to active photovoltaic layers. The era of physicists corresponds to the historical identification of synthetic or organic metals, and parallel to the development of mixed valence crystals [81, 82]. Recently, the domain of electro conductive polymers has driven the interest of physicists in the semiconducting or conducting or even superconducting conductivity transition. It is evident that the field of electrically conducting polymers is still full of controversies, despite years of intensive research on the basic aspects of electropolymerization and the charging or discharging mechanism [47, 48, 75, 83–85].

Electropolymerization could be generally achieved through potentiostatic (constant potential) or galvanostatic (constant current) methods. These techniques are easier to describe quantitatively and have been therefore commonly utilized to investigate the nucleation mechanism and the behavior of macroscopic growth. Potentiodynamic techniques, such as cyclic voltammetry correspond to a repetitive triangular potential waveform applied at the surface of the electrode. The latter method has been mainly used to obtain qualitative information about the redox processes involved in the early stages of the polymerization reaction, and to examine the electrochemical behavior of the polymer film after electrodeposition.

Generally, electrochemical polymerization involves the dissolution of organic monomers in an appropriate solvent containing the desired anionic doping salt, which is later oxidized at the surface of an electrode through application of a considerable amount of anodic potential (oxidation). The nature of solvent and electrolyte used are of specific importance in electrochemistry as both must be stable at the oxidation potential of the monomer and provide an ionically conductive medium. In this regard, organic solvents like acetonitrile have very good potential values with relatively high permittivity's, which allow good dissociation of the electrolyte, thereby establishing a good ionic conductivity. In case of application aspects, the lack of stability and performance limitation in terms of long-term operation have questioned the presence of liquid electrolyte in a practical dye sensitized solar cells (DSSC's), for which efforts have been taken to replace the liquid electrolyte with room-temperature molten salts, inorganic p-type semiconductors, ionic conducting polymers, and organic hole transport materials [4–6]. So, considering such factors, progress in establishing an ideal hybrid solid-state type solar cell using wide band gap oxide and polymer material is under intense research, which has been focused in this chapter.

16.6   FABRICATION OF HYBRID STRUCTURES

16.6.1   Hydrothermal Synthesis

The term hydrothermal is purely of geological origin, which was first used to describe the action of water at elevated temperature and pressure, in bringing out changes in the earth's crust. In such processes, a homogeneous or heterogeneous reaction takes place in the presence of aqueous solvents or mineralizers under high pressure and temperature conditions to dissolve and recrystallize the materials that are relatively insoluble under ordinary conditions. In the present chapter, the hydrothermal reactions are carried out to prepare well-defined nanostructures in a closed system in the presence of a solvent like deionized water, whether it is sub-critical or supercritical. In general, the term hydrothermal methodology could be defined using a homogeneous or heterogeneous chemical reaction that takes place in the presence of a solvent (whether aqueous or non-aqueous) above the room temperature at pressure greater than 1atm in a closed system [86]. Processing of nanomaterials through such methodology is commonly referred as green processing or green chemistry.

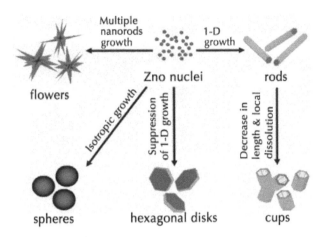

FIGURE 16.1   Schematic representation of the shape-selective synthesis of ZnO nanorods, nanocups, nanodisks, nanoflowers, and nanospheres through hydrothermal technique.

Hydrothermal synthesis is a pliable chemical method with malleability in producing a great class of nanostructured materials in the form of powders as shown in Figure 16.1. The main advantages come from the low crystallization temperature done in one-step process, avoiding thermal treatments usually required in other processes. The nanostructure of bulk and thin films obtained using this approach may open new applications to materials for biosensors, implants, or substrates for growing biological cells. A great variety of materials, such as native elements, metal oxides, hydroxides, silicates, carbonates, phosphates, sulphides, tellurides, nitrides, selenides, and so on, both as particles and nanostructures, such as nanotubes, nanowires, nanorods, could be prepared by the hydrothermal method [87–90].

16.6.2   Oxide Nanorods

Fine rod-like one dimensional oxide nanostructures could be synthesized through a two-step process; including the precipitation reactions at low temperatures and its subsequent hydrothermal treatment [46, 90–92]. Typical ZnO nanorods were prepared from an aqueous solution of zinc acetate prepared from deionized water at room temperature, and then precipitated with sodium hydroxide solution.

FIGURE 16.2 Nanorod-like structures of ZnO observed in the field emission scanning electron microscope for (a) undoped and (b) Cr doped specimens.

Here, the entire precipitation reactions are carried out in an ice bath until it was ensured that the precipitation had ceased. The colloids are then subjected to ultrasonic treatment and introduced into Teflon-lined stainless steel tube-type autoclaves of dimensions 4.5cm inner diameter with fill capacity of 115ml. The percentage of fill was kept constant at 60% during all the experiments. The autoclaves were heated, and maintained at the desired temperature (e.g., 180°C) for 6hr, and then cooled to room temperature naturally. The resulting products were collected and rinsed copiously with deionized water and ethanol and air dried at 40°C, after that the as-prepared samples were vacuum dried at 400°C for 30min. Similar procedures could be carried out for the preparation of Cr, Co, Mn, and Cu doped ZnO nanorods (using selected acetates as dopant source) (Figures 16.2 and 16.3).

FIGURE 16.3 Transmission electron and high resolution transmission electron microscopic images of Cr-doped ZnO nanorod-like structures, along with a typical selected area electron diffraction pattern shown at the inset (Images

reproduced from the reference 90 by permission of The Royal Society of Chemistry).

16.6.3   Fabrication of Oxide Nanocorns

Gd-free and doped ZrO_2 nanocorns synthesized through a cationic surfactant assisted low temperature solution synthesis technique involving zirconyl nitrate (ZN), gadolinium nitrate (GN), and sodium hydroxide as the starting materials are shown in Figure 16.4. A desirable amount of cetyltrimethylammonium bromide (CTAB-cationic surfactant) has been used as the structure-directing agent in all the hydrolysis reactions (solvent or deionized water), prior to the precipitation reactions. Sodium hydroxide was initially added to the ZN stock solution held at 60°C under continuous stirring for 24h to ensure complete precipitation.

FIGURE 16.4   a, b and c Field Emission Scanning Electron Microscope images exhibiting corn like structures of the undoped and Zr0.97Gd0.03O2 specimens. d Formation of corn-like structure over the nanorod forZr0.97Gd0.03O2 nanocorn.

The entire reactions including precipitation were carried out in polypropylene bottles, which were gently heat-treated in an electric oven at 100°C for 48hrs. Later, the precipitates are cooled to room temperature and filtered, washed with absolute ethanol, and deionized water for several times to lower the pH value to 7, followed by air drying at 60°C for 12hr. The resulting white products were gently ground in an agate mortar and calcined subsequently at 450°C for 2hr, for further characterizations.

16.6.4   Electropolymerization of Polymers

Over the last three decades, electronically conducting polymers, such as polyacetylene, polypyrrole, polythiophene, and polyaniline have received considerable attention because of their remarkable electronic, magnetic, and optical properties and their wide range of potential applications in many [66, 93–97]. The conducting polymers have properties that make them suitable for several applications including light-emitting diodes, sensors, batteries, and electrochemical super capacitors [98, 99]. For the majority of these studies, a single polymer was prepared by chemical or electrochemical oxidation of the corresponding monomer [100]. Further, the lack of clear experimental evidence in the formation of radical cations during oxidation reactions could be a major setback. Such a situation is due to the complexity involved in the film formation process and the lack of effective approaches for studying the reaction kinetics and characterizing the insoluble polymers.

In case of pyrrole, it is a well-known conducting material in its polymer form (PPy) while doped with a wide variety of anions [101]. The electrochemical behavior and mechanical properties of the conducting polymers depend significantly on the chemical structure, the conditions by which they are synthesized, and their microstructure [102, 103], which seems to be an advantage for our studies. On the basis of a number of studies carried out on the monodispersed conjugated oligomers, the interrelation among chain length, redox potential, and chemical reactivity of oligomers or oligomeric cations has been well established [104, 105]. This reveals that higher potentials favor the formation of polymers with longer chains and lower redox potentials for the present investigation.

16.7   ELECTROPOLYMERIZATION OF PYRROLE ON OXIDE ELECTRODES

16.7.1 Preparation of Photo Anode or Working Electrodes

To establish a hybrid solar cell structure, it is desirable to have a suitable working electrode, composed of wide band gap oxides. Such electrodes could be constructed on transparent conducting electrodes such as FTO or ITO. The experimental procedures involved in such preparations involve the dispersion of a desirable amount of the oxide nanostructures in 60 ml of deionized water containing 0.5 ml of Conc.HNO$_3$. The sol could be then magnetically stirred and ultrasonically treated for a certain period, to attain homogeneous distribution of the nanostructures. Later the transparent conducting FTO substrates cleansed with acetone, ethanol and deionized water are dip coated in the prepared sol for a period of 10s, and then dried at room temperature for 30–45min. The procedure is usually repeated for 2–3 cycles, in order to attain a considerable thickness of the oxide layers. The final stage involved in the fabrication of the photo anode is their heat treatment (deposited specimens) at 450°C for a period of 30min at a heating rate of 15°C/min [46].

16.7.2 Electropolymerization of Pyrrole

Electropolymerization reactions are carried out in a one-compartment cell provided with a window type arrangement, to which the fabricated photo anode is fixed. The electrochemical processes are usually carried out through a conventional three-electrode set up using Ag/AgCl/KCl ,and Pt sheet as the reference and counter electrodes, respectively. Polypyrrole (PPy) layers were then deposited under cyclic voltammetry conditions (in an applied potential of 100mV s^{-1}) from 0.1M of pyrrole and Tetraethylammonium p-Toleunesulfonate dissolved in deionized water.

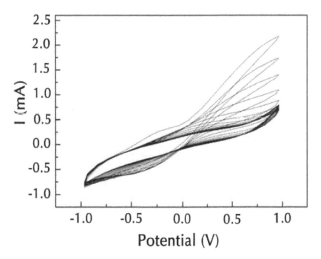

FIGURE 16.5 Cyclic voltammograms measured during the electropolymerization of pyrrole monomers at a scan rate of 100mV s-1 (Image reproduced from [46] by permission of The Royal Society of Chemistry).

Electrochemical synthesis of the polymer material was carried out under UV light irradiations. The ZnO electrodes were subjected to the UV radiation through their transparent backside (for their photo-excitation and also to directly catalyze pyrrole monomers to polymerize at their surface).

A typical cyclic voltammograms (CV) recorded during the electropoly-merization of the pyrrole (PPy/polymer) monomers on the surface of ZnO: Co nanorod made working electrodes has been provided in Figure 16.5. Here, 25 consecutive cycles were carried out at a scan rate of 100mV s-1. During the initial 5 cycles a complete PPy polymer matrix was found to be deposited on the working electrodes. This in fact could also be understood from the gradual decrease in current density values, which has been noted with the increase in number of cycles. This sort of behavior is generally related with the decrease in oxidation process (corresponding to the solvent) at each step. Also, the initially formed PPy deposits might have a limiting effect on the mass transport of water molecules involved in oxidation, thereby slowing down the polymerization rate later. The electropolymerization of pyrrole monomers could be explained using the Diaz's mechanism [106]. During the initial oxidation stage the monomer gets converted to a cation (possessing greater unpaired electron density), which later couples with another cation to form the polymer chain through the release of two protons [107].

Electrochemical synthesis of the polymer material could also be performed using the assistance of UV light radiations, subjected from the backside of ZnO: Co electrodes along with the applied anodic potentials. On the application of a considerable amount of anodic potential the pyrrole monomers undergo oxidation reaction to result in the formation of polypyrrole chains by releasing two protons [106], in accordance to the given equation.

$$2n \ Py-H + m \ A- \longrightarrow [(Py-Py)n-m \ A] + 2n \ H+ + (2n+m)e-$$

During the initial oxidation process in electropolymerization, the radical cation of the monomer is formed, which reacts with the other monomers present in solution to form the oligomeric products and then the polymer. The extended conjugation in the polymer results in a lowering of the oxidation potential compared with the monomer. Therefore, the synthesis and doping of the polymer are generally done simultaneously. The anion is then incorporated into the polymer to ensure the electrical conductivity of the film and, at the end of the reaction; a polymeric film of controllable thickness is formed on the surface of working electrode.

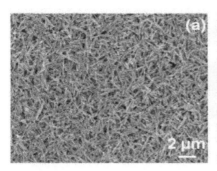 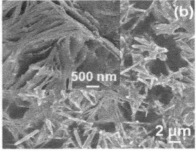

FIGURE 16.6 (a) The homogeneous distribution of ZnCoO nanorods on the surface of working electrodes. (b) Low and high magnification (inset) SEM images of PPy/ZnCoO hybrid systems. (Images reproduced from [46] by permission of The Royal Society of Chemistry).

16.7.3 Surface Analysis on Hybrid Nanostructures

To understand the morphological evolution and influence of a polymer material on the surface of a photoanode at a greater insight, the electropolymerized photoanodes could be examined using FE-SEM or an atomic force microscope (AFM). Figure 16.6a shows the well-dispersed nanorod-like structures deposited on the FTO substrates prior to electropolymerization. Figure 16.6b shows the low magnification image of the PPy/ZnCoO hybrid materials, respectively, with its corresponding high magnification image at the inset. Here, at a microscopic scale the texture of the material appears to suggest a direct adsorption like process to have taken place over the rod like structures, stretching out two dimensionally. Analyzing such areas under high magnification one could assimilate those structures to contain a number of disc shaped particles at a nanoscale regime. This tendency could be due to the higher surface area of the rod like structures, which influences the precipitating polymer molecules to get deposited on them, resulting from the formation of oligomers in solution phase [108]. So, actually at the end of electropolymerization reaction a well-covered polypyrrole layer is thus formed, leading to the formation of multi-particle aggregates, presumably due to the weaker inter-particle interactions.

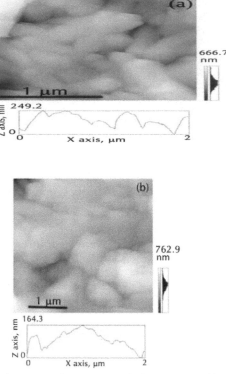

FIGURE 16.7 Top view AFM images and their corresponding line profiles for (a) ZnCoO made working electrodes and (b) PPy/ZnCoO hybrid layers. (Images reproduced from [46] by permission of The Royal Society of Chemistry).

The mean (Ra) and root mean square (Rrms) roughness values obtained from the area shown in Figure 16.7, revealed the Rrms values for PPy deposited films to be increased by 6–10 nm. The line profiles of the corresponding AFM images were compared to notify the PPy deposited areas on the surface of electrode. The small curve like features observed at certain portions in the line scan suggests the polymer material to be deposited on the surface and voids in between the rod like structures, hence forming a complete matrix [46].

The morphological evolution and distribution of the polymer material on the surface of TiO₂ working electrodes examined from the FE-SEM observations are shown in Figure 16.8. Here, the electropolymerized specimens were found to be denser and more compact while compared with that of the pure TiO₂ layers. In accordance with a 3D model explained earlier, the monomer tends to get desorbed during electropolymerization, resulting in the formation of oligomers in solution phase, which subsequently precipitates on the surface of TiO_2 working electrodes [108, 109]. Similar morphological features have been reported to enhance the conductivity of the hybrid material system [110, 111].

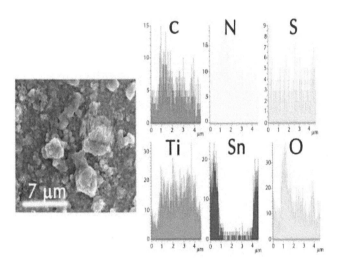

FIGURE 16.8 Energy Dispersive Spectral line analysis corresponding to a PPy/TiO2 aggregate, revealing the distribution of C, N and S elements on their surface. X and Y-axis denoting the scale scanned and counts, respectively.

In addition, the high magnification SEM images reveal the absence of deterioration or alterations in the spherical morphology of the TiO_2 nanoparticulate system. Due to the higher surface area of TiO_2 nanocrystallites, the precipitating polymer molecules could tend to aggregate them into clusters, which in turn make them well covered with polypyrrole layers. This further leads to the formation of multi-particle aggregates, presumably due to the weaker inter-particle interactions, which could help in improving the rate of light absorption [112]. Further to analyze these behavioral aspects, elemental analyses could be carried out in form of a line scan along a micrometer sized aggregate as shown in Figure 16.8, where C, N, and S traces (from PPy) were sorted out to be uniformly distributed on its surface, thereby confirming the aforementioned aspects.

16.7.4 Electrical Property Studies on Hybrid Nanostructures

The electrical property studies could be carried out using a conventional three-electrode set up with Ag/AgCl/KCl, and Pt sheet as the reference and counter electrodes, respectively in a research model potentiostat or galvanostat. IVIUMSTAT based electrochemical interface could be additionally utilized to evaluate the nature of doping type from the Mott–Schottky plots in the nano hybrid systems. The flat band potential (V_{FB}) of the PPy deposited photoanodes evaluated using Mott–Schottky plots, where the variations in $1/C^2$ values are plotted as a function of the applied potential as shown in Figure 16.9. Such plots could be used to determine the nature of doping type, apparent donor density (N_D) and the flat band potential (V_{FB}) though extrapolating the linear part of the $C^{-2}–E$ curves [113].

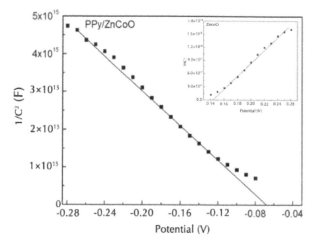

FIGURE 16.9   Mott–Schottky plots of PPy/ZnCoO hybrid systems with the inset showing the plots for ZnCoO photo anodes (Images reproduced from [46] by permission of The Royal Society of Chemistry).

Here, one can evaluate the flat band potential V_{FB} of the photoanodes from a standard phosphate based buffer solution with pH 6, through a potential scan method at a frequency of 1Hz. Nitrogen bubbling is usually preferred to remove the traces of any oxygen present in the electrolyte, which may influence the potential of the redox reaction by acting as an oxidizing agent [114]. The direction of slope is used to reveal the nature of dopant type in the hybrid systems (n or p type), respectively. The p type nature observed in the latter system is a clear indication of the polymer material to be deposited in a well-dispersed manner. However, a small deviation from linearity could be observed that may result from the effect of surface states, recombination effects and non negligible contributions of the Helmholtz layer to the interfacial capacitance. The V_{FB} was observed to shift toward a more negative value (from +143 to -67mV) after the deposition of polymer materials on the surface of photoanodes, presumably due to the presence of the PPy composites, which gives rise to a more negative potential [115]. Preliminary photovoltaic studies were carried out under AM 1.5, 1 sun condition for these photo electrodes. The studies showed a photo voltage of 0.45V.

16.8 INTERFACIAL ELECTRONIC STRUCTURE STUDIES

16.8.1 Ultraviolet Photoelectron Spectroscopy (UPS)

Ultra violet photoelectron spectroscopy is generally used to obtain the information about the density of occupied states and the work function of the material. Figure 16.10a shows a typical valence band spectrum of the PPy/TiO$_2$ hybrid layers fabricated under different photo anodic potentials. Here, the pure TiO$_2$ layers exhibit an unstructured emission with very low intensities, which may be due to the continuous decrease in the vicinity of Fermi energy as observed. This behavior is in contrast to the well-structured emission patterns exhibited by the polypyrrole films with molecular signals corresponding to pyrrole molecules at 6.75eV (Signal II) [116]. Also, a significant shift of this signal could be noted with respect to the applied potential, due to the bands that are formed out of polaronic states [117]. The remarkable difference in the intensities between the T2, T3, and T4 polypyrrole hybrid films (deposited at an anodic potential of 200, 300, and 400mV, respectively) could be due to the difference in thickness of the PPy layers deposited, which is directly dependant on the applied anodic potentials. We further believe the conductivity of the hybrid layers to occur through the aid of the polaronic states described above, which in turn results in the occurrence of a strong electron-phonon coupling to take place [118].

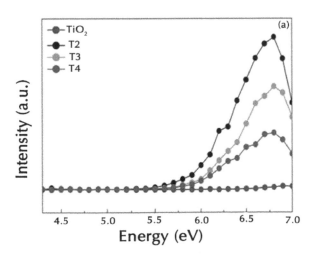

TABLE 16.10 *(Continued)*

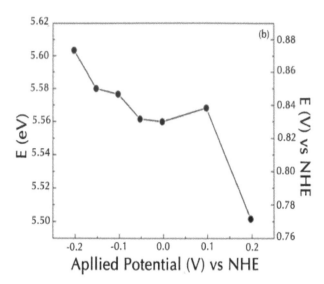

FIGURE 16.10   a UPS spectral images of photo electropolymerized PPy/TiO$_2$ layers fabricated under coulombic-controlled conditions. b The shift in work function of the hybrid system has been found to be decreasing with respect to the increase in applied anodic potential.

The work function (WF) of the hybrid layers were evaluated from the UPS results through the interpolation of the slope of curves along X and Y axis, so as to obtain their point of intersection. The work functions were then plotted as a function of the applied photo anodic potentials as shown in Figure 16.10b. Here, a decreasing tendency could be noted with respect to the increase in anodic potentials, controversially reported in some other systems [119]. The applied potential along with the UV radiation incident on the back side of the working electrode could photo excite the electrons toward the valence level (TiO$_2$), resulting in the movement of electrons from the HOMO level of polypyrrole into the conduction band of inorganic semiconductor. Further, the increasing values of applied potential (increases the value of current density), tends to decrease the net up-take rate of the anionic dopant, resulting in the modification in conductivity of the systems. This in turn results in the increase in overpotential for polarization, thereby lowering the values of work function and henceforth causing the variations in electronic structures. This behavior has been illustrated using the energy diagram as shown in Figure 16.11 and could be clearly understood from the coulombic-controlled experiments, where the time taken to attain the desired charge value (100 mC) varies with respect to the applied potential (smaller time is required for higher potentials).

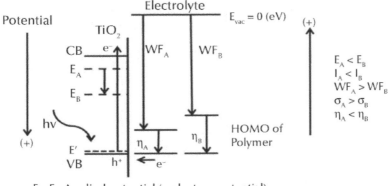

E$_A$, E$_B$: Applied potential (= electron potential)
E': Potential of hole (independent on applied potential)
CB: Conduction band of TiO$_2$
VB: Valence band of TiO$_2$
I$_A$, I$_B$: Current
σ$_A$,σ$_B$: Conductivity
η$_A$, η$_B$: Over potential
WF$_A$, WF$_B$: Work function

FIGURE 16.11   Energy diagrams of the PPy/TiO2 hybrid systems, revealing the methodology for the observed shift to take place with respect to the increase in applied potential.

16.9   SUMMARY

The interest in conducting polymers is steadily growing, driven by the unique electronic properties of these materials. The reactions leading from a suitable monomer to the conducting polymer include a series of oxidation reactions. This has stimulated the application of electrochemical techniques not only for studies of the properties of the polymeric materials, but also for studies of the details of the polymerization process.

The utilization of electropolymerization for energy conversion (i.e., conversion of sunlight to electricity) may offer some advantages compared to more common processing techniques (e.g., spin-coating). Advantages include the ability to use simple precursors without solubilizing groups and subsequent deposition of otherwise non processable materials, and the deposition of pinhole-free thin films. Hybrid organic or inorganic solar cells, composed of an organic donor material, such as a conjugated polymer combined with an inorganic n-type semiconductor nanostructure, are of interest as alternatives to all-organic bulk heterojunction structures because of the potential to control the nanostructure through synthesis, and so achieve high physical as well as chemical stability.

The hybrid matrix discussed in this chapter reveals the preliminary photovoltaic studies carried out under AM 1.5, 1 sun condition to possess a photo voltage of 0.45V, which is really appreciable for such a sort of system at this stage especially. Also the present work could be considered as a major mile stone in the studies related to the adjustments of electronic levels at the hybrid interface.

KEYWORDS

- **Electropolymerization reactions**
- **Hybrid solar cells**
- **Hydrothermal synthesis**
- **Oxide nanorods**
- **Photovoltaic technology**

REFERENCES

1. Chapin, D. M., Fuller, C. S., & Pearson, G. L. (1954). *Journal of Applied Physics, 25,* 676.
2. Paul, W. M. B., Mihailetchi, V. D., Koster, L. J. A., & Markov, D. E. (2007). *Advanced Materials, 19,* 1551–1566.
3. http://www.heliatek.com.
4. Yu, G., Gao, J., Hummelen, J. C., Wudl, F., & Heeger, A. J. (1995). *Science, 270,* 1789.
5. Shaheen, S. E., Brabec, C. J., Padinger, F., Fromherz, T., Hummelen, J. C., & Sariciftci, N. S. (2001). *Applied Physics Letters, Applied Physics Letters, 78,* 841.
6. Wienk, M. M., Kroon, J. M., Verhees, W. J. H., Knol, J., Hummelen, J. C., Van Hal, P. A., & Janssen, R. A. (2003). *Journal of Angewandte Chemie International Edition, 42,* 3371.
7. Schilinsky, P. Waldauf, C., & Brabec, C. (2002). *Applied Physics Letters, Applied Physics Letters, 81,* 3885.
8. Brabec, C. (2004). *Journal of Solar Energy Materials and Solar Cells, 83,* 273.
9. Halls, Walsh, C. A., Greenham, N. C., Marseglia, E. A., Friend, R. H., Moratti, S. C., & Holmes, A. B. (1995). *Nature, 376,* 498.
10. Tian, B., Zheng, X., Kempa, T. J., Fang, Y., Yu, N., Yu, G., Huang, J., & Lieber, C. M. (2007). *Nature,* 449, 885.
11. Veenstra, S. C., Verhees, W. J. H., Kroon, J. M., Koetse, M. M., Sweelssen, J., Bastiaansen, J. J. A. M., Schoo, H. F. M., Yang, X., Alexeev, A., Loos, J., Schubert, U. S., & Wienk, M. M. (2004). *Chemistry of Materials Chemistry of MaterialsChemistry of Materials, 16,* 2503.

12. Schmidt-Mende, L., Frechtenkotter, A., Mullen, K., Moons, E., Friend, R. H., & MacKenzie, J. D. (2001). *Science, 293,* 1119.
13. Ouali, L., Krasnikov, V. V., Stalmach, U., & Hadziioannou, G. (1999). *Advanced Materials, 11,* 1515.
14. Melzer, C., Krasnikov, V. V., & Hadziioannou, G., (2003). *Applied Physics Letters, 82,* 3101.
15. Uchida, S., Xue, J., Rand, B. P., & Forrest, S. R. (2004). *Applied Physics Letters, 84,* 4218.
16. Peumans, P., Uchida, S., & Forrest, S. R. (2003). *Nature, 425,* 158.
17. Xue, J., Uchida, S., Rand, B. P., & Forrest, S. R. (2004). *Applied Physics Letters, 85,* 5757.
18. Greenham, N. C., Peng, X., & Alivisatos, A. P. (1996). *Physical Review B, 54,* 17628.
19. Huynh, W. U., Peng, X., Alivisatos, A. P. (1999). *Advanced Matter, 11,* 923.
20. Huynh, W. U., Dittmer, J. J., & Alivisatos, A. P. (2002). *Science, 295,* 2425.
21. Sun, B., Marx, E., & Greenham, N. C. (2003). *Nano Letters, 3,* 961.
22. Sun, B., Snaith, H. J., Dhoot, A. S., Westenhoff, S., & Greenham, N. C. (2005). *Journal of Applied Physics, 97,* 014941.
23. Kwong, C. Y., Djurisic, A. B., Chui, P. C., Cheng, K. W., & Chan, W. K. (2004). *Chemical Physics Letters, 384,* 372.
24. Chiang, C. K., Fincher, J. C. R., Park, Y. W., Heeger, A. J., Shirakawa, H., & Louis, E. J. (1977). *Physical Review Letters, 39,* 1098.
25. Shirakawa, H., Louis, E. J., MacDiarmid, A. G., Chiang, C. K., & Heeger, A. J. (1977). *Journal of the Chemical Society and Chemical Communications,* 578.
26. Tomozawa, H., Braun, D., Phillips, S., Heeger, A. J., & Kroemer, H. (1987). *Synthetic Metals, 22,* 63.
27. Tang, C. W. & VanSlyke, S. A. (1987). *Applied Physics Letters, 51,* 913.
28. Burroughes, J. H., Bradley, D. D. C., Brown, A. R., Marks, R. N., Mackay, K., Friend, R. H., Burns, P. L., & Holmes, A. B. (1990). *Nature, 347,* 539.
29. Gustafsson, G., Cao Y., Treacy, G. M., Klavetter, F., Colaneri, N., & Heeger, A. J. (1992). *Nature, 357,* 477.
30. Tang, C. W. (1986) *Applied Physics Letters, 48,* 183.
31. Halls, J. J. M., Walsh, C. A., Marseglia, E. A., Friend, R. H., Moratti, S. C., & Holmes A. B. (1995). *Nature, 376,* 498.
32. Chen, H.-Y., Lo, M. K. F., Yang, G., Monbouquette, H. G., & Yang, Y. (2008) *Nature Nanotechnology, 3,* 543.
33. Burroughes, J. H., Jones, C. A., & Friend, R. H. (1988) *Nature, 335,* 137.
34. Garnier, F., Hajlaoui, R., Yasser, A., & Srivastava, P. (1994) *Science, 265,* 1684.
35. Yang, Y. & Heeger, A. J. (1994) *Nature, 372,* 344.
36. Torsi, L., Dodabalapur, A., Rothberg, L. J., Fung, A. W. P., & Katz, H. E. (1996). *Science, 272,* 1462.
37. Ma, L., Liu, J., & Yang, Y. (2002). *Applied Physics Letters, 80,* 2997.
38. Ouyang, J., Chu, C.-W., Szmanda, C. R., Ma, L., & Yang, Y. (2004). *Nature Materials, 3,* 918.
39. Yang, Y., Ma, L., & Wu, J. (2004). *MRS Bulletin, 29,* 833.
40. Tseng, R. J.-H., Huang, J., Ouyang, J., Kaner, R. B., & Yang, Y. (2005). *Nano Letters, 5,* 1077.

41. Yu, G. & Heeger, A. J. (1995). *Journal of Applied Physics, 78,* 4510.
42. Dittmer, J. J., Lazzaroni, R., Leclere, P., Moretti, P., Granstrom, M., Petritsch, K., Marseglia, E. A., Friend, R. H., Bredas, J. L., Rost, H., & Holmes, A. B. (2000). *Solar Energy Materials and Solar Cells, 61,* 53.
43. Dittmer, J. J., Marseglia, E. A., & Friend, R. H. (2000). *Advanced Materials, 12,* 1270.
44. Huynh, W. U., Dittmer, J. J., & Alivisatos, A. P. (2002). *Science, 295,* 2425.
45. Peir, o A. M., Ravirajan, P., Govender, K., Boyle, D. S., Brien, Bradley, D. D. C., Nelson, J., & Durrant, J. R. (2006). *Journal of Materials Chemistry, 16,* 2088.
46. Mohan Kumar, G., Raman, V., Kawakita, J., Ilanchezhiyan, & Jayavel, R. (2010). *Dalton Transactions, 39,* 8325.
47. Martinson, A. B. F., Goes, M. S., Fabregat-Santiago, F., Bisquert, J., Pellin, M. J., & Hupp, J. T. (2009). *The Journal of Physical Chemistry, 113,* 4015.
48. Law, M., Greene, L. E., Johnson, J. C., Saykally, R., & Yang, P. (2005). *Nature Materials, 4,* 455.
49. Huang, M. H., Mao, S., Feick, H., Yan, H., Wu, Y., Kind, H., Weber, E., Russo, R., & Yang, P. (2001). *Science, 292,* 1897.
50. Yang, P., Yan, H., Mao, S., Russo, R., Johnson, J., Saykally, R., Morris, N., Pham, J., He, R., & Choi, H.-J. (2002). *Advanced Functional Materials, 12,* 323.
51. Pan, A. L., Liu, R. B., Wang, S. Q., Wu, Z. Y., Cao, L., Xie, S. S., & Zou, B. S. (2005). *Journal of Crystal Growth, 282,* 125.
52. Choy, J. H., Jang, E.-S., Won, J. H., Chung, J. H., Jang, D. J., & Kim, Y. W. (2003). *Advanced Materials, 15,* 1911.
53. Govender, K., Boyle, D. S., O'Brien, P., Brinks, D., West, D., & Coleman, D. (2002). *Advanced Materials, 14,* 1221.
54. Greene, L. E., Law, M., Goldberger, J., Kim, F., Johnson, J. C., Zhang, Y., Saykally, R. J., & Yang, P. (2003). *Angewandte Chemie International Edition, 42,* 3031.
55. Zhu, Y. W., Zhang, H. Z., Sun, X. C., Feng, S. Q., Xu, J., Zhao, Q., Xiang, B., Wang, R. M., & Yu, D. P. (2003). *Applied Physics Letters, 83,* 144.
56. Hung, C.H., & Whang, W. T. (2004). *Journal of Crystal Growth, 268,* 242.
57. Tseng, Y. K., Huang, C. J., Cheng, H. M., Lin, I. N., Liu, K. S., & Chen, I. C. (2003). *Advanced Functional Materials, 13,* 811.
58. Lee, C. J., Lee, T. J., Lyu, S. C., Zhang, Y., Ruh, H., & Lee, H. J. (2002). *Applied Physics Letters, 81,* 3648.
59. Park, W. I., Yi, G. C., Kim, M., & Pennycook, S. J. (2002). *Advanced Materials, 14,* 1841.
60. Park, W. I., Yi, G. C., Kim, J. W., & Park, S. M. (2003). *Applied Physics Letters, 82,* 4358.
61. Park, W. I. & Yi, G. C. (2004). *Advanced Materials, 16,* 87.
62. Jeong, M. C., Oh, B.Y., Lee, W., & Myoung, J. M. (2005). *Applied Physics Letters, 86,* 103105.
63. Trivikrama, Rao G. S., Tarakarama, & Rao, D. (1999). *Sensors and Actuators B: Chemical, 55,* 166.
64. Skotheim, T. A., Elsenbaumer, R. L., & Reynolds, J. R. (Eds.). (1998). *Handbook of Conducting Polymers.* New York: Marcel Dekker.
65. Roncali, J. (1992). *Chemical Reviews, 92,* 711.

66. Diaz, A. F., Crowley, J., Bargon, J., Gardini, G. P., & Torrance, J. B. (1981). *Journal of Electroanalytical Chemistry, 121,* 355.
67. Wei, Y., Chan, C.-C., Tian, J., Jang, G.-W., & Hsueh, K. F. (1991). *Chemistry of Materials, 3,* 888.
68. Asavapiriyanont, S., Chandler, G. K., Gunawardena, G. A., Pletcher, D. J. (1984). *Electroanalytical Chemistry, 177,* 229.
69. Kuwabata, S., Nakamura, J., & Yoneyama, H. J. (1988). *Chemical Society, Chemical Communications,* 779.
70. Yamaura, M., Hagiwara, T., & Iwata, K. (1988). *Synthetic Metals, 26,* 209.
71. Chung, T. C., Moraes, F., Flood, J. D., & Heeger, A. (1984). *Journal of Physical Review B: Condensed Matter, 29,* 2341.
72. Bredas, J. L. & Street, G. B. (1985). *Accounts of Chemical Research, 18,* 309.
73. Sun, Z. W. & Frank, A. J. (1991). *Journal of Chemical Physics, 94,* 4600.
74. Christensen, P. A., Hamnett, A., Hillman, A. R., Swann, M. J., & Higgins, S. J. (1992). *Journal of Chemical Society, Faraday Transactions, 88,* 595.
75. Batz P., Schmeisser D., & Gopel W. (1991), *Physics Review B: Condensations Matter, 43,* 9178.
76. Zhou, M. & Heinze. (1999). *Journal of Electrochimica Acta, 44,* 1733.
77. Murray, R. W. (1980). *Accounts of Chemical Research, 13,* 135.
78. Murray, R. W., Ewing, A. G., & Durst, R. A. (1987). *Analytical Chemistry, 59,* 379.
79. Anson, F. C., Ni, C.-L., & Saveant, J.-M. (1985). *Journal of American Chemical Society, 107,* 3442.
80. Skotheim, T. A. & Reynolds, J. (2007). *Handbook of Conducting Polymers,* (3rd edn.). US: CRC Press.
81. Torrance, J. (1979). *Accounts of Chemical Research, 12,* 79.
82. Jerome, D. (2004). *Chemical Reviews, 104,* 5565.
83. Papageorgiou, N., Athanassov, Y., Armand, M., Bonhote, P., Pettersson, H., Azam A., &. Gratzel, M. (1996). *Journal of Electrochemical Society, 143,* 3099.
84. O'Regan, B., & Schwartz, D. T. (1996). *Journal of Applied Physics, 80,* 4749.
85. Bach, U., Lupo, D., Comte, P., Morse, J. E., Weissortel, F., Salbeck, J., Spreitzer, H., & Gratzel, M. (1998). *Nature, 395,* 583.
86. Byrappa, K. & Yoshimura, M. (2001), *"Handbook of hydrothermal technology"*. USA: Noyes Publications.
87. Bin Liu, and Hua Chun Zeng. (2003). *Journal of American Chemical Society, 125,* 4430.
88. Zhao, H., Qu, Z. R., Ye, H. Y., & Xiong, R. G. (2008). *Chemical Society Review, 37,* 84.
89. Wang, X., Zhuang, J., Peng, Q., & Li, Y. (2005). *Nature, 437,* 121.
90. Mohan Kumar, G., Ilanchezhiyan, P., Kawakita, J., Subramanian, M., & Jayavel R. (2010). *Crystal Engineering Communication, 12,* 1887.
91. Ilanchezhiyan, P., Mohan Kumar, G., Vinu, A., Salem Al-Deyab, S., & Jayavel, R. (2010). *International Journal of Nanotechnology, 7,* 1087.
92. Ilanchezhiyan, P., Mohan Kumar, G., Subramanian, M., & Jayavel, R. (2010). *Materials Science and Engineering B, 175,* 238.
93. Andrieux, C. P., Audebert, P., Hapiot, P., & Saveant, J. M. (1991). *Journal of Physical Chemistry, 95,* 10158.

94. Genies, E. M., Bidan, G., & Diaz, A. F. (1983), *Analytical Chemistry Electroanalytical Chemistry, 100*, 101.

95. Tourillon, G. & Garnier, F. (1982). *Journal of Electroanalytical Chemistry, 135*, 173.

96. Roncali, J., Garnier, F., Lemaire, M., & Garreau, R. (1986), *Synthetic Metals, 15*, 323.

97. Ming Zhou & Jurgen Heinze. (1999). *Journal of Physical Chemistry B, 103*, 8443.

98. Halls, J. J. M., Walsh, C. A., Greenham, N. C., Marseglia, E. A., Friend, R. H., Moratti, S. C., & Holmes, A. B. (1995). *Nature, 376*, 498.

99. Bassler, H. (1994). *Molecular Crystals Liquid Crystals Science Technology Section A, 252*, 11.

100. Fusalba, & Daniel. (1999). *Journal of Physical Chemistry B, 103*, 9044.

101. Kuwabata, S., & Charles Martin, R. (1994). *Analytical Chemistry, 66*, 2757.

102. Brian Saunders, R., Keith Murray, S., & Robert Fleming, J. (1992). *Synthetic Metals, 47*, 167.

103. Beck, F., Michaelis, R., Schloten, F., & Zinger, B. (1991). *Electrochimica Acta, 39*, 229.

104. Brédas, J. L., Scott, J. C., Yakushi, K., & Street, G. B. (1984). *Physical Review B, 30*, 1023.

105. Bauerle, P. (1992). *Advanced Materials, 4*, 102.

106. Sadki, S., Schottland, P., Brodie N., & Sabouraud, G. (2000). *Chemical Society Review, 29*, 283.

107. Liu, L., Zhao, C., Zhao, Y., Jia, N., Zhou, Q., Yan, M., & Jiang, Z. (2005). *European Polymer Journal, 41*, 2117.

108. Hwang, B. J., Santhanam, R., & Lin, Y. L. (2000). *Journal of Electrochemical Society, 147*, 2252.

109. Liu, Y. C., Huang, J. M., Tsai, C. E.,Chuang, T. C., & Wang, C. C. (2004). *Chemical Physics Letters, 387*, 155.

110. Liu, Y. C. & Hwang, B. J. (2000). *Thin Solid Films, 360*, 1.

111. Neoh, K. G., Young, T. T., Kang, E. T., & Tan, K. L. (1997), *Journal of Applied Polymer Science, 64*, 519.

112. Li, G., Shrotriya, V., Yao, Y., Huang, J., & Yang, Y. (2007). *Journal of Materials Chemistry, 17*, 3126.

123. Liu, H., Piret, G., Sieber, B., Laureyns, J., Roussel, P., Xu, W., Boukherroub R., & Szunerits, S. (2009). *Electrochemistry Communications, 11*, 945.

124. Kang, S. H. & Sung, Y. E. (2006). *Electrochimica Acta, 51*, 4433.

125. Yin, X., Zhao, H., Chen, L., Tan, W., Zhang, J., Weng, Y., Shuai, Z., Xiao, X., Zhou, X., Li, X., & Lin, Y. (2007). *Surface and Interface Analysis, 39*, 809.

126. Batz, P., Schmeisser, D., & Gopel, W. (1991). *Physical Review B, 43*, 9178.

127. Schmeisser, D. & Gopel, W. (1993). *Bericht der Bunsengesellschaft für physikalische Chemie, 97*, 372.

128. Heeger, J., Kivelson, S., Schrieffer, J. R., & Su, W. P. (1988). *Reviews of Modern Physics, 60*, 781.

129. Petr, F., Zhang, H., Peisert, M., Knupfer, & Dunsch, L. (2004). *Chemical Physics Letters, 385*, 140.

CHAPTER 17

BIOCOMPATIBLE AND BIODEGRADABLE POLYMERS FOR BIOMEDICAL APPLICATION

RAJANGAM THANAVEL and AN A. SEONG SOO

CONTENTS

17.1 INTRODUCTION

Biocompatible polymers can be defined as a material meant to interface traditional organic polymers with living systems to assess, deal, augment or substitute preconceived (or designed) tissues and organs [49, 167, 192]. Such biocompatible polymers may play a crucial role in human body, especially, when interfaced with biopolymers of many types of biomaterials for implementing diverse functional tissues and organs in living body. Based on their biocompatible properties, the biocompatible-biopolymers can be further classified into biocompatible and non compatible polymers, in addition to their degradation properties. In many typical applications, such as wound dressings and bandages, and implants such as vascular graft, bone fixing materials, bone substitution materials and dental materials, the biocompatible-biopolymers should provide biocompatibility with short or long term stabilities for consistent performances along with biodegradation property, as needed. For more than two decades, biologically derived natural biomaterials were significantly explored for their applications in controlled drug delivery, gene therapy, cell delivery, biomolecules delivery, tissue engineering, and regenerative areas because of their biocompatibility and biodegradability [48, 162, 200]. Usually, biodegradation occurred through the temperature, chemical or structural deteriorations, and/or enzymatic reaction associated within human body. This type of degradations occurred in two steps; initially, polymers can be fragmented into lower molecular mass species by means of biotic (degradation by microorganisms) or abiotic reactions (oxidation, hydrolysis, and photo degradation), followed by bio-assimilations of microorganisms and their mineralization. Although, the degradation of biocompatible-biopolymers may not only depends upon the source of the biomaterials, but also based on its chemical compositions, their structures, methods of production, processing and storage conditions, and environmental factors (pH and temperature) [185].

17.2 NATURALLY OCCURRING BIOPOLYMERS

Natural biomaterials such as collagens, gelatin, fibrinogen, casein, alginate, chitosan, and polyesters were frequently considered and widely used as biomaterials of choices in various biomedical applications due to their original involvements in the native extracellular matrix (ECM). These ECM proteins could render structures and mechanical supports to tissues, and also could help in communication between neighboring cells and cellular components in support of facilitating cells adhesion, proliferation, regulation of daily cellular metabolic process, and wound healings. Furthermore, natural ECM could perform a different set of functions in their living system. For example, protein based polymers could act as central components of living system because they provided an architectural environment in nurturing and supporting interactions between cell-cell, cell-ECM and cell-growth factors for creating regenerative niches. Polysaccharides based biomaterials may also provide functions in membranes and intracellular communication. Advantages of the natural biomaterials or ECM could be their degradation by their respective enzymes in active transitions of ECM or tissues, such as wound healing and tissue regeneration. Without the degradation properties, formation of the ECM or tissue would be very limited in their sizes and functions. The ideal bioma-terials should reproduce the structure of their ECM, in addition to regenerat-ing their functions and mechanical strength as performed by ECM. However, most of the natural biomaterials were observed with less mechanical strength. Malafaya et al. [107] classified the natural–origin polymers with their detailed biomedical applications, categorizing natural biomaterials based on their chemical structures.

- **Protein-Origin Polymers**: Collagen, gelatin, fibrinogen, silk fibroin, elastin, soybean, and wool
- **Polysaccharides**: Alginate, chitosan, starch, cellulose, chondroitin sul-phate, and hyaluronan
- **Polyesters**: Polyhdroxyalkanoates

17.2.1   Collagen

It is a primary protein with the triple helix widely distributed in connective tissues. As one of the key components in tissue architecture, it could pro-vide mechanical strength in cell-matrix and matrix-matrix interactions. So far, 28 types of collagen were identified, but among these, type I collagen was the most abundant and widely investigated for its biomedical applications. Mostly, predominant amino acid compositions in collagen were glycine, ly-sine, proline, and hydroxyl proline. The direct correlation of flexibility was observed in collagen with increased number of glycine [185]. Since collagen is the major natural protein component of the ECM for providing major sup-port in connective tissues such as tendons, cartilage, ligaments, bones, blood vessels, and skin, it was considered as an ideal biomaterial for creating di-verse scaffolds in tissue engineering. Because of biocompatibility and multi-functional role of collegen, it could easily interact with cells in connective tissues, even transducing important signals for the regulation of cell adhesion, migration, proliferation, differentiation, and survival. Importantly, collagen could be enzymatically degraded through collagenase in restructuring scaf-folds, ECM, or tissues.

17.2.2   Gelatin

Gelatin is a denatured protein, derived from collagen either by acid or alkaline treatments. As a result, two different types of gelatin could be manufactured, depending on the pre-treatment methods (alkaline and acidic) of collagen prior to the extraction process [36]. For example, the alkaline treatment of collagen through hydrolytic process yielded gelatin with a high density of carboxyl groups, which reduced the isoelectric point (IEP) of gelatin due to the negative charges. In contrast, collagen treated through the acid process yielded low density of carboxyl groups from its limited invasive reaction to amide groups of collagen, resulting similar IEP to the original collagen [197]. However, both types of gelatin were commonly used for pharmaceutical, cosmetic manufacturing and medical applications because of their biodegradability and biocompatibility in physiological environments [74]. The unique electrical properties of gelatin made it as ideal scaffolds, because positively or negatively charged proteins or peptides or drugs could electrostatically interact with an oppositely charged gelatin by forming polyionic complexes. Moreover, gelatin revealed relatively less antigenicity in comparison to collagen.

17.2.3   Fibrinogen

Fibrinogen is a 340kDa blood plasma protein, made up of a pair of three polypeptide chains: $2A\alpha$, $2B\beta$, and 2γ, involved in blood coagulation, platelet aggregation, wound healing and tissue regeneration. Fibrinogen synthesized by liver is freely flowing in the circulatory system and plays a most important role in hemostasis and wound healing. When fibrinogen reacts with thrombin in the presence of Ca^{2+}, it polymerize into fibrin clot in the coagulation process. After the wound healing, the fibrin clot is cleaved by proteolytic enzyme, plasmin, in a process called fibrinolysis [14]. Similar to collagen, fibrin(ogen) scaffolds could attain high cell seeding efficiency and uniform cell distribution, then cells could proliferate, migrate, and differentiate into the specific tissue or organ by secreting ECM in the process of constructing complete tissue or organ [23]. Fibrin(ogen) has major advantages, as it provides a favorable surface for cellular attachment and proliferation, because of 3-D fibrous structural support and nanotextured surfaces, comprising of a fibrous network for cell signaling, cell-matrix, and cell-cell interaction [52]. Especially, fibrin (ogen) based scaffolds could induce ECM production that render support to connective tissues such as cartilages, ligaments, bones, tendons, nerves, blood vessels, and skin in tissue engineering [52, 128, 157].

17.2.4   Chitosan

Chitosan as a copolymer of glucosamine and N-acetyl glucosamine has mucopolysaccharides structural characteristics similar to glycosamines. Chitosan is apolycationic, biocompatible, and biodegradable polymer derived from chitin by alkaline deacetylation method [107]. It was sought as an ideal scaffold for range of biomaterial and industrial applications [75]. In addition, chitosan contains different functional groups for variety offunctions with diverse biomolecules and ligands either by adsorption, affinity/charge, or conjugation methods [29]. Usually, chitosan is not soluble in water, but it is more readily soluble in diluted acidic solution with pH near 6.0, because chitosan can be considered as strong base due to the high content of primary-NH$_2$ groups with a pKa value of 6.3, which affected considerably changes in the charged state and properties of chitosan from pH [195]. At low pH, these amines get protonated and turn positively charged, making it water-soluble in the cationic polyelectrolyte. Because of these unique physicochemical properties, chitosan was widely used in tissue engineering and biomolecules for non viral gene delivery, drug delivery, and enzyme immobilization. Chitosan also could be easily blended with other natural biomaterials like collagen, which showed an increased proliferation of endothelial and smooth muscle cells [21]. Since chitosan is soluble in most acidic solutions, it was shown to form nano fibers in different sizes easily by electrospinning [38].

17.2.5   Alginate

Alginate is a linear unbranched polysaccharide containing repeated units of (1–4)-linked β-mannuronic acid and α-guluronic acid monomers. Generally, alginate referred as a family of polyanionic copolymers, derived from brown sea weeds [150]. Alginate became most important in recent years for biomedical applications, since it uses one of the most versatile natural biomaterials in synthesizing hydrogels [31, 73]. Alginates contained reversible gelling properties in aqueous solutions, where the ionic cross-linking of alginates could occur between positively charged cations (barium, calcium, and magnesium) and negatively charged carboxylic acid moieties on the guluronic acid residues [28]. This cross-linking property was employed to immobilize different microorganisms for diverse industrial applications, such as ethanol production from yeast. And in the food, beverage and pharmaceutical industry, alginates remained as one of the most important ingredients [54]. Moreover, the ionic cross-linked alginate did not cause any harm to microorganism and other cells during the cross-linking process, revealing their biocompatibility and since their properties were similar to natural tissue with high biocompatible and biodegradable properties, hence, alginate hydrogel was demonstrated as great potential candidate biomaterial for wound dressing [37], tissue regeneration [30], drug delivery [170], anti-adhesion materials [79], and encapsulation materials for immunoisolation-based cell therapeutics [143].

17.2.6   Starch

Starch is a biopolymer derived as minute granules from variety of plants, such as corn, rice, wheat, millet, potatoes, and barely including roots, stems, and seeds. Starch is entirely composed with two of D-glucose (homopolymers) as unit; amylose, mostly organized into the essentially linear α-(1, 4)-glucose residue (20–30%), and branched amylopectin containing α-(1,6)-linkages (70–80%) [109]. Among natural biopolymers, starch is also one of the most promising natural polymers for its availability, economical feasibility, and inherent biodegradability. Unfortunately, the physicochemical properties, including mechanical strength of native starch were shown to be very poor, and the aqueous solubility also restricted for its long term usage [13]. These demerits could be overcome either by grafting or blending with other synthetic hydrophobic polymers or by converting into a thermoplastic, which could make it as an ideal biomaterial for different biomedical applications with their sustained biodegradation [148]. The recent report suggested that starch or synthetic polymer composites revealed a great potential to use in biomedical applications, especially, starch based composite for bone and cartilage tissue engineering [59], bone fixation and replacement, and fillings for bone defects [45]. Also, starch ornatural and/or synthetic polymer composites were used as a carrier for various biomedical applications. For example, starch hybrid with chitosan is used as biodegradable scaffold for bone tissue engineering *in situ* [112] with polycaprolactone, as micro composite used for angiogenesis [155] and with poly(L-lactic acid), as bi-layered scaffold for tissue engineering of osteochondral defects [56]. Other synthetic or natural polymers blended with starch were also found in literature [185].

17.2.7 Polyhydroxyalkanoates

Polyhydroxyalkanoates (PHAs) are a family of polyhydroxyesters of 3-(R)-hydroxyalkanoic acids synthesized by microorganism (*Alcaligeneseutrophus*) as an intracellular carbon and energy storage compound under nutrient-limiting conditions with excess carbon sources [139]. The 3-(R)-hydroxy fatty acids were bound together by an ester linked between the hydroxyl group and carboxyl group of an adjacent monomer [169]. Since PHAs could be synthesized from a variety of renewable resources, they received much consideration in addition to their other properties, highly biocompatibility, biodegradability, highly thermo stability, and optically active phenomenon [198]. Interestingly, PHAs could also be synthesized by transgenic plants [142]. PHAs were divided into two major groups; medium chain length PHAs with composition of 3-(R)-hydroxyhexanoate/3-(R)-hydroxytetradecanoate and short chain length PHAs with Polyhydroxybutyrate (PHB) and the copolymer polyhydroxy-co-valerate (PHBV) [183]. Depending on the necessities of biomedical applications, PHAs could be either surface modified, blended or grafted with other biopolymers, enzymes or even minerals, such as hydroxyl apatite, in order to improve their biocompatibility and mechanical properties (Chen et al. 2005). Therefore, PHAs are becoming a promising class of emerging biopolymers in the fields of medicine, tissue engineering, nanocomposites, and polymer blends [139].

17.3 BIODEGRADABLE SYNTHETIC POLYMERS

Synthetic polymers are also playing significant role in medical, surgical sutures, drug delivery, protein/peptide delivery and gene therapy. The properties of the synthetic polymers mainly depended on their chemical compositions, structure and molar mass etc. Based on the polymer functional groups, which presented on the polymers surface, and their internal structures can contribute either in various therapeutic applications or toxic effects. For example, the presence of carboxylic or amine groups on the polymers surface could induce therapeutic activity of many pharmaceuticals, in addition to helping cells and biomolecules for binding or anchoring. In recent years, synthetic biodegradable polymers became highly useful in biomedical applications through their pre-determined properties, since their porosity, mechanical strength, and degradation time could be tailored for the specific applications. Most of commercially available synthetic polymers were superior in both mechanical and physicochemical properties in comparison to biological tissues, which required continuous regeneration [64]. Production of synthetic polymers in large quantities under controlled conditions could be a major advantage. Moreover, they could exhibit predictable and reproducible physical properties, such as tensile strength, elastic modulus, and degradation rate [65]. On the other hand, the biocompatible and biodegradable synthetic polymers could facilitate the restoration of structures and functions of injured or damaged or diseased tissues, and they could be continuously replaced by native compositions. The PEG, PVA, PLA, PGA, and PLGA copolymers were among the most commonly used biodegradable synthetic polymers for biomedical applications. These synthetic polymers were also shown to be degraded often by chemical and mechanical degradations; in addition, few synthetic biopolymers could also be depolymerized by microbial enzymes [178], where the monomers were absorbed into microbial cells and biodegraded [57]. But abiotic hydrolysis was the most important reaction for initiating the environmental degradation of synthetic polymers [3]. The major advantage of hybrid-biomaterials was their ability to be designed and manipulated in order to improve the desired mechanical strength and degradation time and rates [141]. However, most of the synthetic scaffold inherently lacked in binding sites for cell attachment, proliferation, and signaling. Therefore, for making an ideal scaffold, hybrid composite of natural and synthetic polymers by blending or grafting may ascertain cells attachment and proliferation with good mechanical properties [148]. Here are some examples for biocompatible and biodegradable synthetic polymers with their physicochemical properties.

- Polyalkylene esters
- Polyamide esters
- Polyvinyl esters
- Polyanhydrides
- Poly(dioxanone)
- Poly(caprolactone)
- Polylactic acid(PLA)
- Polyglycolic acid (PGA)
- Polyglycolic-co-lactic acid (PLGA)
- Polypropylene fumarates
- Polyethylene glycol (PEG)

Additional biocompatible and biodegradable synthetic polymers for biomedical applicationswere listed [64].

17.3.1 Poly-Lactic Acid (PLA)

The PLA is commonly prepared by ring opening polymerization of lactide or polycondensation of D-or L-lactic acid. Higher molecular weight PLAs could be obtained through ring opening polymerization with better mechanical properties [135, 164]. PLA existed in three isomeric forms; D-lactide, L-lactide, and racemic (D, L) mixtures, where the chirality was usually abbreviated. Poly(D)-lactic acid and poly(L)-lactic acid were semicrystalline solids with similar rates of hydrolytic degradation as with polyglycolic acid (PGA). With the presence of methyl ($-CH_3$) side groups, PLA was more hydrophobic than PGA, and was more resistant to hydrolytic attack than PGA due to the steric shielding effect of the $-CH_3$ side groups [64, 185]. For the most of the biomedical applications, the (L)-lactic acid was chosen due to its preferential metabolism in body. PLA mechanical properties could be synthesized to accommodate large degrees of mechanical properties; from soft and elastic materials to stiff and high-strength materials, by adjusting the molecular weight, degree of crystalline and formation of stereo complexation of enantiomeric lactic acids [46, 62]. The commercial PLA represented the typical glass transition temperature and the tensile strength of 63.8°C and 32.22MPa, respectively [15]. Regulation of the biodegradability and other physical properties of PLA could be achieved by racemization of D- and L-isomers or by employing a hydroxy acids comonomer component [164]. The solubility of PLA depended mostly on the degree of crystallinity, molar mass, and the presence of other co-monomer units in the polymer. Poly(racemic) lactic acid and poly(meso) lactic acid were readily soluble in many organic solvents, such as acetone, ethyl acetate, pyridine, dimethylformamide, and dimethyl sulfoxide. In addition, fluorinated or chlorinated solvents, furane, dioxane, dioxalane were also good solvents for enantiomerically pure PLA.

17.3.2 Poly Glycolic Acid

The PGA is simplest linear aliphatic polyester and is a rigid thermoplastic polymer with high crystallinity in 45–55%. Due to the high crystalline nature, PGA was not soluble in many organic solvents except highly fluorinated organic solvents, such as hexafluoroisopropanol [122]. The PGA could be obtained by ring opening polymerization of a cyclic lactone and glycolide. On the other hand, PGA revealed excellent mechanical properties due to its high crystallinity. PGA was known to lose its strength in 1–2 months during hydrolysis, and its mass within 6–12 months. However, PGA's biomedical applications would be limited by its low solubility and its high rate of degradation, yielding acidic byproducts. But the copolymers of PGA with PLA, caprolactone or trimethylene carbonate became interesting in several biomedical applications [117, 122].

17.3.3 Poly Glycolic-Co-Lactic Acid (PLGA)

The PLGA is a copolymer prepared from both L-lactide and DL-lactide (LA) with glycolic acid (GA) monomers by co-polymerization. Different ratios of lactic acid and glycolic acid for the PLGA were commercially developed [185]. The degraded monomeric acids, lactic acid, glycine, and glycolic acids from PLA, PGA, and PLGA, respectively, could be further eliminated from body through urine or converted into carbon dioxide and water by the citric acid cycle [113, 163]. Furthermore, these polymers could also be degraded by certain enzymes, especially with esterase activity, where their respective intermediate or byproducts could be excreted by urine [190].

17.3.4 Polyvinyl Alcohol (PVA)

PVA is a water soluble synthetic polymer. It revealed good chemical stability, biocompatibility, non-carcinogenity, and achieve high performance in drug delivery and tissue engineering applications without known toxicity. PVA also presented good physical properties, such as elasticity, compliance, and resistance to mechanical stress. Above unique properties made PVA to be prepared in different types of scaffolds; gels, membranes, films, and adhesion protection sheets [87, 103, 134, 196]. PVA was extensively studied synthetic polymer than others, and drew many interests from pharmaceuticals, medicine, food chemistry, biotechnology, and biomedical applications [91, 129, 172]. Interestingly, it was approved by the Food and Drug Administration (FDA). PVA microspheres [176], nanofibers [102], and hydrogel were effective in drug delivery and tissue engineering applications. Especially, PVA hydrogels revealed to be biocompatible, biodegradable, and are non irritating to soft tissues during contact with PVA, which made it as a suitable carrier for wound healing and other biomedical applications [121]. The PVA hydrogel also presented several other useful properties including hydrophilicity, permeability, and low frictional function. Although, PVA showed good physicochemical properties, the usage of PVA hydrogel in tissue engineering including bone tissue engineering and orthopedic surgery, was limited due to its low mechanical strength and durability.

17.3.5 Polyethylene Glycol (PEG)

The PEG is the most widely used water soluble synthetic polymer for diverse biomedical applications. Its biocompatibility, biodegradability, flexibility, and stealth properties made PEG as ideal material for applications in tissue engineering and drug delivery [93, 125]. PEG-based hydrogels demonstrated their high biocompatibility and absence of toxic influence on surrounding tissues, suggesting their potential usages as good biomaterials for drug delivery systems [180, 202]. PEG hydrogel was mainly used as a carrier for drug release, because it had ability to control the release of drugs, protein, growth factor, and other biomolecules. To improve their durability, chemically cross-linked PEG hydrogels were also developed biologically active hydrogel scaffolds in functional tissue development [60, 100].

17.4 BIODEGRADABLE NATURAL AND SYNTHETIC POLYMERS AS DELIVERY SYSTEMS

17.4.1 Drug Delivery Systems

Biodegradable and biocompatible (natural or synthetic) polymer carriers with responsive characteristics or reactive functional groups were widely studied for drug delivery. The drugs and biomolecules incorporated by biomaterials could be released through diffusion, swelling or degradation of polymer carrier. Biomaterials made from natural sources presented several advantages of being highly biocompatible, biodegradable, and nontoxic with good functional properties and well characterized structures. Because of these properties, natural polymer-based carriers in forms of microspheres, nanospheres or nanoparticles, microfibers or nanofibers, and hydrogels were studied in depth for the drug delivery system or tissue replacements. Collagen carriers were tested for the local delivery of aminoglycosides like gentamicin [153] and tobramycin [101, 123]. The controlled deliveries of highly sensitive drugs or proteins from gelatin carriers were also studied through gelatin carrier-mediated pharmaceutical drug delivery system; cancer chemotherapy [18, 120] and sustained antibiotic delivery for bone infection and repair [34, 89]. Gelatin-based carriers needed to be cross-linked to release the drug in controlled manner. Commonly aldehydes, such as glutaraldehyde [173] and formaldehyde [35] or bifunctional agents [1] were widely used to cross-link the gelatin.

Fibrinogen based microspheres and nanoparticles were mainly used for the anti-cancer treatment [149], expanding from their regular tissue engineering role. A preliminary data suggested that the adsorbed fibrinogen into docetaxel-loaded oil droplets helped the retention of the droplets within the fibrin-rich tumor microenvironment [42]. Fibrinogen nanoparticles loaded with 5-Fluorouracil [149] demonstrated the controlled release for anti-cancer therapy *in vitro*. Lu da et al. [105] also reported that fibrinogen functionalized anti-cancer drug could treat solid tumors.

Duarte et al. [39] showed the release of dexamethasone from chitosan scaffolds *in vitro*, which directed the proliferation of stem cells toward the osteogenic lineage. Chitosan were also used as a drug delivery carrier in various studies [145, 162], whereas Shanmuganathan et al. [159] fabricated the chitosan into microspheres for delivering doxycycline as antibiotic. In one study, due to the excellent muco-adhesive properties, the cationic chitosan polymer was used as carrier for muco-adhesive drug delivery system [166].

Starch was another widely used biopolymeras drug delivery carrier with good biocompatibility *in vitro* and *in vivo* [154]. But, it was difficult to process the starch by itself without plasticizer. Recently, several concerns of using starch based biodegradable blends for fabricating scaffolds were reported [133].

The PLGA-based microspheres became one of the most popular drug delivery system among biodegradable synthetic polymers, since it showed favorable degradation characteristics with possibility of controlled drug release. Dexamethasone PLGA loaded microspheres were embedded within PVA hydrogels system, which suggested a new potential approach for variety of pharmaceuticals for the sustained drug delivery applications in local tissues [53]. In another study, dexamethasone loaded PVA hydrogel provided soft, permeable, and hydrophilic interfaces with body tissues, and the drug release from the scaffold suppressed or controlled the inflammatory responses and tissue injury processes at the implant site [70].

17.4.2 Approaches for Biomolecules or Protein Delivery

Biomolecules and growth factors or proteins are susceptible to chemical changes, proteolysis and to easy denaturation during storage or shortly after the administration into the body. In contrast, biomaterials in biopolymer were shown to have better stabilization of proteins over a sufficiently long storage time, to prolong the biological activity of biomolecules after the administration, and achieved controlled release at the site of action over a long period of time. It was highly possible that biomolecules were protected against proteolysis and other changes, when incorporated into an appropriate polymer matrix. The delivery of growth factors through biopolymer improved the healing response by host at the site of injury by helping the neighboring cells to renovate; in addition biopolymer manipulated and facilitated the growth *in vitro* into fully active engineered tissues. In tissue engineering and regenerative medicine, single or multiple growth factors delivery system was much designed because the growth factors could stimulate specific single or multiple tissues by providing the additional necessary information or signals for cell attachment, multiplication, and differentiation by meeting the requirement of dynamic reciprocity for tissue engineering. For example, basic fibroblast growth factor (bFGF) could induce proliferation and differentiation of a variety of cells, and played crucial roles in an early stage of wound healing, stimulating angiogenesis and neural outgrowth [2, 140]. Next, collagen was also extensively used in protein or peptide deliveries in the field of biomedical applications. When VEGF was incorporated into collagen gel, covalently enhanced angiogenesis was observed [88]. Shen et al. [160] reported that VEGF immobilized by collagen matrix promoted the penetration and proliferation of endothelial cells for vascular development. Recently, collagen sponges loaded with bone morphogenic protein (BMP)-7 and BMP-12 revealed the improved periodontal ligament regeneration [119]. The gelatin matrix could form polyion complexes by mixing with biomolecules for releasing them as biologically active form [197]. Usually, the release was controlled by matrix degradation; hence, the time period for the release of biomolecules could be adjusted by tailoring the matrix degradation. In other study, aqueous protein solution was dropped onto the freeze-dried gelatin matrix allowing for sorption of the protein to the matrix and its subsequent sustained release by degradation of the matrix *in vivo*. Recently, it was established that such utility of silk fibroin might be further expanded by chemically cross-linking or coupling with bioactive proteins and/or peptides for various biomedical applications. For example, silk fibroin was chemically conjugated with BMP-2 [81] and with peptides of RGD [115, 165]. The other important protein derived from blood fibrin(ogen) scaffold could be immobilized with growth factors, such as VEGF [41], epidermal growth factors (EGF) [107], nerve growth factors (NGF) (Bhang SH) and BMP [128] for stimulating blood vessel, skin, nerve and bone growth respectively. Fibrin(ogen) scaffold were extensively studied with various tissue inductive growth factors, such as transforming growth factor (TGF) [138], FGF [194], fibronectin (FN) [25], and keratinocyte [69], for various tissues growth. Furthermore, the controlled release of growth factors from fibrin(ogen) provided number of vital advantages, since growth factors could be transported into the desirable environments, where they may locally govern multiple processes of the cell chemotaxis, attachment, proliferation, differentiation and morphogenesis [23, 128]. Study by Yu et al. [199] revealed the beneficial effects of RGD functionalized alginate matrix on angiogenesis and left ventricular function in a chronic rat infarct model.

Biodegradable synthetic polymers, such as PLA, PGA, PLGA, and PEG were also used to deliver the biomolecules in an active form. Encapsulated PLGA microspheres with recombinant human vascular endothelial growth factor (rhVEGF) showed the continuous release of growth factor with initial burst of release, followed by steady-state controlled release up to 14 days [82]. King and Patrick [86] developed PLGA-PEG scaffold for growth factor release. The VEGF loaded PLGA scaffold degraded over time with controlled release of VEGF in an active form, and demonstrated the increased proliferation of HUVECs than the controls *in vitro*. Recently, the VEGF loaded PLGA microspheres showed angiogenesis in tissue engineered intestine [152]. When adhesive biomolecule, FN, was chemically cross-linked with 3-D PVA scaffold, improved cell attachment, proliferation, and infiltration depth were observed in comparison with control scaffold (without FN) [40]. Here is the list of biodegradable polymers in making scaffold system for biomolecules delivery in Table 17.1.

TABLE 17.1   List of polymer carriers or scaffolds and the biomolecules for tissue engineering application.

Polymer carriers or scaffold	Biomolecules delivery	Tissue engineering application	References
Collagen scaffold	VEGF	Vascularization	Shen et al. [160]
Collagen sponge	bFGF	Cartilage	Fujisato et al. [50]
Chondroitin sulfate A	VEGF	Vascularization	Liu et al. [103]
Collagen/hydroxylapatite	NGF	Bone	Letic-Gavrilovic et al. [96]
Gelatin hydrogel	bFGF	Bone	Ikada and Tabata [74]
Gelatin sponge	BMG-2	Bone	Okamoto et al. [127]
Gelatin microspheres/collagen sponge	bFGF	Adipose	Kimura et al. [85]
Fibrin gel	NT-3	Spinal cord injury	Taylor et al. [174]
Fibrin gel	VEGF	Vascularization	Ehrbar et al. [41]

TABLE 17.1   *(Continued)*

Silk fibroin nanofibrous scaffolds	BMG-2	Bone	Li et al. [97]
Chitosan hydrogels	FGF-2	Vascularization	Fujita et al. [51]
Chitosan/chitin tubular scaffold incorporated with PLGA microspheres	EGF	Peripheral nerve regeneration	Goraltchouk et al. [61]
Chitosan hydrogels	hEGF	Skin	Alemdaroglu et al. [5]
Alginate beads	BMG-2	Bone	Grunder et al. [63]
Alginate hydrogel	VEGF and bFGF	Vascularization	Lee et al. [95]
Starch-based microparticles	Non-steroid anti-inflammatory agent	Bone	Malafaya et al. [106]
Starch-based porous scaffolds	Non-steroid anti-inflammatory agent	Bone	Gomes et al. [59]

17.4.3   Cell Delivery

Combining stem cells with biomaterials may render a promising strategy for cellular delivery. The encapsulated stem cells could be cultured *in vitro* or transplanted *in vivo* for producing growth factors and other tissue promoting proteins and for restoring defective sites [179]. For example, for the artificial conduct or regeneration of nerve, stem cells, such as Schwann cells (SCs) and mesenchymal stem cells (MSCs) were shown to enhance the growth by secreting nerve stimulating factors [176]. MSCs could be induced to differentiate into SCs, promoting the regeneration of injured sciatic nerves [16]. Natural ECM based 3D scaffolds were also effective in culturing a wide range of stem cells for cell-based therapies. Batorsky et al. [11] developed a system to encapsulate adult human MSC within collagen (type I) scaffold. The 3D microenvironment scaffold-cell composition extended a way to control cell-matrix interactions, thereby leading the differentiation of human MSC. Especially, cell delivery strategies provided a novel approaches for bone tissue engineering. Collagen (type I) sponge was seeded with human alveolar osteoblasts cells, then implanted into osseous injure or damaged sites. After 4 weeks from the implantation, the newly formed bone tissue from transplanted alveolar osteoblasts area was observed suggesting the utilization of ECM scaffolds with incorporated cells derived from human alveolar bone for the effective induction of new bone formation [191]. In another study, chondrocytes were seeded onto collagen membrane and improved the cartilage formation in rabbits [32]. When human embryonic stem cells (hES) combined with appropriate cues, hES could be developed into blood vessels with *in vitro* model [55]. Gelatin scaffold also demonstrated its promising usage as a material for cell delivery. Another supporting result was obtained, where encapsulated porous gelatin scaffold with human adipose-derived adult stem cells was effective ingenerating cartilage [9]. Encapsulated adult human MSCs into porous gelatin sponge was also successful in therapy for cartilage regeneration [144]. Several other cells including bovine chondrocytes [130], human chondrocytes [108], human pre-adipocytes [85], and mesenchymal stem cells [104] were delivered with gelatin scaffolds and yielded successful implantation.

In other study, osteoblasts cells were encapsulated by silk fibroin hydrogel and showed promising results of bone formation with rabbit distal femurs [48]. MSCs and endothelial cells were also loaded into silk fibrin scaffold and observed stimulations in cartilage and angiogenesis, respectively [181, 186]. Cultured ES onto fibrin scaffold seemed to stimulate neurons, revealing its potential usage as a platform for neural tissue engineering [189]. In addition, cultured mesenchymal progenitor cells onto fibrin(ogen) 3D microbeads secreted the mineralized ECM around the implantation area and promoted bone formation by secretion of osteogenesis (ascorbic acid, β-glycerophosphate, and dexamethasone) and osteoblast growth factors [67]. Seeded endothelial cells onto fibrin(ogen) hydrogel archived a good mechanical strength with significant production of collagen and elastin, and revealed considerable vasoreactivity within 2 weeks from the implantation into 12 weeks old lambs [171].

When other natural carbohydrate polymer scaffolds were developed from hyaluronan for tissue regeneration including nerve, skin, and cartilage, the promising results with hyaluronan suggested it as a material of choice for the culture and differentiation of various adult stem cells. Cultured MSCs onto hyaluronan matrix seemed to provide the directionality of regenerating or fixing cartilages both *in vitro* and *in vivo* [7, 114].

Synthetic polymers such as poly (ε-caprolactone), PGA, and PLGA also may prove as alternatives for their uses in stem cell cultures in comparison with natural polymers. Based on their defined chemical composition or blending, these synthetic polymers could offer several advantages of reproducibility, stronger mechanical properties, and independent shape designing [43]. Neubauer et al. [124]

demonstrated that MSCs could be seeded onto PLGA matrix for the generation of adipose tissue. Their result suggested that the added bFGFin the PLGA scaffolds enhanced the differentiation of these stem cells into adiposities. Teng et al. [175] also reported that cultured neural stem cells onto PLGA scaffolds revealed an increase in functional recovery after traumatic spinal cord injury in preclinical testing, demonstrating its applicability for further investigation into clinical usage. Some examples of biodegradable polymers as a carrier system for cell delivery were listed in Table 17.2.

TABLE 17.2   List of biodegradable polymers and incorporated cells in various tissue engineering applications.

Polymer carriers or scaffold	Encapsulated or seeded cell type	Tissue engineering application	References
Collagen gel	Bone marrow stromal cells	Bone or cartilage	Xu et al. [192]
Transglutaminase cross-linked gelatin hydrogel	Fibroblast-like NIH/3T3 cell line	Bone or cartilage	Ito et al. [78]
Gelatin microspheres encapsulated in an hydrogel matrix	Rat marrow stromal osteoblasts	Bone	Payne et al. [132]
Silk fibroin hydrogel	Osteoblasts	Bone	Fini et al. [48]
Silk fibroin Electrospun fiber scaffold	Keratinocytes and fibroblasts	Wound dressing	Fuchs et al. [49]
Silk fibroin net	Endothelial cells	Angiogenesis	Min et al. [118]
Fibrin scaffold	Murine embryonic stem cells	Spinal cord injury	Willerth et al. [189]
Fibrin–collagen gel	Embryonic chondrogenic cells	Cartilage	Perka et al. [136]
Fibrin tubes (autologous)	Outgrowth endothelial cells	Vascularization	Aper et al. [8]
Chitosan scaffolds	Human adipose-derived adult stem cells	Bone/cartilage	Malafaya et al. [107]

TABLE 17.2 *(Continued)*

Chondroitin sulfate–collagen scaffolds	Chondroitin sulfate–collagen scaffolds	Cartilage	van Susante et al. [184]
Chondroitin sulfate–PVA hydrogels	Baby-hamster kidney(BHK) cells	Kidney	Lee et al. [92]
Hyaluronic acid (Hyaff®11)	Chondrocytes	Cartilage	Aigner et al. [4]
Hyaluronic acid membrane	Preadypocytes	Skin	Hemmrich et al. [72]

17.4.4 Gene Delivery

Various growth factors were studied as therapeutic agents for the regeneration of various tissues including chronic dermal wound. Unfortunately, they failed some times in producing substantial improvements in tissue regeneration, in part due to an ineffective delivery approach and poor retention in the defected tissues. Chandler et al. [19] suggested that gene therapy might overcome the growth factor or protein delivery by ongoing transcriptions and translations, thus prolonging the availability of the therapeutic proteins.

Here, several advantages of gene delivery system will be mentioned in comparison to the protein delivery. First, the inherent stability of plasmid DNA could allow for an extended shelf-life in relative to its corresponding proteins [83]. Secondly, diverse natural biomaterials from nanoparticulates to three-dimensional scaffolds could be applied for the wide range of gene therapeutics. Hence, the combination of gene therapy and tissue engineering may offer a powerful synergistic treatment option for the tissue growth. For example, polymer scaffold incorporated with DNA could release plasmid DNA encoding a growth factor gene to be taken up by the cells and the transfected cells could release a growth factor upon induction, differentiation, and growth [168]. Xu et al. [192] studied the effects of natural polymer scaffold for BMP-2 *in vivo* using genetic tissue engineering elaborately. Several natural polymer gels including alginate, agarose, collagen, fibrin, and hyaluronate were loaded with transduced bone marrow stromal cells (BMSCs) with adenovirus-mediated human BMP-2 gene (Adv-hBMP-2) yielding good results of its expression, which suggested the potential approach in the orthopedic tissue engineering applications. When collagen gel was mixed with therapeutic transgenes (platelet-derived growth factor-A or -B) encoded adenoviral or plasmid gene vectors, this potential gene therapy revealed the improved wound healing performance with increased granulation tissue formation, re-epithelialization, and vascularization in comparison to the controls, treated with collagen alone [19].

Positively charged gelatin matrix was also used to facilitate delivery of small interference RNA (siRNA) to silence TGF-h receptor gene as a target in the interstitium of renal tissue upon delivery through the ureter [10]. Most commonly, cationic polymers were employed as gene transfer substance, since these cationic polymers could form complexes with anionic DNA through electrostatic interaction [58]. The cationic polymer-DNA complexes could be easily taken up by living cells through the electrostatic interaction or endocytosis, since the cell surface was negatively charged. Among the natural polymers, chitosan had a significant role in gene therapy, since its polycation nature revealed a strong affinity for DNA. Hence, the transfection efficiency of chitosan polymer was investigated by using a chitosan or pLacZ nanoparticle delivery (CH150kDa, 10.0g DNA) [27] and chitosan or pGL3-Luc (luciferase plasmid) complex for gene therapy [77]. DNA-chitosan complexes were also investigated against tumor cells and showed the easy uptake of DNA-chitosan complex by tumor cells [156]. Erbacher et al. [44] demonstrated the effective ability to transfect luciferase plasmid through chitosan as vehicle. Other natural polysaccharide based polymers, such as cyclodextrin were also studied as potential carriers for gene delivery [146].

Past few years, synthetic polymers, such as PLA, PGA, PLGA, and PEG were used as non viral gene carrier in gene therapy. Among them, the PEG-based carriers were more widely investigated. Polyethyleneimine immobilized by PEG (PEI-g-PEG) was developed and investigated their biological activities, especially as gene delivery systems [137]. After repeated intrathecal injections of PEI-g-PEG/ DNA complexes *in vivo*, PEI-g-PEG revealed a prolonged gene expression in the spinal cord without any decrease of gene expression [161]. The degradable analogue of PLA namely, Poly (-[4-aminobutyl]-l-glycolic acid) (PAGA) could be used as safe gene delivery system, which could protect DNA from a DNAse I attack by forming polyelectrolyte complexes similar to PLA [99]. Recently, dedifferentiated rabbit chondrocytes, transduced from PLGA matrix with cultured baculovirus expressing recombinant BMP-2couldproved their efficiency in repairing osteochondral defect *sex vivo* [20]. De Laporte et al. [33] showed that ECM-coated multiple channel bridged PLGA could be used as gene delivery system for spinal cord injury. Commonly, scaffold could be considered in two ways for the effective and controlled gene delivery; by encapsulating the uncondensed or condensed naked DNA with polycationic [111]. The natural and synthetic polymers based scaffolds aimed at tissue engineering were summarized in Table 17.3.

TABLE 17.3 List of biodegradable polymers and encapsulated gene in various tissue engineering applications.

Polymer carriers or scaffold	Encapsulated gene	Tissue engineering application	References
Collagen gel	BMP-2 gene	BMP-2 gene	Xu et al. [192]
Collagen scaffolds	IGF-1 gene	Articular cartilage	Capito and Spector [17]
Collagen gel	PDGF-A gene, PDGF-B gene	Skin	Xu et al. [192]
Porous chitosan-gelatin scaffold	DNA encodingTGF-β1	Cartilage	Guoet al. [66]
Hyaluronic acid microspheres	DNA	Vascularization	Yun et al. [200]
Chitosan or collagen scaffolds	TGFβ1 plasmid	Periodontal bone	Zhang et al. [201]
Porous PLG scaffold	Plasmid encoding FGF-2	Angiogenesis	Rives et al. [151]
PEG-PLGA-PEG triblock copolymer scaffold	TGF-β1	Wound healing	Lee et al. [94]

17.5 BIOCOMPATIBLE AND BIODEGRADABLE POLYMERS SCAFFOLD FOR TISSUE ENGINEERING APPLICATION

Several important requirements must be satisfied in developing polymer scaffolds for tissue engineering with biocompatibility, biodegradability, designed physical and adhesive strengthen, elasticity, sterilization, stability, and controlled release of necessary chemicals. Biocompatibility and biodegradability require materials for scaffolds to avoid evidences of extreme immunogenicity or cytotoxicity, or elicit or unresolved inflammatory responses during their interfaces and degradations. The degradable scaffolds should be designed to degrade at pre-determined time-period, and the initial space occupied by the scaffolds should be replaced by grown cell or tissue. Scaffolds must mimic the functions of ECM and molecules naturally found in tissues, where scaffolds should be recognized and interact with the endogenous substances from the surrounding cells providing biochemical signals in aiding cellular development and morphogenesis.

Next, scaffolds should have mechanical properties matching those of the tissue at the implantation sites or those of the tissue that are needed to be regenerated. Particularly, in the reconstruction of hard, load-bearing tissues, such as cartilages and bone, the mechanical properties in retaining the scaffold's structure after implantation are must.

In terms of the adhesive interfaces, they should provide and allow other biomolecules or cells to attach on the surface of scaffold through flow or exchange of growth factors and/or other growth enhancing agents. In addition, the scaffold should support cell adhesion, proliferation, cell-cell, and growth factor interactions, and cell migration.

Viscosity and elasticity are very important properties, especially for skin and vascular tissue engineering. Vascular graft must withstand the tensile or compressive stress with elastic property for blood to flow freely in pressure.

Electrical properties may be necessary for nerve regeneration and connecting the electrical impulses between cells and tissues.

All scaffold materials must be easily sterilized to prevent infection before any procedures. Sterilization methods should not interfere with bioactivity of scaffold materials and the loaded biomolecules on the scaffold or modifying their chemical compositions, which consecutively could affect their biocompatibility or degradation properties.

For the complex design and tailoring of the scaffolds' functions, biomolecules or drugs may be pre-loaded onto the scaffold, and the kinetic of controlled release would be necessary to maintain or regenerate the surrounding tissues over a given period of time.

17.5.1   Vascular Tissue Engineering

Cardiovascular disease is the leading cause of mortality in the world wide, exceeding 300 billion dollars for the postoperative cares [76]. Vascular bypass or coronary artery bypass graft procedures are performed in approximately 600,000 patients annually in USA [147]. Vessels for the graft are typically harvested either from the patient or other mammary arteries for use in reconstructive surgeries for angina and other vascular diseases. However, supplementing with native vessels may not be adequate for multiple bypass or repeated procedures [26]. To overcome above limitations, vascular grafts were fabricated either from natural or synthetic biomaterials. Biodegradable polymer scaffolds with seeded vascular cells revealed promising results of arterial regeneration *in situ* experiments [171]. Various methodologies have emerged for fabricating the blood vessel substitutes with biological functionality for flexibility and biocompatibility. Most common systems included the cell-seeded gels and cell-seeded biodegradable vascular grafts [158]. In 1986, first attempt was made to construct blood vessels by tissue engineering strategy with collagen gel, which was cultured with bovine endothelial cells (ECs), smooth muscle cells (SMCs) and fibroblasts [187]. In the same year, van Buul-Wortelboer et al. [182] developed a vascular wall with SMCs and ECs grown on a collagen network. However, these approaches were not suitable for surgical implantation due to the mechanical weakness of the collagen gel. Then, improved cell scaffold technology was developed for the generation of blood vessels. In this approach, the respective cells were seeded with a suitable biodegradable scaffold. The seeded cells proliferated well; in addition, they also released growth factors and other necessary matrix proteins, resulting in the formation of complete respective tissue [84, 188]. This type of scaffolds offered a temporary template rendering a defined architecture to guide cellular adhesion, tissue growth, and devolvement [80]. Several synthetic (PGA, PLA, and their co-polymers of PLGA) and natural (collagen, fibrin, silk fibroin, and chitosan) biomaterials are being investigated as potential scaffolds for vascular tissue engineering applications. Among the several synthetic biodegradable polymers, PGA based scaffold is the most commonly examined due to its simple handling and manipulations into different size and shapes of scaffolds with highly porous structures for helping to provide nutrient diffusions and differentiations upon subsequent neovascularization [80]. Niklason et al. [126] also developed small-diameter vascular biomimetic perfusion system using PGA, where bovine aortic SMCs were seeded PGA scaffolds. Their results showed 100% patency after 4 weeks of implantation into porcine saphenous arteries, but endothelialization occurred only in pockets of the graft surface. This failure of endothelialization could be due to the lack of cell signaling support or cytotoxic degradation products of PGA. Recently, the naturally available matrix with fibrin with many unique properties of cell adhesion and interactions, angiogenesis, and tissue repair may prove to be a good biological polymer for making scaffolds for vascular tissue engineering applications [6, 12]. In addition, fibrin contains natural binding sites for endothelial and fibroblast cells. Although, the development of engineered autologous tissue or vascular grafts with few successes still are in need for potential improvements in order to overcome currently used synthetic grafts. Many challenges for the well-organized endothelialization on the surface of the luminal graft and its long-term patency are remaining to be solved. Unfortunately, none of above scaffolds in vascular tissue engineering met the characteristics of an ideal scaffold.

17.5.2 Bone Tissue Engineering

Bone tissue engineering is a rapidly expanding area of research due to better results of biocompatibility and biodegradability of polymers. Especially, the problems of autografts and transplantation were improved. But the ideal biomaterials should have properties such as, osteoconductivity, osteogenicity, and osteoinductivity. Since collagen is the chief component of the ECM providing support to connective tissues such as, cartilage and bones, it was considered as an ideal composite material with other natural and synthetic scaffolds for bone tissue engineering applications. Collagen sponge with immobilized bFGF showed the cartilage development from subcutaneous implantation experiments with nude mice [50]. Other collagen based scaffolds were developed, commercialized, and marketed by Medtronic Sofamor Danek in USA under the name of FUSE® Bone Graft, where the bone graft from collagen sponges used an osteoconductive carrier of BMP-2 for spinal cord fusion. Other examples of commercialized collagen and other scaffolds from natural polymers for various tissue engineering applications are being reported in the literature [22]. For additional example, Collagraft® is another commercialized collagen product by Angiotech Pharmaceuticals Inc. in Canada. Gelatin microspheres with incorporated bFGF demonstrated the highest rate of cell attachment and growth, which could be a promising material for cartilage tissue engineering [107]. Interestingly, Patel et al. [131] investigated the release of dual growth factors, BMP-2 and VEGF through delivery from gelatin microspheres *in vivo*, and the results suggested that delivery of dual growth factors could regenerate the bone defect into critical size. Several commercially available gelatin based scaffolds for drug and protein delivery utilized for their specific biomedical applications [108, 144]. The most commonly used gelatin product is Gelfoam® as a sterile and workable surgical sponge from Pfizer Inc. in USA.

Recently, PLGA-based scaffold became applicable in bone regeneration or repair, which showed biocompatibility without any toxic effect and inflammation, and with good biodegradability and sustained release of BMP-2 [94]. Other studies also demonstrated that utilization of PLGA and polypropylene fumarate (PPF) microspheres in scaffold composite could be effective for sustained release of growth factor for bone tissue engineering [47, 71].

17.5.3 Skin Tissue Engineering

Successful wound healing is a complex process, requiring interactions between dermal cells, epidermal, angiogenesis, and ECM; all of which are regulated by cytokines and other growth factors to govern the cellular proliferation and skin restoration locally [68]. Many growth factors are involved in skin regeneration, such as epidermal growth factor (EGF), fibroblast growth factor (FGF), keratinocyte growth factor (KGF), PDGF, and TGF-β, where they encourage surrounding areas for the enhanced wound healings [111]. EGF is a powerful growth factor for the growth of epidermal cells, which became a therapeutic agent for epidermal cell migration, proliferation, and skin stimulant for re-epithelialization [110]. Cross-linked gelatin scaffolds with tissue stimulating factor, such as fibronectin, vitronectin and RGD peptides were further developed to be used as artificial skin [78]. Recently, a wide variety of biomedical products with collagen based matrix from animal-sources were formulated and commercialized. For example, double-layered collagen gels with seeded human keratinocytes in the upper part and human fibroblasts in the lower layer were used as artificial commercial skin products in USA under the brand name of Apligraf® by Organogenesis Inc. They has also developed other collagen based scaffolds; Revitix™ for topical cosmetic, Forta-Derm™ Antimicrobial as antimicrobial wound dressing, and VCTO1™ as bilayered bio-engineered skin [24]. Other detailed information regarding the growth factor, cell types, and their delivery design for the skin tissue engineering can be found in the literature [116]. Recently, synthetic polymer scaffolds were studied for re-epithlialization. Electrospun PLGA nanofibersscaffolds were proposed as good skin substitutes through *in vitro* experiments [90]. Bandages with the desired elasticity and physiochemical properties were developed with different concentrations of synthetic polymers, such as microfiber, films, electrospun scaffold (nanofiber), and sponges, which showed significant improvements in designed specific tissue regenerations [98, 185].

17.6　CONCLUSIONS AND FUTURE PERSPECTIVES

The biocompatible and biodegradable natural and synthetic biomaterials received considerable interests for drug delivery, tissue engineering, and other biomedical applications for the last two decades. Both natural and synthetic polymers were effective in delivering biomolecules, genes, and cells. However, the success rate depended on the utilization of both advantages and disadvantages of each material. For example, natural polymers were more cell friendly with less mechanical properties. On the other hand, the synthetic polymers revealed strengthened mechanical properties without biological cues for cells bindings than natural polymers. Recently, it became clear that mechanical strength of the matrix could influence cell multiplication and differentiation, and would be necessary to take it into an account during the scaffold fabrications. The scaffold constructed in micron size showed better mechanical strength in comparison to submicron carriers. But still, the micro-scaffolds lacked a nanotextured surface, while the nanotextured scaffolds further lacked a 3-D microenvironment. It was again understood that both the nanotextured surface with 3-D microenvironment properties should be considered to be combined together. Here, we discussed the biocompatible and biodegradable (natural and synthetic) polymers for diverse biomedical applications from cancer therapy to bone constructions. Brief explanations of scaffold designs for drug, biomolecules, genes, and cell delivery for the regeneration of various tissues such as, bone, vascular, and skin were discussed. In summary, ideal biomaterial scaffolds should provide structural, mechanical, chemical, and biochemical signals, which may intend for the new tissue formations. It will be good to make collaborative efforts from material scientists, biologists, chemists and clinical surgeons for planning and constructing various functional scaffolds in clinical studies with economic feasibility.

KEYWORDS

- **Biocompatible polymer**
- **Biodegradable synthetic polymers**
- **Extracellular matrix**
- **Homopolymer**
- **Vascular graft**

REFERENCES

1. Adhirajan, N.; Shanmugasundaram, N.; Babu, M. J. Microencapsul. 2007, 24, 647–659.
2. Aebischer, P.; Salessiotis, A. N.; Winn, S. R. J. Neurosci. Res. 1989, 23, 282–289.
3. Agarwal, M.; Koelling, K. W.; Chalmers, J. J. Biotechnol. Progr. 1998, 14, 517–526.
4. Aigner, J.; Tegeler, J.; Hutzler, P.; Campoccia, D.; Pavesio, A.; Hammer, C.; Kastenbauer, E.; Naumann, A. Biomed. Mater. Res. 1998, 42, 172–181.
5. Alemdaroglu, C.; Degim, Z.; Celebi, N.; Zor, F.; Ozturk, S.; Erdogan, D. Burns 2006, 32, 319–327.
6. Amrani, D. L.; Diorio, J. P.; Delmotte, Y. Ann. N. Y. Acad. Sci. 2001, 936, 566–579.
7. Angele, P.; Johnstone, B.; Kujat, R.; Zellner, J.; Nerlich, M.; Goldberg, V.; Yoo, J. J. Biomed. Mater. Res A. 2008, 85, 445–455.
8. Aper, T.; Schmidt, A.; Duchrow, M.; Bruch, H. P. Eur. J. Vasc. Endovasc. Surg. 2007, 33, 33–39.
9. Awad, H. A.; Wickham, M. Q.; Leddy, H. A.; Gimble, J. M.; Guilak, F. Biomaterials. 2004, 25, 3211–3222.
10. Barquinero, J.; Eixarch, H.; Perez–Melgosa, M. Gene. Ther. 2004, 11 Suppl 1, S3–9.

11. Batorsky, A.; Liao, J.; Lund, A. W.; Plopper, G. E.; Stegemann, J. P. Biotechnol. Bioeng. 2005, 92, 492–500.
12. Bensaid, W.; Triffitt, J. T.; Blanchat, C.; Oudina, K.; Sedel, L.; Petite, H. Biomaterials. 2003, 24, 2497–2502.
13. Bertz, A.; Wohl–Bruhn, S.; Miethe, S.; Tiersch, B.; Koetz, J.; Hust, M.; Bunjes, H.; Menzel, H. J. Biotechnol. 2012.
14. Blomback, B.; Hessel, B.; Hogg, D.; Therkildsen, L. Nature. 1978, 275, 501–505.
15. Briassoulis, D. J. Polym. Environ. 2004, 12, 65–81.
16. Caddick, J.; Kingham, P. J.; Gardiner, N. J.; Wiberg, M.; Terenghi, G. Glia. 2006, 54, 840–849.
17. Capito, R. M.; Spector, M. Gene. Ther. 2007, 14, 721–732.
18. Cascone, M. G.; Lazzeri, L.; Carmignani, C.; Zhu, Z. J. Mater. Sci. Mater. Med. 2002, 13, 523–526.
19. Chandler, L. A.; Gu, D. L.; Ma, C.; Gonzalez, A. M.; Doukas, J.; Nguyen, T.; Pierce, G. F.; Phillips, M. L. Wound. Repair. Regen. 2000, 8, 473–479.
20. Chen, H. C.; Chang, Y. H.; Chuang, C. K.; Lin, C. Y.; Sung, L. Y.; Wang, Y. H.; Hu, Y. C. Biomaterials. 2009, 30, 674–681.
21. Chen, Z. G.; Wang, P. W.; Wei, B.; Mo, X. M.; Cui, F. Z. Acta. Biomater. 2010, 6, 372–382.
22. Chevallay, B.; Herbage, D. Med. Biol. Eng. Comput. 2000, 38, 211–218.
23. Chung, Y. I.; Kim, S. K.; Lee, Y. K.; Park, S. J.; Cho, K. O.; Yuk, S. H.; Tae, G.; Kim, Y. H. J. Control. Release. 2010, 143, 282–289.
24. Chunlin, Y.; Hillas, P.J.; Buez, J.A.; Nokelainen, M.; Balan, J.;Tang, J.;Spiro, R.; Polarek, J.W.The application of recombinant human collagen intissue engineering. Bio. Drugs. 2004, 18, 103–119.
25. Clark, R. A.; Lanigan, J. M.; DellaPelle, P.; Manseau, E.; Dvorak, H. F.; Colvin, R. B. J. Invest. Dermatol. 1982, 79, 264–269.
26. Cooper, G. J.; Underwood, M. J.; Deverall, P. B. Eur. J. Cardiothorac. Surg. 1996, 10, 129–140.
27. Corsi, K.; Chellat, F.; Yahia, L.; Fernandes, J. C. Biomaterials. 2003, 24, 1255–1264.
28. Coviello, T.; Matricardi, P.; Marianecci, C.; Alhaique, F. J. Control. Release. 2007, 119, 5–24.
29. Custodio, C. A.; Alves, C. M.; Reis, R. L.; Mano, J. F. J. Tissue. Eng. Regen. Med. 2010, 4, 316–323.
30. Dadsetan, M.; Szatkowski, J. P.; Yaszemski, M. J.; Lu, L. Biomacromolecules. 2007, 8, 1702–1709.
31. Dang, J. M.; Leong, K. W. Adv. Drug. Deliv. Rev. 2006, 58, 487–499.
32. De Franceschi, L.; Grigolo, B.; Roseti, L.; Facchini, A.; Fini, M.; Giavaresi, G.; Tschon, M.; Giardino, R. J. Biomed. Mater. Res A. 2005, 75, 612–622.
33. De Laporte, L.; Yan, A. L.; Shea, L. D. Biomaterials. 2009, 30, 2361–2368.
34. Di Silvio, L.; Bonfield, W. J. Mater. Sci. Mater. Med. 1999, 10, 653–658.
35. Digenis, G. A.; Gold, T. B.; Shah, V. P. J. Pharm. Sci. 1994, 83, 915–921.
36. Djagny, V. B.; Wang, Z.; Xu, S. Crit. Rev. Food Sci. Nutr. 2001, 41, 481–492.
37. Draye, J. P.; Delaey, B.; Van de Voorde, A.; Van Den Bulcke, A.; De Reu, B.; Schacht, E. Biomaterials. 1998, 19, 1677–1687.
38. Duan, B.; Dong, C.; Yuan, X.; Yao, K. J. Biomater. Sci. Polym. Ed. 2004, 15, 797–811.

39. Duarte, A. R. C.; Mano, J. F.; Reis, R. L. Eur. Polym. J. 2009, 45, 141–148.
40. Dubey, G.; Mequanint, K. Acta. Biomater. 2011, 7, 1114–1125.
41. Ehrbar, M.; Metters, A.; Zammaretti, P.; Hubbell, J. A.; Zisch, A. H. J. Control. Release. 2005, 101, 93–109.
42. Einhaus, C. M.; Retzinger, A. C.; Perrotta, A. O.; Dentler, M. D.; Jakate, A. S.; Desai, P. B.; Retzinger, G. S. Clin. Cancer. Res. 2004, 10, 7001–7010.
43. Engler, A. J.; Sen, S.; Sweeney, H. L.; Discher, D. E. Cell. 2006, 126, 677–689.
44. Erbacher, P.; Zou, S.; Bettinger, T.; Steffan, A. M.; Remy, J. S. Pharm. Res. 1998, 15, 1332–1339.
45. Espigares, I.; Elvira, C.; Mano, J. F.; Vazquez, B.; San, R. J.; Reis, R. L. Biomaterials. 2002, 23, 1883–1895.
46. Fambri, L.; Pegoretti, A.; Fenner, R.; Incardona, S. D.; Migliaresi, C. Polymer. 1997, 38, 79–85.
47. Felix Lanao, R. P.; Leeuwenburgh, S. C.; Wolke, J. G.; Jansen, J. A. Biomaterials. 2011, 32, 8839–8847.
48. Fini, M.; Motta, A.; Torricelli, P.; Giavaresi, G.; Nicoli Aldini, N.; Tschon, M.; Giardino, R.; Migliaresi, C. Biomaterials. 2005, 26, 3527–3536.
49. Fuchs, S.; Motta, A.; Migliaresi, C.; Kirkpatrick, C. J. Biomaterials. 2006, 27, 5399–5408.
50. Fujisato, T.; Sajiki, T.; Liu, Q.; Ikada, Y. Biomaterials. 1996, 17, 155–162.
51. Fujita, M.; Ishihara, M.; Morimoto, Y.; Simizu, M.; Saito, Y.; Yura, H.; Matsui, T.; Takase, B.; Hattori, H.; Kanatani, Y.; Kikuchi, M.; Maehara, T. J. Surg. Res. 2005, 126, 27–33.
52. Gailit, J.; Clarke, C.; Newman, D.; Tonnesen, M. G.; Mosesson, M. W.; Clark, R. A. Exp. Cell Res. 1997, 232, 118–126.

53. Galeska, I.; Kim, T. K.; Patil, S. D.; Bhardwaj, U.; Chatttopadhyay, D.; Papadimitrakopoulos, F.; Burgess, D. J. AAPS J. 2005, 7, E231–240.
54. George, M.; Abraham, T. E. J. Control. Release. 2006, 114, 1–14.
55. Gerecht–Nir, S.; Ziskind, A.; Cohen, S.; Itskovitz–Eldor, J. Lab. Invest. 2003, 83, 1811–1820.
56. Ghosh, S.; Viana, J. C.; Reis, R. L.; Mano, J. F. Materials. Science. and Engineering. C. 2008, 28, 80–86.
57. Goldberg, D. J. Polym. Environ. 1995, 3, 61–67.
58. Goldman, C. K.; Soroceanu, L.; Smith, N.; Gillespie, G. Y.; Shaw, W.; Burgess, S.; Bilbao, G.; Curiel, D. T. Nat. Biotechnol. 1997, 15, 462–466.
59. Gomes, M. E.; Ribeiro, A. S.; Malafaya, P. B.; Reis, R. L.; Cunha, A. M. Biomaterials. 2001, 22, 883–889.
60. Gong, C.; Shi, S.; Dong, P.; Kan, B.; Gou, M.; Wang, X.; Li, X.; Luo, F.; Zhao, X.; Wei, Y.; Qian, Z. Int. J. Pharm. 2009, 365, 89–99.
61. Goraltchouk, A.; Scanga, V.; Morshead, C. M.; Shoichet, M. S. J. Control. Release. 2006, 110, 400–407.
62. Grijpma, D. W.; Nijenhuis, A. J.; Wijk, P. G. T.; Pennings, A. J. Polym. Bull. 1992, 29, 571–578.
63. Grunder, T.; Gaissmaier, C.; Fritz, J.; Stoop, R.; Hortschansky, P.; Mollenhauer, J.; Aicher, W. K. Osteoarthritis. Cartilage. 2004, 12, 559–567.

64. Gunatillake, P.; Mayadunne, R.; Adhikari, R. Biotechnol. Annu. Rev. 2006, 12, 301–347.
65. Gunatillake, P. A.; Adhikari, R. Eur. Cell. Mater. 2003, 5, 1–16.
66. Guo, T.; Zhao, J.; Chang, J.; Ding, Z.; Hong, H.; Chen, J.; Zhang, J. Biomaterials. 2006, 27, 1095–1103.
67. Gurevich, O.; Vexler, A.; Marx, G.; Prigozhina, T.; Levdansky, L.; Slavin, S.; Shimeliovich, I.; Gorodetsky, R. Tissue. Eng. 2002, 8, 661–672.
68. Harding, K. G.; Morris, H. L.; Patel, G. K. BMJ. 2002, 324, 160–163.
69. Hartmann, A.; Quist, J.; Hamm, H.; Brocker, E. B.; Friedl, P. Dermatol. Surg. 2008, 34, 922–929.
70. Hassan, C. M.; Stewart, J. E.; Peppas, N. A. Eur. J. Pharm. Biopharm. 2000, 49, 161–165.
71. Hedberg, E. L.; Tang, A.; Crowther, R. S.; Carney, D. H.; Mikos, A. G. J. Control. Release. 2002, 84, 137–150.
72. Hemmrich, K.; von Heimburg, D.; Rendchen, R.; Di Bartolo, C.; Milella, E.; Pallua, N. Biomaterials. 2005, 26, 7025–7037.
73. Hsiong, S. X.; Huebsch, N.; Fischbach, C.; Kong, H. J.; Mooney, D. J. Biomacromolecules. 2008, 9, 1843–1851.
74. Ikada, Y.; Tabata, Y. Adv. Drug. Deliv. Rev. 1998, 31, 287–301.
75. Illum, L. Pharm. Res. 1998, 15, 1326–1331.
76. Isenberg, B. C.; Williams, C.; Tranquillo, R. T. Circ. Res. 2006, 98, 25–35.
77. Ishii, T.; Okahata, Y.; Sato, T. Biochim. Biophys. Acta. 2001, 1514, 51–64.
78. Ito, A.; Mase, A.; Takizawa, Y.; Shinkai, M.; Honda, H.; Hata, K.; Ueda, M.; Kobayashi, T. J. Biosci. Bioeng. 2003, 95, 196–199.
79. Ito, T.; Yeo, Y.; Highley, C. B.; Bellas, E.; Kohane, D. S. Biomaterials. 2007, 28, 3418–3426.
80. Kakisis, J. D.; Liapis, C. D.; Breuer, C.; Sumpio, B. E. J. Vasc. Surg. 2005, 41, 349–354.
81. Karageorgiou, V.; Meinel, L.; Hofmann, S.; Malhotra, A.; Volloch, V.; Kaplan, D. J. Biomed. Mater. Res. A. 2004, 71, 528–537.
82. Karal–Yilmaz, O.; Serhatli, M.; Baysal, K.; Baysal, B. M. J. Microencapsul. 2011, 28, 46–54.
83. Kasper, F. K.; Mikos, A. G.Molecular and Cellular Foundations of Biomaterials; Elsevier: San Diego, 2004, pp 131–168.
84. Khademhosseini, A.; Langer, R.; Borenstein, J.; Vacanti, J. P. Proc. Natl. Acad. Sci. U. S. A. 2006, 103, 2480–2487.
85. Kimura, Y.; Ozeki, M.; Inamoto, T.; Tabata, Y. Biomaterials. 2003, 24, 2513–2521.
86. King, T. W.; Patrick, C. W., Jr. J. Biomed. Mater. Res. 2000, 51, 383–390.
87. Kobayashi, M.; Toguchida, J.; Oka, M. J Hand. Surg. Br. 2001, 26, 436–440.
88. Koch, S.; Yao, C.; Grieb, G.; Prevel, P.; Noah, E. M.; Steffens, G. C. J. Mater. Sci. Mater. Med. 2006, 17, 735–741.
89. Kuijpers, A. J.; Engbers, G. H.; van Wachem, P. B.; Krijgsveld, J.; Zaat, S. A.; Dankert, J.; Feijen, J. J. Control. Release. 1998, 53, 235–247.
90. Kumbar, S. G.; Nukavarapu, S. P.; James, R.; Nair, L. S.; Laurencin, C. T. Biomaterials. 2008, 29, 4100–4107.

91. Laurent, A.; Wassef, M.; Saint Maurice, J. P.; Namur, J.; Pelage, J. P.; Seron, A.; Chapot, R.; Merland, J. J. Invest. Radiol. 2006, 41, 8–14.
92. Lee, C. T.; Kung, P. H.; Lee, Y. D. Carbohydr. Polym. 2005,61, 348–354.
93. Lee, H. J.; Bae, Y. Biomacromolecules. 2011, 12, 2686–2696.
94. Lee, J. W.; Kang, K. S.; Lee, S. H.; Kim, J. Y.; Lee, B. K.; Cho, D. W. Biomaterials. 2011, 32, 744–752.
95. Lee, K. Y.; Peters, M. C.; Mooney, D. J. J. Control. Release. 2003a, 87, 49–56.
96. Letic–Gavrilovic, A.; Piattelli, A.; Abe, K. J. Mater. Sci. Mater. Med. 2003, 14, 95–102.
97. Li, C.; Vepari, C.; Jin, H. J.; Kim, H. J.; Kaplan, D. L. Biomaterials. 2006, 27, 3115–3124.
98. Li, W. J.; Laurencin, C. T.; Caterson, E. J.; Tuan, R. S.; Ko, F. K. J. Biomed. Mater. Res. 2002, 60, 613–621.
99. Lim, Y.B.; Kim, C.H.; Kim, K.; Kim, S.W.; Park, J.S. Developement of asafe gene delivery system using biodegradable polymer, poly [alpha–(4–aminobutyl)–l–glycolic acid]. J. Am. Chem. Soc.2000, 122, 6524–6525.
100. Lin, C. C.; Anseth, K. S. Pharm. Res. 2009, 26, 631–643.
101. Lindsey, R. W.; Probe, R.; Miclau, T.; Alexander, J. W.; Perren, S. M. Clin. Orthop. Relat. Res. 1993, 303–312.
102. Linh, N. T.; Min, Y. K.; Song, H. Y.; Lee, B. T. J. Biomed. Mater. Res. B. Appl. Biomater. 2010, 95, 184–191.
103. Liu, Y.; Shu, X. Z.; Prestwich, G. D. Tissue Eng. 2006, 12, 3405–3416.
104. Liu, Y.; Yang, H.; Otaka, K.; Takatsuki, H.; Sakanishi, A. Colloids Surf B Biointerfaces 2005, 43, 216–220.
105. Lu da, Y.; Chen, X. L.; Ding, J. Med. Hypotheses. 2007, 68, 188–193.
106. Malafaya, P. B.; Silva, G. A.; Reis, R. L. Adv. Drug. Deliv. Rev. 2007, 59, 207–233.
107. Malafaya, P. B.; Stappers, F.; Reis, R. L. J. Mater. Sci. Mater. Med. 2006, 17, 371–377.
108. Malda, J.; Kreijveld, E.; Temenoff, J. S.; van Blitterswijk, C. A.; Riesle, J. Biomaterials. 2003, 24, 5153–5161.
109. Mano, J. F.; Koniarova, D.; Reis, R. L. J. Mater. Sci. Mater. Med. 2003, 14, 127–135.
110. Marcantonio, N. A.; Boehm, C. A.; Rozic, R. J.; Au, A.; Wells, A.; Muschler, G. F.; Griffith, L. G. Biomaterials. 2009, 30, 4629–4638.
111. Marimuthu, M.; Kim, S. Current. Nanoscience 5, 189–203.
112. Martins, A. M.; Santos, M. I.; Azevedo, H. S.; Malafaya, P. B.; Reis, R. L. Acta. Biomater. 2008, 4, 1637–1645.
113. Maurus, P. B.; Kaeding, C. C. Operative Techniques in Sports Medicine. 2004, 12, 158–160.
114. Mehlhorn, A. T.; Niemeyer, P.; Kaschte, K.; Muller, L.; Finkenzeller, G.; Hartl, D.; Sudkamp, N. P.; Schmal, H. Cell. Prolif. 2007, 40, 809–823.
115. Meinel, L.; Karageorgiou, V.; Hofmann, S.; Fajardo, R.; Snyder, B.; Li, C.; Zichner, L.; Langer, R.; Vunjak–Novakovic, G.; Kaplan, D. L. J. Biomed. Mater. Res. A. 2004, 71, 25–34.
116. Metcalfe, A. D.; Ferguson, M. W. Biomaterials. 2007, 28, 5100–5113.
117. Middleton, J. C.; Tipton, A. J. Biomaterials. 2000, 21, 2335–2346.

118. Min, B. M.; Lee, G.; Kim, S. H.; Nam, Y. S.; Lee, T. S.; Park, W. H. Biomaterials. 2004, 25, 1289–1297.

119. Moioli, E. K.; Clark, P. A.; Xin, X.; Lal, S.; Mao, J. J. Adv. Drug. Deliv. Rev. 2007, 59, 308–324.

120. Muvaffak, A.; Gurhan, I.; Hasirci, N. J. Biomed. Mater. Res. B. Appl. Biomater. 2004, 71, 295–304.

121. Nacer Khodja, A.; Mahlous, M.; Tahtat, D.; Benamer, S.; Larbi Youcef, S.; Chader, H.; Mouhoub, L.; Sedgelmaci, M.; Ammi, N.; Mansouri, M. B.; Mameri, S. Burns. 2012.

122. Nair, L. S.; Laurencin, C. T. Prog. Polym. Sci. 2007, 32, 762–798.

123. Nelson, C. L.; McLaren, S. G.; Skinner, R. A.; Smeltzer, M. S.; Thomas, J. R.; Olsen, K. M. J. Orthop. Res. 2002, 20, 643–647.

124. Neubauer, M.; Hacker, M.; Bauer–Kreisel, P.; Weiser, B.; Fischbach, C.; Schulz, M. B.; Goepferich, A.; Blunk, T. Tissue. Eng. 2005, 11, 1840–1851.

125. Ni, P.; Fu, S.; Fan, M.; Guo, G.; Shi, S.; Peng, J.; Luo, F.; Qian, Z. Int. J. Nanomedicine. 2011, 6, 3065–3075.

126. Niklason, L. E.; Abbott, W.; Gao, J.; Klagges, B.; Hirschi, K. K.; Ulubayram, K.; Conroy, N. Jones, R.; Vasanawala, A.; Sanzgiri, S.; Langer, R. L. Morphologic and mechanical characteristics of bovine engineered arteries. J. Vasc. Surg. 2001,33, 628–638.

127. Okamoto, T.; Yamamoto, Y.; Gotoh, M.; Huang, C. L.; Nakamura, T.; Shimizu, Y.; Tabata, Y.; Yokomise, H. J. Thorac. Cardiovasc. Surg. 2004, 127, 329–334.

128. Osathanon, T.; Linnes, M. L.; Rajachar, R. M.; Ratner, B. D.; Somerman, M. J.; Giachelli, C. M. Biomaterials. 2008, 29, 4091–4099.

129. Pal, K.; Banthia, A. K.; Majumdar, D. K. AAPS. Pharm. Sci. Tech. 2007, 8, 21.

130. Park, H.; Temenoff, J. S.; Holland, T. A.; Tabata, Y.; Mikos, A. G. Biomaterials. 2005, 26, 7095–7103.

131. Patel, Z. S.; Young, S.; Tabata, Y.; Jansen, J. A.; Wong, M. E.; Mikos, A. G. Bone. 2008, 43, 931–940.

132. Payne, R. G.; Yaszemski, M. J.; Yasko, A. W.; Mikos, A. G. Biomaterials. 2002, 23, 4359–4371.

133. Pedroso, A. G.; Rosa, D. S. Carbohydr. Polym. 2005, 59, 1–9.

134. Peppas, N. A.; Merrill, E. W. J. Biomed. Mater. Res. 1977, 11, 423–434.

135. Perego, G.; Cella, G. D.; Bastioli, C. J. Appl. Polym. Sci. 1996, 59, 37–43.

136. Perka, C.; Schultz, O.; Lindenhayn, K.; Spitzer, R. S.; Muschik, M.; Sittinger, M.; Burmester, G. R. Clin. Exp. Rheumatol. 2000, 18, 13–22.

137. Petersen, H.; Fechner, P. M.; Martin, A. L.; Kunath, K.; Stolnik, S.; Roberts, C. J.; Fischer, D.; Davies, M. C.; Kissel, T. Bioconjug. Chem. 2002, 13, 845–854.

138. Petratos, P. B.; Felsen, D.; Trierweiler, G.; Pratt, B.; McPherson, J. M.; Poppas, D. P. Wound Repair Regen. 2002, 10, 252–258.

139. Philip, S.; Keshavarz, T.; Roy, I. J. Chem. Technol. Biotechnol. 2007, 82, 233–247.

140. Pierce, G. F.; Tarpley, J. E.; Yanagihara, D.; Mustoe, T. A.; Fox, G. M.; Thomason, A. Am. J. Pathol. 1992, 140, 1375–1388.

141. Place, E. S.; George, J. H.; Williams, C. K.; Stevens, M. M. Chem. Soc. Rev. 2009, 38, 1139–1151.

142. Poirier, Y. Prog. Lipid. Res. 2002, 41, 131–155.
143. Ponce, S.; Orive, G.; Hernandez, R.; Gascon, A. R.; Pedraz, J. L.; de Haan, B. J.; Faas, M. M.; Mathieu, H. J.; de Vos, P. Biomaterials. 2006, 27, 4831–4839.
144. Ponticiello, M. S.; Schinagl, R. M.; Kadiyala, S.; Barry, F. P. J. Biomed. Mater. Res. 2000, 52, 246–255.
145. Prabaharan, M.; Rodriguez–Perez, M. A.; de Saja, J. A.; Mano, J. F. J. Biomed. Mater. Res. B. Appl. Biomater. 2007, 81, 427–434.
146. Pun, S. H.; Bellocq, N. C.; Liu, A.; Jensen, G.; Machemer, T.; Quijano, E.; Schluep, T.; Wen, S.; Engler, H.; Heidel, J.; Davis, M. E. Bioconjug. Chem. 2004, 15, 831–840.
147. Ravi, S.; Chaikof, E. L. Regen. Med. 2010, 5, 107–120.
148. Reis, R. L.; Cunha, A. M. J. Mater. Sci. Mater. Med. 1995, 6, 786–792.
149. Rejinold, N. S.; Muthunarayanan, M.; Chennazhi, K. P.; Nair, S. V.; Jayakumar, R. J. Biomed. Nanotechnol. 2011, 7, 521–534.
150. Ribeiro, C. C.; Barrias, C. C.; Barbosa, M. A. Biomaterials. 2004, 25, 4363–4373.
151. Rives, C. B.; des Rieux, A.; Zelivyanskaya, M.; Stock, S. R.; Lowe, W. L., Jr.; Shea, L. D. Biomaterials. 2009, 30, 394–401.
152. Rocha, F. G.; Sundback, C. A.; Krebs, N. J.; Leach, J. K.; Mooney, D. J.; Ashley, S. W.; Vacanti, J. P.; Whang, E. E. Biomaterials. 2008, 29, 2884–2890.
153. Rutten, H. J.; Nijhuis, P. H. Eur. J. Surg. Suppl. 1997, 31–35.
154. Salgado, A. J.; Coutinho, O. P.; Reis, R. L.; Davies, J. E. J. Biomed. Mater. Res. A. 2007, 80, 983–989.
155. Santos, M. I.; Tuzlakoglu, K.; Fuchs, S.; Gomes, M. E.; Peters, K.; Unger, R. E.; Piskin, E.; Reis, R. L.; Kirkpatrick, C. J. Biomaterials. 2008, 29, 4306–4313.
156. Sato, T.; Ishii, T.; Okahata, Y. Biomaterials. 2001, 22, 2075–2080.
157. Schmoekel, H. G.; Weber, F. E.; Schense, J. C.; Gratz, K. W.; Schawalder, P.; Hubbell, J. A. Biotechnol. Bioeng. 2005, 89, 253–262.
158. Shaikh, F. M.; Callanan, A.; Kavanagh, E. G.; Burke, P. E.; Grace, P. A.; McGloughlin, T. M. Cells. Tissues. Organs. 2008, 188, 333–346.
159. Shanmuganathan, S.; Shanumugasundaram, N.; Adhirajan, N.; Ramyaa Lakshmi, T. S.; Babu, M. Carbohydr. Polym. 2008, 73, 201–211.
160. Shen, Y. H.; Shoichet, M. S.; Radisic, M. Acta. Biomater. 2008, 4, 477–489.
161. Shi, J.; Alves, N. M.; Mano, J. F. J. Biomed. Mater. Res. B. Appl. Biomater. 2008, 84, 595–603.
162. Shi, L.; Tang, G. P.; Gao, S. J.; Ma, Y. X.; Liu, B. H.; Li, Y.; Zeng, J. M.; Ng, Y. K.; Leong, K. W.; Wang, S. Gene. Ther. 2003, 10, 1179–1188.
163. Sinha, V. R.; Trehan, A. J. Control. Release. 2003, 90, 261–280.
164. Södergård, A.; Stolt, M. Prog. Polym. Sci. 2002, 27, 1123–1163.
165. Sofia, S.; McCarthy, M. B.; Gronowicz, G.; Kaplan, D. L. J. Biomed. Mater. Res. 2001, 54, 139–148.
166. Sogias, I. A.; Williams, A. C.; Khutoryanskiy, V. V. Int. J. Pharm. 2012, 436, 602–610.
167. Song, Y.; Kamphuis, M. M. J.; Zhang, Z.; Sterk, L. M. T.; Vermes, I.; Poot, A. A.; Feijen, J.; Grijpma, D. W. Acta. Biomater. 2010, 6, 1269–1277.
168. Storrie, H.; Mooney, D. J. Adv. Drug. Deliv. Rev. 2006, 58, 500–514.
169. Sudesh, K.; Abe, H.; Doi, Y. Prog. Polym. Sci. 2000, 25, 1503–1555.
170. Swain, S.; Behera, A.; Beg, S.; Patra, C. N.; Dinda, S. C.; Sruti, J.; Rao, M. E. Recent patents on drug delivery & formulation 2012.

171. Swartz, D. D.; Russell, J. A.; Andreadis, S. T. Am. J. Physiol. Heart. Circ. Physiol. 2005, 288, H1451–1460.
172. Takeuchi, H.; Kojima, H.; Yamamoto, H.; Kawashima, Y. J. Control. Release. 2000, 68, 195–205.
173. Tanaka, N.; Takino, S.; Utsumi, I. J. Pharm. Sci. 1963, 52, 664–667.
174. Taylor, S. J.; McDonald, J. W., 3rd; Sakiyama–Elbert, S. E. J. Control. Release. 2004, 98, 281–294.
175. Teng, Y. D.; Lavik, E. B.; Qu, X.; Park, K. I.; Ourednik, J.; Zurakowski, D.; Langer, R.; Snyder, E. Y. Proc. Natl. Acad. Sci. U. S. A. 2002, 99, 3024–3029.
176. Thanoo, B. C.; Sunny, M. C.; Jayakrishnan, A. J. Pharm. Pharmacol. 1993, 45, 16–20.
177. Tohill, M.; Terenghi, G. Biotechnol. Appl. Biochem. 2004, 40, 17–24.
178. Toncheva, V.; Van Den Bulcke, A.; Schacht, E.; Mergaert, J.; Swings, J. J. Polym. Environ. 1996, 4, 71–83.
179. Uludag, H.; De Vos, P.; Tresco, P. A. Adv. Drug. Deliv. Rev. 2000, 42, 29–64.
180. Underhill, G. H.; Chen, A. A.; Albrecht, D. R.; Bhatia, S. N. Biomaterials. 2007, 28, 256–270.
181. Unger, R. E.; Peters, K.; Wolf, M.; Motta, A.; Migliaresi, C.; Kirkpatrick, C. J. Biomaterials. 2004, 25, 5137–5146.
182. van Buul–Wortelboer, M. F.; Brinkman, H. J.; Dingemans, K. P.; de Groot, P. G.; van Aken, W. G.; van Mourik, J. A. Exp. Cell Res. 1986, 162, 151–158.
183. van der Walle, G. A.; de Koning, G. J.; Weusthuis, R. A.; Eggink, G. Adv. Biochem. Eng. Biotechnol. 2001, 71, 263–291.
184. van Susante, J. L. C.; Pieper, J.; Buma, P.; van Kuppevelt, T. H.; van Beuningen, H.; van Der Kraan, P. M.; Veerkamp, J. H.; van den Berg, W. B.; Veth, R. P. H. Biomaterials. 2001, 22, 2359–2369.
185. Vroman, I.; Tighzert, L. Materials. 2009, 2, 307–344.
186. Wang, Y.; Kim, U. J.; Blasioli, D. J.; Kim, H. J.; Kaplan, D. L. Biomaterials. 2005, 26, 7082–7094.
187. Weinberg, C. B.; Bell, E. Science. 1986, 231, 397–400.
188. Whitaker, M. J.; Quirk, R. A.; Howdle, S. M.; Shakesheff, K. M. J. Pharm. Pharmacol. 2001, 53, 1427–1437.
189. Willerth, S. M.; Arendas, K. J.; Gottlieb, D. I.; Sakiyama–Elbert, S. E. Biomaterials. 2006, 27, 5990–6003.
190. Williams, D. F.; Mort, E. J. Bioeng. 1977, 1, 231–238.
191. Xiao, Y.; Qian, H.; Young, W. G.; Bartold, P. M. Tissue Eng. 2003, 9, 1167–1177.
192. Xu, X. L.; Lou, J.; Tang, T.; Ng, K. W.; Zhang, J.; Yu, C.; Dai, K. J. Biomed. Mater. Res. B. Appl. Biomater. 2005, 75, 289–303.
193. Yang, C.; Hillas, P. J.; Baez, J. A.; Nokelainen, M.; Balan, J.; Tang, J.; Spiro, R.; Polarek, J. W. Bio.Drugs. 2004, 18, 103–119.
194. Yang, H. S.; Bhang, S. H.; Hwang, J. W.; Kim, D. I.; Kim, B. S. Tissue. Eng. Part. A. 2010, 16, 2113–2119.
195. Yi, H.; Wu, L. Q.; Bentley, W. E.; Ghodssi, R.; Rubloff, G. W.; Culver, J. N.; Payne, G. F. Biomacromolecules. 2005, 6, 2881–2894.
196. Young, S.; Wong, M.; Tabata, Y.; Mikos, A. G. J. Control. Release. 2005, 109, 256–274.

197. Young, T. H.; Yao, N. K.; Chang, R. F.; Chen, L. W. Biomaterials. 1996, 17, 2139–2145.
198. Yu, J.; Gu, Y.; Du, K. T.; Mihardja, S.; Sievers, R. E.; Lee, R. J. Biomaterials. 2009, 30, 751–756.
199. Yu, L.; Dean, K.; Li, L. Prog. Polym. Sci. 2006, 31, 576–602.
200. Yun, Y. H.; Goetz, D. J.; Yellen, P.; Chen, W. Biomaterials. 2004, 25, 147–157.
201. Zhang, Y.; Cheng, X.; Wang, J.; Wang, Y.; Shi, B.; Huang, C.; Yang, X.; Liu, T. Biochem. Biophys. Res. Commun. 2006, 344, 362–369.
202. Zhao, X.; Harris, J. M. J. Pharm. Sci. 1998, 87, 1450–1458.

CHAPTER 18

NEW PERSPECTIVES FOR NANOTECHNOLOGY FROM AYURVEDA AND SIDDHA MEDICINES: MERGE TO EMERGE

SREEVIDYA NARASIMHAN

CONTENTS

18.1   INTRODUCTION

Nanotechnology needs no introduction as is nano-medicine because much is written and known in these fields. The aim of nano-medicine is multi-fold like therapeutics, diagnostics, and targeted drug delivery and like. For achieving this many ideas and methodologies like nanobots, tubes, pores, magnetic probes, denrimers, liposomes, shells and so on, are being synthesized using a variety of synthetic methodologies, physical, chemical, and biological. Even though much success has been achieved, the safety and toxicological aspects of these are really questioned by many and the time taken to evaluate these parameters is more with greater success rate. Thus to bridge the time gap with higher success rate some proven methods which have been used will help.

Ayurvedic system of medicine is now well known to the western world as is known in the country of origin India, and *Siddha* system of medicine is not known that well, though from India. Both these systems of medicine have well written treatises in Sanskrit and Tamil respectively, and have a rich traditional system of practice in India, and dates back to several centuries. One wonders what these old systems of medicines have to offer to a very highly emerging and technically advanced nano science. In this manuscript the connection between these two fields is revealed with scientific evidence and the new methods and ideas which can be borrowed to merge the gap in time and emerge as a much stronger field with much less effort. Help to two fields of nano-medicine, targeted drug delivery and therapeutics are readily available from these old systems of medicine with a variety of green synthetic methodologies in physical, chemical and biological modes. Quality controlled and relatively less toxic medicines with high shelf life can be made available if these methods are followed [14, 15, 23, 24, 36].

18.2   PREPARATION OF NANO-MATERIALS

The general methods of preparation of modern nano materials are (i) top-down approach which involves mainly physical or mechanical reaction of bulk material to get nano particles and (ii) bottom up approach which involves chemical or biological reactions of atoms or molecules or clusters to build up nano particles. Here it is looked into two methods one each in these two methods namely mechano-chemical reactions and photo-synthesis of nano-materials *vis a vis* it analogous counterparts in *Ayurveda* and *Siddha* preparations [8].

Mechanochemical method of preparation of nano materials involves attrition or shearing or grinding or milling of bulk materials in different type of milling equipments like ball mills, vibration mills, attrition mills, and so on, by supplying suitable energy and controlled parameters to get desired nano forms [37].

Analogous method of synthesis in *Ayurveda* is called *khalviya* [39], and *mardhana* where the medicine is prepared by mixing the ingredients in a mortar and pestle. The number of times the pestle is run and hours the mixing is done alone is taken in to account for the preparation and is recorded in the *rasashastra* texts. Apart from these two, we have *bhavana*, wet grinding of materials [2] which is not a method adopted by modern nanotechnologists. In the wet grinding method of *bhavana* in Ayurvedic medicine, different plant juices are used for preparation of variety of metal bhasmas and are well recorded in the classical texts. For the preparation of a particular bhasma, definite quantity and quality of plant material and specific methods are prescribed.

Phytosynthesis is the green or biological method of synthesis of nano material using plant parts. In modern nano-technological evolution, people are trying to find different plants which can be used to synthesis different metal nano particles. Different plants like *Ocimum tenuiflorum*, *Morinda citrifolia*, *Rhizophora mucronata* were used for synthesis of silver nano particles of different size distribution, shapes, and crystallinity [4, 7, 11,12, 17, 28]. Here there is no systemic approach to the deciding factor by the researchers in choosing the plants for the synthesis and the final characteristics of the product are unpredictable.

Here a small review of the scientific validation and findings of some of the traditional bhasmas or parpams or chendurums of gold, silver, copper, iron, tin, zinc, mercury and sulfur is given to understand the way how nanotechnology is involved in these traditional medicines and what one can learn from these findings.

18.2.1 Gold Preparations [9, 22]

Swarna Bhasma

Gold based preparation from Ayurvedic medicine

- **Use**: Bronchial asthma, rheumatoid arthritis, diabetes mellitus and nervous system diseases, tuberculosis, schizophrenia, anemia, dyspnoea, anorexia, ophthalmic disorders usually given orally with honey, ghee or milk, powder of *Embelica officinalis* or *Acorus calamus*.
- **Raw materials used**: Pure gold, mercury, sulfur, sesame oil, butter milk, kanji, decoction of *Dolichos biflorus*.
- **Characteristics**: Ninety-two percent gold, no mercury detected, no organic compounds present, crystalline nano gold forms with crystallite in the range of 23–37nm as calculated using Scherrer equation and TEM indicated the average particle size to be 57nm with globular morphology.

Thanga Parpam

Gold based preparation from *Siddha* medicine

- **Use**: Tuberculosis, chlorosis and nervous debility, aphrodisiac, hyperpyrexia, ulcerative wounds, oedema, usually given with ghee, butter, milk or honey twice daily.
- **Raw materials used**: Gold, water soluble ash of adina, *Euphorbia hirta* herb juice.
- **Characteristics**: No reports on chemical characterization.

18.2.2 Silver Preparations [18, 26]

Rajatha Bhasma

Silver based preparation from Ayurvedic medicine

- **Use**: Diabetes, vitiligo, tuberculosis, anemia, piles, emaciation dyspnoea, ophthalmic disorders, thirst, usually given with sugar, ghee, meat juice, juice of *Adhatoda vasica* leaves.
- **Raw materials used**: Silver, *Dolichos biflorus* decotion, sesame oil, butter milk, lemon juice, sulfur, asafotedia, aloe pulp, mercury, cow's urine.
- **Characteristics**: Ag_2S was the major phase with Ag_5SiO_4 as minor phase through XRD.

Velliparpam

Silver based preparation from *Siddha* medicine

- **Use**: Cough haemoptysis, tuberculosis, colic, piles, leucorrhoea and gonorrhoea, spermatorrhoea usually with butter twice daily.
- **Raw materials used**: Silver, mercury, egg shell, *Pergularia daemia* leaf juice.
- **Characteristics**: The ICP-OES analysis shows large silver in major portions and also Ca, it may be in the form of silver oxide and calcium oxide. No organic matter detected as confirmed by IR studies. Agglomeration of particles with 10% of particles below 188.7nm.

18.2.3 Copper Preparations [13, 40]

Tamara Bhasma

Copper based preparation from ayurvedic medicine

- **Use**: Anemia, worm infestation, abdominal tumor, spleenic disorder, liver disorders, spasmodic pain, peptic ulcer, piles, rhinorrhoea, acidity, usually with milk, ghee, honey, sugar candy, powder of *Piper longum* fruit, juice of *Bombax malabaricum* root.

- **Raw materials used**: Copper, mercury, sulfur, butter milk, sesame oil, cows urine, juice of *Vitex negundo, Dolichos biflorus* decotion, turmeric, aloe juice.
- **Characteristics**: Copper is predominantly present as copper sulphide as revealed by XRD studies. In another study, it was found that copper was in the form of its oxide and PES showed Cu in $^2P_{3/2}$ state. The IR spectra show absence of organic matter.

Thamira Parpam

A copper based *Siddha* preparation

- **Use**: In vitated khapha given with honey, in tuberculosis given with ghee, skin diseases given with sugar, in veneral and delirum disease it is given with tender coconut water.
- **Raw materials used**: Copper, rock salt, *Citrus reticulate* juice, *Aristolochia bracteolate* juice, *Alangium salvifolium* root decoction.
- **Characteristics**: No work reported on chemical characterization.

18.2.4   Mercury Preparations [3, 5, 35]

Ras Sindoor

Mercury based preparation from Ayurvedic medicine

- **Use**: Syphilis, genital disorders, rejuvenation usually with sugar, powder of *Embilica officinalis* fruit, ginger juice, lemon juice.
- **Raw materials used**: Mercury, sulfur, cow's milk, ghee, juice of aerial root of *Ficus benghalensis*.
- **Characteristics**: Sharper and intense peaks of mercury sulphide in XRD, and the crystallite size as determined using Scherrer equation was 25–50nm. Has a spongy like structure with several crystallites agglomerated in a single particle with loss of grain boundaries as determined from TEM studies. The XPS analysis shows additional peaks for C and O apart from Hg ($^4f_{5/2}$ and $^4f_{7/2}$) and S ($^2P_{3/2}$). Presence of organic substances confirmed through IR spectra.

Lingachenduram, Poornachandrodayachenduram

Mercury based preparations from *Siddha* medicine

Lingachenduram

- **Use:** Used to treat fevers, skin diseases and veneral diseases, given along with honey.
- **Raw materials used:** Method (a) cinnabar, *Citrus colocynthes* plant juice or method (b) cinnabar, gum benzoin, camphor, *Morinda tinctoria* bark powder.
- **Characteristics:** The ICP-OES studies indicate presence of mercury, and XRD studies indicate presence of HgS. Dynamic light scattering for particle size studies indicate 10% particles below 69.2nm.

Poorna Chandrodaya Chenduram

- **Use:** Used for treating tuberculosis, jaundice, fever and bronchitis, rat bites cancerous ulcer and myalgia. Generally administered along with karpooraadichooranam or honey or betel leaf juice.
- **Raw materials used:** Gold, mercury, sulfur, flower juice of *Gossypium herbaceum*, *Musa paradiasica* rhizome juice.
- **Characteristics:** The ICP-OES studies indicate presence of 0.8mg/g gold and mercury in the form of HgS and the size close to nano with 10% particles below 139.9nm which was confirmed by TEM-EDAX and the nature of gold was uncertain. In another study, ICP-OES analysis revealed the presence of mercury (141.76mg g^{-1}), Calcium (11.68mg g^{-1}) and gold (0.84mg g^{-1}). The TEM particles size in the range of 60–70nm. HgS was the main component and a small percentage of Hg as oxide with a small percentage of gold was revealed by EDAX and IR. Absence of organic compounds was confirmed by IR studies.

18.2.5   Iron Preparations [1, 2, 6, 20, 21, 25, 31, 32]

Lahua Bhasma

Iron based preparations from Ayurvedic medicine

- **Use:** Anemia, diabetes, skin disorders, cachexia, obesity, bowel syndrome, spleenic disorders, hyperlipidemia, usually with honey, ghee, powder of Triphala.
- **Raw materials used:** Iron, triphala, cinnabar, aloe pulp.
- **Characteristics:** Mossbauer spectrum shows single sextets of α-Fe_2O_3 and the same was observed in XRD. In another study, Fe_3O_4 was also

reported along with Fe_2O_3. In yet another study, it was found to contain particles with around 28nm size in aggregates of distorted spheres.

Kshayakulathaga Chenduram

Iron based preparation from *Siddha* medicine

- **Use**: Respiratory diseases along with honey and thirikaduguchooranam for 10 days.
- **Raw materials used**: Yellow orpiment, calomel, red orpiment, *Calotropis gigantea* latex, *Ficus racemosus* yellowed leaves, lode stone, coral, asbestos, gold, cinnabar, sulfur, aloe juice.
- **Characteristics**: Many elements mainly Ca, Fe, Mg and As, 7.524mg/g gold, and Hg was absent. The IR and XRD showed the presence of Fe_3O_4 and the crystallite size being 397.1nm.

18.2.6   Zinc Preparations [16, 19, 29, 33, 38]

Jashada or Yashada Bhasma

Zinc based preparation from Ayurvedic medicine

- **Use**: Emaciation, depression, tremor, ophthalmic disorders, diabetes, anemia usually with ghee, betel leaf juice, rice water, sugar candy, *Phoenix sylvestris* fruit.
- **Raw materials used**: Zinc, butter milk, rice khanji, *Dolichus biflrous* decoction, cows urine, sesame oil, lime water, *Achyranthes aspera* powder, mercury, sulfur, aloe juice, lemon juice.
- **Characteristics**: The XRD showed characteristic peaks of ZnO, $ZnCO_3$, ZnS in intermediate stages of preparation and the final product contained only ZnO. In yet another study the same was confirmed and was reported to be non stoichiometric with particles in nano meter size.

Nagachenduram, or Nagaparpam

Zinc based preparation from *Siddha* medicine

- **Use**: Used to treat ascities, piles, skin disease, leprosy and white discharge. Usually 100–200mg is given along with honey or ghee.
- **Raw materials used**: Zinc, potassium nitrate, *Zingiber officinale, Anethum graveolens, Curcuma longa, Achyranthes aspera, Citrus reticulata* juice.

- **Characteristics**: It has 10% of particles less than 357.8nm, and XRD shows the presence of ZnO. The ICP-OES studies show the presence of Ca and Mg too.

Nagaparpam
- **Use**: Used to treat piles fistula, diarrhoea, cough asthma and dysentery along with butter or ghee or ththankottailehyam 100–200mg twice a day.
- **Raw materials used**: Zinc, *Wedelia sinensis* juice, aloe pulp.
- **Characteristics**: It has 10% of particles less than 233.9nm and in the form of ZnO. The IR spectrum revealed no organic components to be present.

18.2.7   Tin Preparations [27, 30, 34]

Vanga Bhasma

Tin based preparation from Ayurvedic medicine
- **Use**: Hyperlipidemia, emaciatin, flatulence, oligospermia usually with ghee, borax, camphor, juice of *Ocimium sanctum* leaves.
- **Raw materials used**: Tin, sesame oil, cows urine, rice kanji, *Dolichos biflorus* decotion, butter milk, churnodaka, Aloe juice.
- **Characteristics**: Contains Sn-81%, Fe-5378.09ppm, As <5ppm, S < 1ppm, with mean particle size of 9.95µm, and shows triclinical structure and P1 space group. In another study, it was reported that an intermediate of tin hydroxide was formed.

Vangaparpam

Tin based preparation from *Siddha* medicine
- **Use**: Used for the treatment of urino genital infections and dyspepsia, arthritis, diabetes, asthma, arthritis, gonorrhea, blood disorders, with a dosage of 65–130mg/day for 7 consecutive days when taken along with an adjuvant such as honey, milk or ghee.
- **Raw materials used**: Tin, sesame oil, turmeric powder, aloe juice.
- **Characteristics**: The particles were in the range of 50nm to 100micorns. The XRD confirm the presence of highly crystalline single tetragonal rutile phase of SnO_2, Characteristic peaks of Sn=O and Sn–O, and oxygen species, intermediate between O_2^- and O_2^{2-} was observed through IR. A bigger spongy microcrystalline aggregate of SnO_2 was observed in SEM.

Sn (^{3}d$_{5/2}$) and Sn (^{3}d$_{3/2}$) photoelectron transitions, respectively, which are characteristic of Sn as SnO or SnO2 were observed in XPS indicating of two types of oxygen. The formation of oxides of Calcium, Silicon at the surface was also noticed.

18.2.8 Lead Preparations [10]

Naga Bhasma

Lead based preparation from Ayurvedic medicine

- **Use:** Rheumatoid arthritis, tetani, oedema, abdominal tumor, piles, ulcer, diarrhea and for all water borne disease, usually with milk, rice, wheat, sugar candy, and sesame oil.
- **Materials used:** Lead, butter milk, rice khanji, *Dolichus biflrous* decoction, cow's urine, sesame oil, orpiment, bark of *Ficus religiosa*, Lime water.
- **Characteristics:** Found to be crystalline and a mixture of PbO, Pb$_3$O$_4$. The IR spectra revealed presence of carbonate $(CO_3)^{-2}$ in yet another study, it was shown to contain particles with varied sizes from <1μm to 20μm and agglomerated. They report the peaks to be related to Pb$_3$Ge$_2$O$_7$ and RbTIPb$_8$(PO$_4$)$_6$ with lead oxide, lead phosphate and some other un-identified peaks. It is also reported that Pb is only 58.4% for the end product and C, S and O to be present through XPS and ICP-AES analysis. In yet another study, where *Vitex nigundo* decoction and *Curcuma longa* powder, As$_2$S$_2$, *Plectranthus coesta, Azadirachta indiaca Adhatoda vasica* leave juice was used, nano-crystalline (60nm) PbS were obtained as observed from XRD studies. The TEM showed that it is irregular spongy structure with sub-micron particles. The XPS studies show Pb(^{4}f$_{7/2}$) and Pb(^{4}f$_{5/2}$) and S (^{2}P$_{3/2}$)respectively for PbS phase. The IR shows association with organic matter which is from the plants used.

18.3 QUALITY CONTROL OF THE PREPARATIONS ACCORDING TO TRADITIONAL METHODS

Two methods, physical and chemical are prescribed for the quality control of the above metal based preparations in *Ayurveda*. Of the physical characteristics, floating in water, freeness of particles from adhesiveness to each other, reduced particle of the size of pollen grains of *Pandanus odoratissimus* flower, color of the preparation, tastelessness, lustreless, and of the chemical characteristics, irreversible to metallic form, are important points for the modern scientists to take in to consideration.

18.4 INFERENCES AND PERSPECTIVE FROM THE STUDIES ON THE PREPARATION OF TRADITIONAL MEDICINES

From the scientific validation studies on the traditional preparations the following points can be inferred:

- Both *Ayurveda* and *Siddha* medicines have preparations equivalent to the metals and their chemical forms seem to be almost the same. Also for the same metals, there are many different preparations.
- They are used for different diseases and the raw materials used for their preparation differ considerably.
- Even though there are many metals and organic compounds used, the final preparation seems to have almost no organic content and few of the metals used are eliminated or reduced in composition.
- Almost all the preparations are nano-particles or near to nano and crystalline and agglomerated.
- The ways these preparations are administered are different for different disease.
- All the preparations confirm to the traditional quality control parameters. But in some of the validation studies, one finds it difficult to find the exact nature of the chemical entity of these preparations and the intermediates differ in different type of preparations.

When one tries to compare with the modern methods of preparations of nano particles and their quality control with that of the traditional methods, one can see the following points.

- The modern methods have their equivalent methods of preparation in the traditional methods of preparation and have many more methods to adopt from the traditional methods.
- The use of plants and their juices are generally more prevalent in traditional systems and they are specific for each preparation, though the reason for the same is not explained. A thorough understanding will help us in easier methods for choosing the plants for green synthesis.
- For targeted drug therapy, the use of anupana or the vehicles is changed for the same drug. For nano medicines which can be prescribed for oral intake, no such vehicles are there and the approach is altogether different.
- The quality parameters like floating on the water is possible only if the density, surface tension, interfacial forces, surface energy, hydrophobic nature, contact angle of the powder with the liquid are taken into con-

sideration. The modern preparations do not take these parameters in to consideration.

- The important quality of the traditional preparation is irreversible nature of the preparation to metallic form and absence of any metallic content. In the modern methods of preparation such a quality is not considered and is not analyzed at all.

18.5 CONCLUSION

The studies of traditional medicinal preparations indicate that quality control needs to be taken care of in the modern preparations. Possibly if these parameters are taken care of, then safe, less toxic more bio-available nano-medicines can be made available. Also, the advance in the analytical technologies for the characterization of nano-materials has enabled the traditional medicines to unravel the chemical identities of its preparation. Thus both these fields can grow by complimenting each other with their knowledge.

KEYWORDS

- **Ayurveda**
- **Magnetic Probes**
- **Nanobots**
- **Phyto-Synthesis**
- **Siddha**

REFERENCES

1. Tripathi, A., Joshi, B., Singh, H. S., Rathore, J. S., & Sharma, G. (2010). Chemical phases of some of the Ayurvedic Heamatinic medicines. *International Journal Engineering Science Technology, 2*(8), 25–32.
2. Chaudhary, A. & Singh, N. (2010). Herbo mineral formulations (Rasaoushadhies) of Ayurveda an amazing inheritance of Ayurvedic Pharmaceutics. *Ancient Science of Life, 30*(1), 18–26.
3. Austin, A. (2012). Chemical characterization of a gold and mercury based Siddha Sasthric preparation—Poorna Chandrodayam. *American Journal of Drug Discovery and Development, 2*(3), 110–123.

4. Leela, A., & Vivekanandan, M. (2009). Tapping the unexploited plant resources for the synthesis of silver nanoparticles. *African Journal of Biotechnology,* 3162–3165.

5. Sudha, A., Murty, V. S., & Chandra, T. S. (2009). Standardization of metal-based herbal medicines. *American Journal of Infectious Diseases,* 5(3), 193–199.

6. Krishnamachary, B., Pemiah, B., Krishnaswamy, S., Krishnan, U. M., Sethuraman, S., & Rajan, K. S. (2012). Elucidation of a core-shell model for lauhabhasma through physico- chemical characterization. *International Journal of Pharmacy and Pharmaceutical Sciences,* 4(2), 644–649.

7. Bhattacharjee, C. R., Sharon, M., & Nath, A. (2012). Synthesis of nano composites from plant-based sources. *Research Journal of Chemical Sciences,* 2(2), 75–78.

8. Raab, C., Simkó, M., Fiedeler, U., Nentwich, M., & Gazso, A. (2011). *Production of nanoparticles and nanomaterials.* Institute of Technology Assessment of the Austrian Academy of Sciences, The Nanotrust Dossiers No. 006.

9. Brown, C. L., Bushell, G., Whitehouse, Mw, Agarwal, D. S., & Tiekink, E. R. T. (2007). Nanogold-Pharmaceutics (I) The use of colloidal gold to treat experimentally induced arthritis in Rat Models; (Ii) Characterization of the gold in Swarna Bhasma, a Microparticulate used in traditional Indian medicine. *Gold Bulletin,* 40(3), 245–250.

10. Smita, D. & Chandrashekhar, B. (2012). Naga Bhasma: An overview. *International Journal Pharmaceutical Invention,* 2(1), 36–47.

11. Sathishkumar, G., Gobinath, C., Karpagam, K., Hemamalini, V., Premkumar, K., & Sivaramakrishnan, S. (2012). Phyto-synthesis of silver nanoscale particles using morindacitrifolia l. and its inhibitory activity against human pathogens. *Colloids and Surfaces B: Biointerfaces, 95,* 235–240.

12. Umashankari, J., Inbakandan, D., Ajithkumar, T. T., & Balasubramanian, T. (2012). Mangrove Plant, RhizophoraMucronata (Lamk, 1804) mediated one pot green synthesis of silver nanoparticles and its antibacterial activity against aquatic pathogens. *Aquatic Biosystems,* 8(1), 11.

13. Chitnis, K. S. & Stanley, A. (2011) Chemical evaluation of Tamra Bhasma. *International Journal of Pharma and Bio Sciences,* 2(2), 160–168.

14. Kapoor, R. C. (2010). Some observations on the metal-based preparations in the Indian systems of medicine. *Indian Journal of Traditional Knowledge,* 9(3), 562–575.

15. Kumar, A., Nair, A. G., Reddy, A. V., & Garg, A. N. (2006). Bhasmas: unique ayurvedic metallic-herbal preparations, chemical characterization. *Biological Trace Elements Research, 109,* 231–254.

16. Lagad, C. E., Sawant, R. S., & Bhange, P. V. (2012). Study of standard operating procedure of Naag Bhasma in relation to its physico-chemical properties. *International Research Journal of Pharmacy,* 3(3), 162–167.

17. Rai, M., Yadav, A., & Gade, A. (2008). Current trends in phytosynthesis of metal nanoparticles. *Critical Reviews in Biotechnology,* 28(4), 277–284.

18. Tanna, M. (2006–2007). A comparative pharmaceutico-pharmaco-clinical study of Rajata Bhasma and RajataSindura W. S. R. to Depression (Avasad). Project report, submitted to Institute of Post Graduate Teaching & Research in Ayurved, From Annual Report 2006–2007, Gujarat Ayurved University Jamnagar.

19. Wadekar, M., Gogte, V., Khandagale, P., & Prabhune, A. (2004) Comparative study of some commercial samples of Naga Bhasma. *Ancient Science of Life,* 23(4), 48–58.

20. Singh, N. & Reddy, K. R. C. (2010). Pharmaceutical study of Lauha Bhasma. *Ayurveda, 31*(3), 387–390.
21. Singh, N. & Reddy, K. R. C. (2011). Particle size estimation and elemental analysis of Lauha Bhasma. *International Journal of Research in Ayurveda and Pharmacy, 2*(1), 30–35.
22. Paul, W. & Sharma, C. P. (2011). Blood compatibility studies of Swarna Bhasma (Gold Bhasma), an Ayurvedic drug. *International Journal of Ayurveda Research, 2*, 14–22.
23. Sarkar, P. K. & Chaudhary, A. K. (2010). Ayurvedic Bhasma: The most ancient application of Nano medicine. *Journal of Scientific and Industrial Research, 69*, 901–905.
24. Sarkar, P. K., Das, S., & Prajapati, P. K. (2010). Ancient concept of metal pharmacology based on Ayurvedic literature. *Ancient Science of Life, 29*(4), 1–6.
25. Rajendraprasad, M. L., Shekhar, S., & Subramanya, A. R. (2010). Pharmaceutical and analytical study on Loha Bhasma. *International Journal of Ayurvedic Medicine, 1*, 47–59.
26. Chaturvedi, R., & Jha, C. B. (2011). Standard manufacturing procedure of Rajata Bhasma. *Ayurveda, 32*(4), 566–571.
27. Hiremath, R., Jha, C. B., & Narang, K. K. (2010). Vanga Bhasma and its XRD analysis. *Ancient Science of Life, 29*(4), 24–28.
28. Patil, R. S., Kokate, M. R., & Kolekar, S. S. (2012). Bio inspired synthesis of highly stabilized silver nanoparticles using ocimum tenuiflorum leaf extract and their antibacterial activity. *Spectrochim Acta: A Molecular and Biomolecular Spectroscopy, 91*, 234–238.
29. Santosh, B., Raghuveer, Jadar, P. G., & Rao, N. (2012). X-ray diffraction analysis of Yashada Bhasma: An Ayurvedic metallic preparation. *International Journal of Research in Ayurveda and Pharmacy, 3*(2), 165–167.
30. Ullagaddi, S. S. (2010). Screening of free radical scavenging activity and immunomodulatory effect of Vanga Bhasma. Dissertation submitted to the Rajiv Gandhi University of Health Sciences, Bangalore.
31. Singh, N., Reddy, K. R.C., Prasad, N. K., & Singh, M. (2010). Chemical characterization of Lauha Bhasma by X-ray diffraction and vibrating sample magnetometry. *International Journal of Ayurvedic Medicine, 1*, 143–149.
32. Singh, N. & Reddy, K. R. C. (2011). Particle size estimation and elemental analysis of Lauha Bhasma. *International Journal of Research in Ayurveda and Pharmacy, 2*, 30–35.
33. Singh, S. K., Gautam, D., Kumar, M., & Rai, S. B. (2010). Synthesis, characterization and histopathological study of a lead-based Indian traditional drug: Naga Bhasma. *Indian Journal of Pharmaceutical Sciences, 72*, 24–30.
34. Sudha, S. & Parimala, A. (2012). Gnanamani, Asit Baran Mandal. Bulk and surface properties of tin based herbal drug during its preparation: Fingerprinting of the active pharmaceutical constituent. *International Journal of Pharmaceutical Sciences and Research, 3*(4), 1037–1042.
35. Singh, S. K., Chaudhary, A., Rai, D. K., & Rai, S. B. (2009). Preparation and characterization of a mercury based Indian traditional drug-Ras-Sindoor. *Indian Journal of Traditional Knowledge, 8*(3), 346–351.
36. Surendiran, A., Sandhiya, S., Pradhan, S. C., & Adithan, C. (2009). Novel applications of Nanotechnology in medicine. *Indian Journal of Medical Research, 130*, 689–701.

37. Yadav, T. P., Yadav, R. M., & Singh, D. P. (2012). Mechanical milling: a top down approach for the synthesis of nanomaterials and Nanocomposites. *Nanoscience and Nanotechnology, 2*(3), 22–48.

38. Bhowmick, T. K., Suresh, A. K., Kane, S. G., Joshi, A. C., & Bellare, J. R. (2009). Physico chemical characterization of Indian traditional medicine Jasada Bhasma: Detection of Nano-particles containing non-stoichiometric zinc oxide. *Journal of Nanoparticle Research, 11,* 655–664.

39. Gupta, K. V., Pallavi, G., Patgiri, B. J., & Naveena, K. (2011). Relevance of Rasa Shastra in 21st Century with special reference to life style disorders. *International Journal of Research in Ayurveda and Pharmacy, 2*(6), 1628–1632.

40. Wadekar, M. P., Rode, C. V., Bendale, Y. N., Patil, K. R., & Prabhune, A. (2005). Preparation and Characterization of a copper based Indian traditional drug: Tamra Bhasma. *Journal of Pharmaceutical and Biomedical Analysis, 39,* 951–955.

INDEX